CONQUERORS OF TIME

Other books by the author

WALES AND THE WELSH
TALKING OF WALES
AMERICANS AND NOTHING ELSE
INDIA FILE
THE STATE OF AMERICA
OUT OF RED DARKNESS
MY FOREIGN COUNTRY
COBRA ROAD

CONQUERORS OF TIME

Exploration and Invention in the Age of Daring

TREVOR FISHLOCK

For David and Karen
— conquering —
with love and best
wishes

Trevor
2006.

JOHN MURRAY

First published in 2004 by John Murray (Publishers)
A division of Hodder Headline

1 3 5 7 9 10 8 6 4 2

A CIP catalogue record for this title is
available from the British Library

ISBN 0-7195-5517 5

Typeset in Monotype Garamond by
Palimpsest Book Production Limited,
Polmont, Stirlingshire

Printed and bound in Great Britain by
Clays Ltd, St Ives plc

John Murray (Publishers)
338 Euston Road
London NW1 3BH

For Penny
and
Ella and Iris

Contents

Illustrations

The author and publishers would like to thank the following for permission to reproduce illustrations: Plate 1, Eric Newby; 2, 9, 12, 20, 28 and 39, Bettman/CORBIS; 3, 26, 31, 33, 34, 38, 40 and 41, Hulton-Deutsch Collection/CORBIS; 4, 6, 16 and 17, National Portrait Gallery, London; 5, The Royal College of Surgeons of England; 7 and 25, Michael Maslan Historic Photographs/CORBIS; 8, British Library; 10, 21, 22, 24, 29, 32 and 37, Hulton Archive; 11 and 36, Mary Evans Picture Library; 13, 14 and 30, John Murray Archive; 15, Stapleton Collection/CORBIS; 18, Royal Geographical Society, London; 19 and 27, CORBIS; 23, Brunel University Library; 35, Underwood & Underwood/CORBIS.

Preface

It is possible that the seed of this book was a boyhood curiosity about the discovery and growth of the modern world. I grew up by the sea and was stirred by the passage of warships through Spithead and the departure of regal Atlantic liners for New York and of Union Castle ships for Cape Town. More certainly, the stimulus was my reporting as a foreign correspondent for *The Times* and *The Daily Telegraph*. Assignments led to the roots of events. The achievements, failures and legacies of discoverers, engineers and scientists were aspects of the stories. History was part of the extended adventure of reporting.

My interest in the North West Passage, for example, grew during two Arctic journeys. On Cornwallis Island, which Sir John Franklin circumnavigated as he voyaged to his doom, I contemplated the spectacle that had filled his imagination, the opal universe of ice rinsed in brilliant light before the darkness fell. A king's coronation in Uganda took me to the Africa of the strangely fated John Speke, the red and chocolate-scented earth and the source of the Nile. The tumult of South African politics drew me to the Great Trek, the gold and diamond trails and colonial scramble. In the West Indies a breadfruit tree marked the pitiful passage of slaves and also of William Bligh. On the stony flinders of Australian plains I learnt a little of the ache of distance and the seduction of deserts. Close to the forty-ninth parallel I followed the tracks of the westering Mounties who helped to make Canada. I travelled the dynamited spiral of the railroad through the Rockies. Shanghai's Bund and the panorama from Kowloon told of money made with gusto. In years of criss-crossing India and America I saw in the railways, bridges, dams and canals the determination and energy of men who sought to subjugate formidable landscapes. Almost everywhere stood the monumental buildings of western civilization, government houses and courts, and legislatures with seats of Westminster green and Speakers' chairs and maces, the gift of the maternal parliament.

I went to battlefields like Quebec, Lucknow, Delhi, Isandhlwana, Little Big Horn, Rorke's Drift, Ladysmith and Magersfontein. But for the purposes of this book my interest was less in wars and statesmen, more in exploration in the broadest sense, in the pathfinders who slowly prised open the shutters and let light fall into the mysterious places, including the mind. This is a story of the search for the shape and nature of the world, principally in the long nineteenth century, and the development of the nerves, sinews and musculature of technology. The technique of empire depended on rivets, rails, roads, wires, cables, pistons and girders, and the genius and perseverance of engineers. It required the financial propellant of business, of great London institutions like the East India Company, the Massachusetts Bay Company, the Hudson's Bay Company and the Imperial East Africa Company. My story deals with the progress that to many minds seemed divine, advances in communications, fuels, science, medicine and art. Since it is about human struggle it also chronicles disaster, horror, suffering, cruelty and, here and there, sex and cannibalism.

One of the themes is the incessant battle with the sea, the distances, the fury of oceans, the relentless demands of navigation. While I was writing this book I made two voyages under sail across the North Atlantic and I very much valued the perspective on history that these humbling experiences gave me, day in, day out. Surveying all of the extraordinary journeys and endeavours of our predecessors I am astonished by the sheer grinding and bloody hardship endured by so many of them and remain mightily moved by their spirit.

Maps

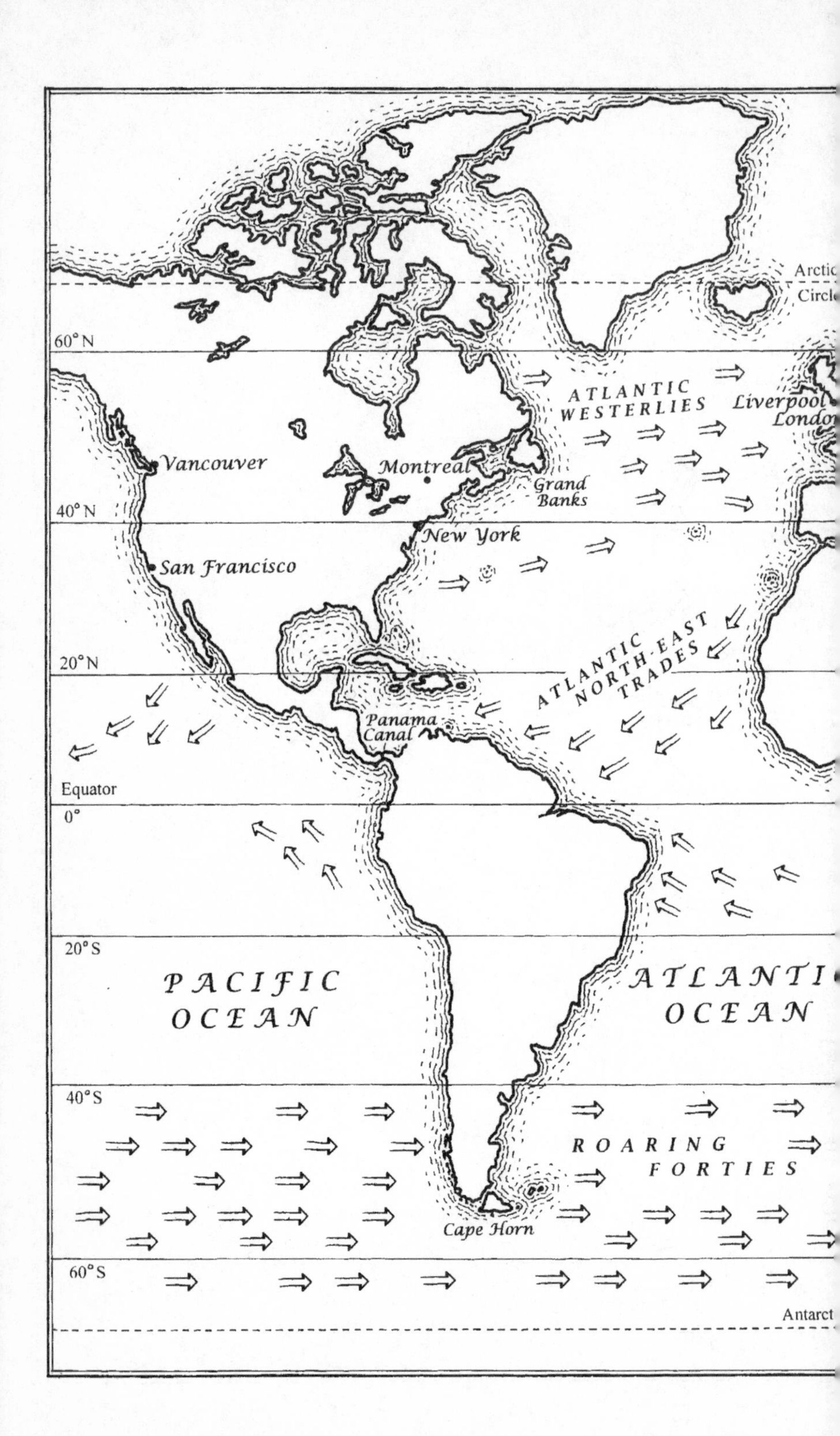

Arctic
Circle
60° N
ATLANTIC WESTERLIES
Liverpool
London
Vancouver
Montreal
Grand Banks
40° N
New York
San Francisco
ATLANTIC NORTH-EAST TRADES
20° N
Panama Canal
Equator
0°
20° S
PACIFIC OCEAN
ATLANTIC OCEAN
40° S
ROARING FORTIES
Cape Horn
60° S
Antarct

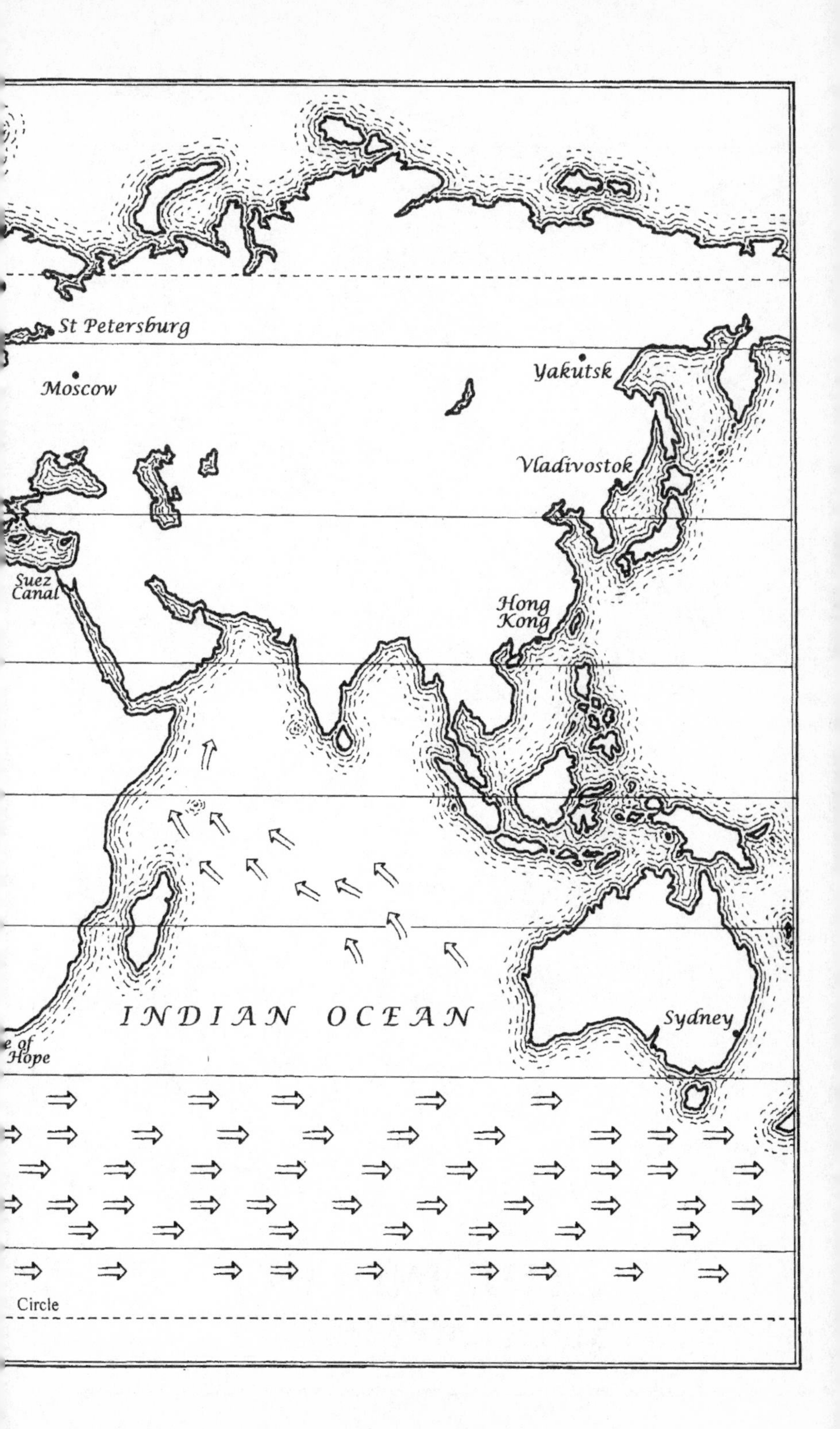
St Petersburg
Moscow
Yakutsk
Vladivostok
Suez Canal
Hong Kong
INDIAN OCEAN
Sydney
e of Hope
Circle

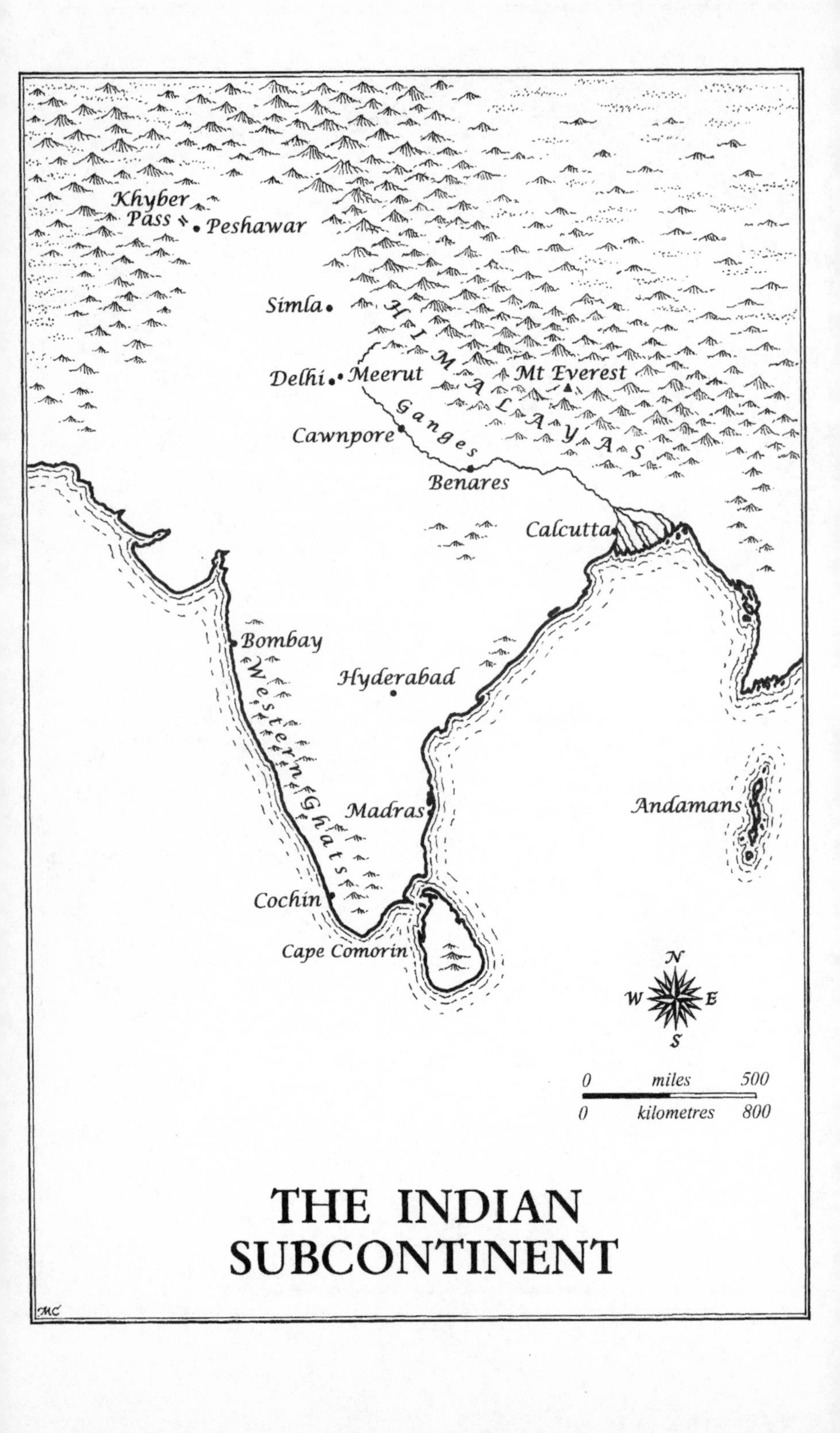

THE INDIAN SUBCONTINENT

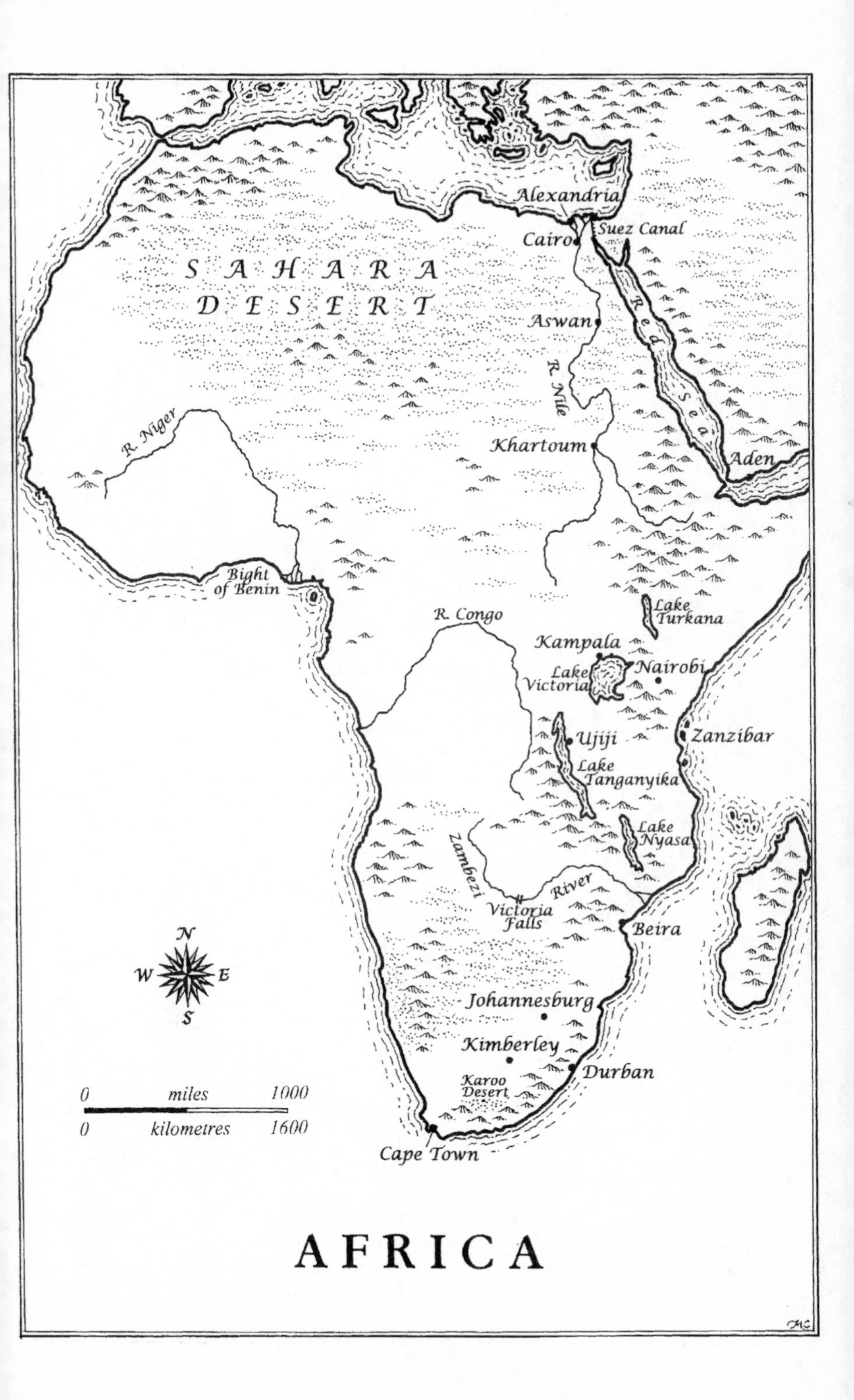

Alexandria
Suez Canal
Cairo
S A H A R A
D E S E R T
Aswan
R. Nile
Red Sea
Khartoum
Aden
R. Niger
Bight of Benin
R. Congo
Lake Turkana
Kampala
Lake Victoria
Nairobi
Ujiji
Zanzibar
Lake Tanganyika
Lake Nyasa
Zambezi River
Victoria Falls
Beira
N
W
E
S
Johannesburg
Kimberley
Durban
Karoo Desert
0 miles 1000
0 kilometres 1600
Cape Town
AFRICA

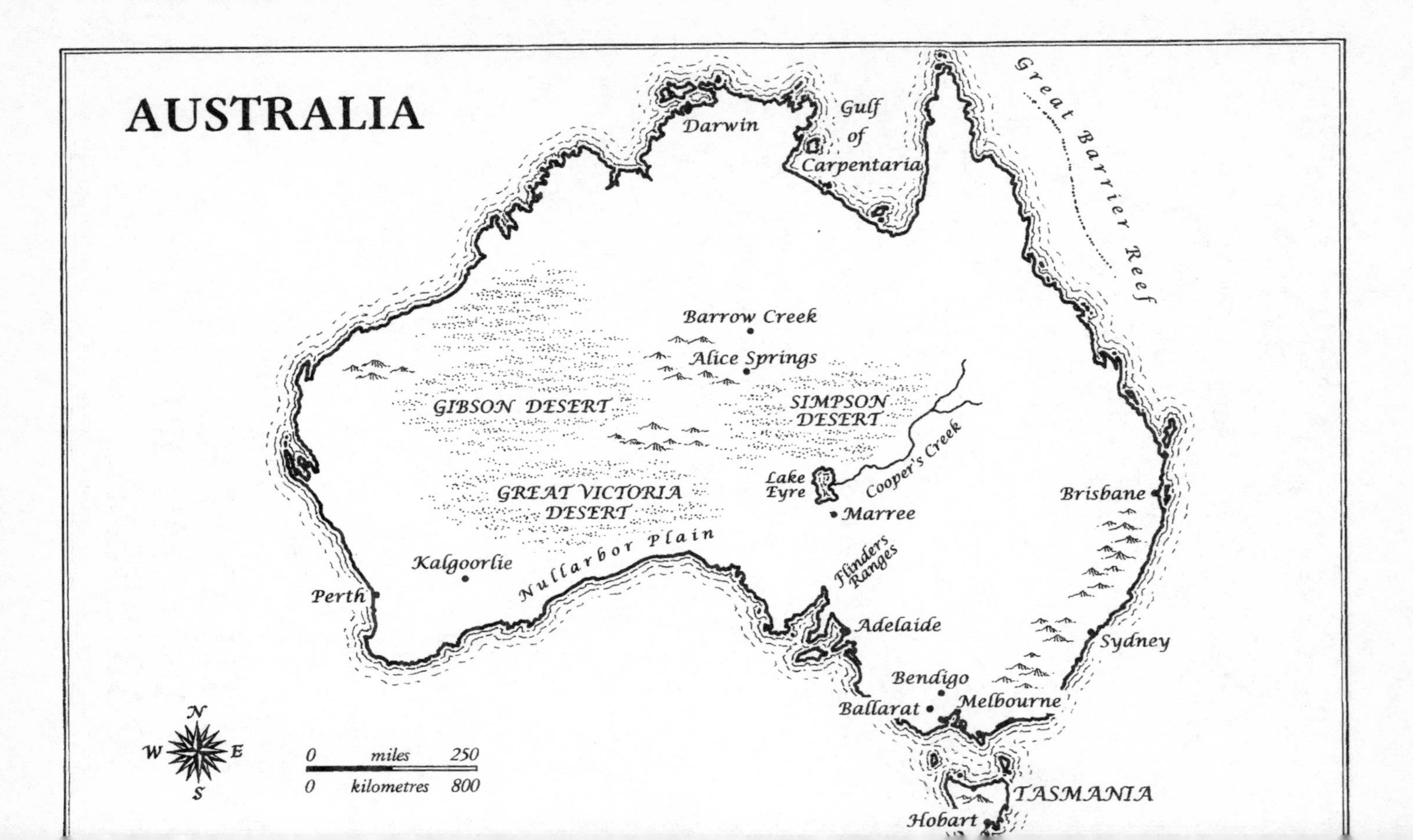
AUSTRALIA
Darwin
Gulf of Carpentaria
Great Barrier Reef
Barrow Creek
Alice Springs
GIBSON DESERT
SIMPSON DESERT
Cooper's Creek
Lake Eyre
Marree
GREAT VICTORIA DESERT
Nullarbor Plain
Kalgoorlie
Perth
Flinders Ranges
Brisbane
Adelaide
Sydney
Bendigo
Ballarat
Melbourne
TASMANIA
Hobart
N
W
E
S
0 miles 250
0 kilometres 800

Cornwallis Island
Baffin Island
King William Island
Mackenzie R.
Great Slave Lake
Hudson Bay
Lake Athabaska
ROCKY
MTS
Vancouver
Regina
Winnipeg
Newfound-land
St Lawrence R.
GREAT LAKES
Bismarck
Ottawa
Montreal
Rideau Canal
Erie Canal
San Francisco
Chicago
Missouri R.
New York
St Louis
Mississippi R.
New Orleans
JAMAICA
N
W
E
S
0 miles 500
0 kilometres 800
NORTH AMERICA AND THE CARIBBEAN
MC

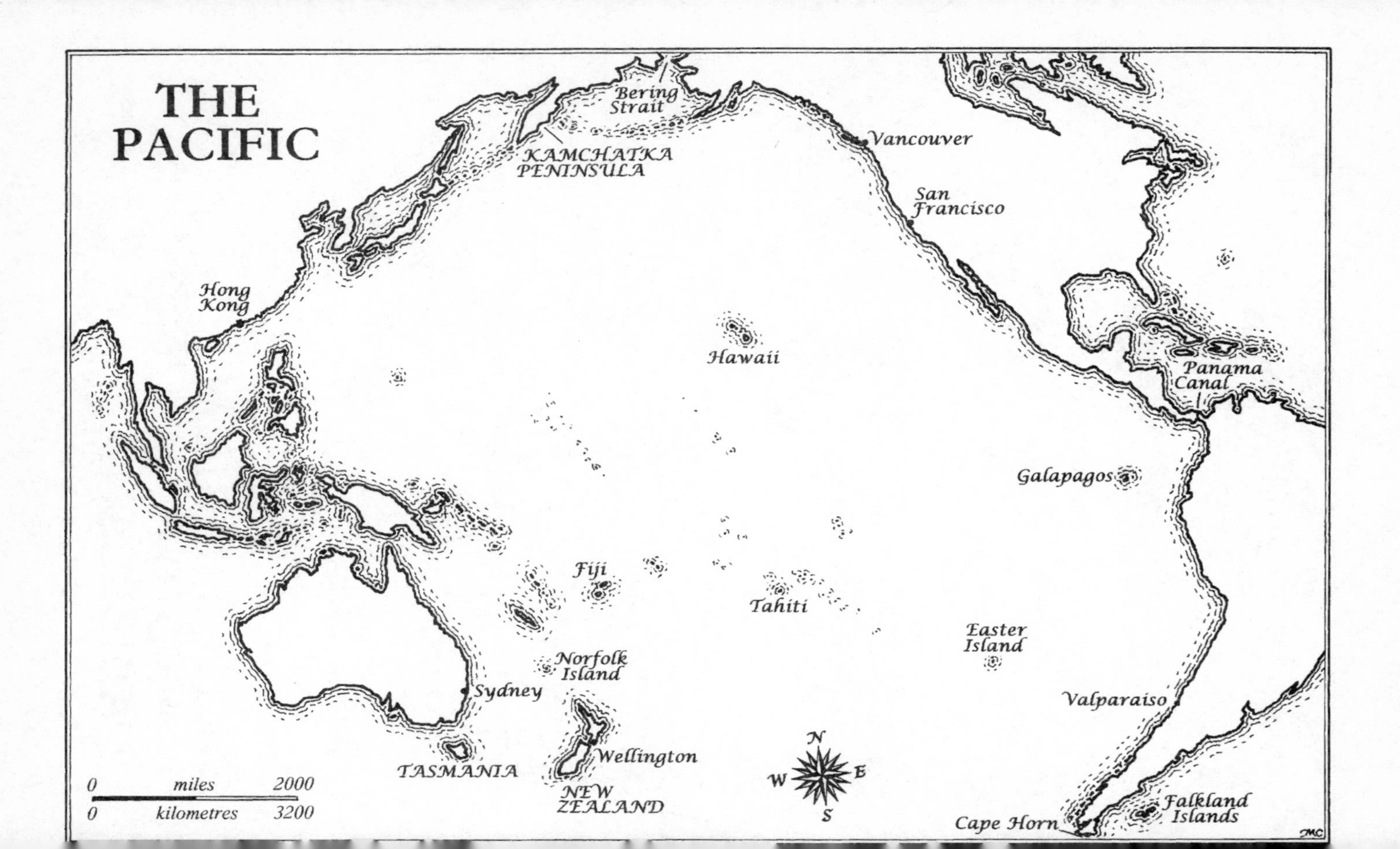
THE PACIFIC
Bering Strait
KAMCHATKA PENINSULA
Vancouver
San Francisco
Hong Kong
Hawaii
Panama Canal
Galapagos
Fiji
Tahiti
Easter Island
Norfolk Island
Sydney
Valparaiso
Wellington
TASMANIA
NEW ZEALAND
N
W
E
S
Falkland Islands
Cape Horn
0 miles 2000
0 kilometres 3200
MC

Introduction

'There's a magic in the distance,
where the sea-line meets the sky.'
Alfred Noyes, 'Forty Singing Seamen'

Far out in the North Atlantic, taut sails like sabres as we rolled, we struggled towards America.

For almost all of human history this is how it was. The sea was where the dreams began. The world took shape in the sailor's eye.

Years from home and adrift in time beneath strange constellations the seafarers were the first to hear surf booming on the uttermost reefs and dare their feet on fabled shores. In ships clogged and slow, often lost, frequently ambushed by adversity, they submitted to the dark unmeasured waters and pricked the horizon in the mysterious hemispheres. As students of the stars and the set of the swells, counting miles with sandglasses and sounding fathoms with knotted ropes, they delineated the coasts of continents and fixed the whereabouts of straits and promontories in the territories of the future.

Seafarers were men apart, their experience distinct. In the sixth century BC the Scythian sage Anacharsis, inventor of the fluked anchor, noted three kinds of people: the living, the dead, and those who sailed in ships. Distant-water sailormen stumbled ashore as Argonauts, the glittering-eyed survivors of wonderful and terrible events. 'All extraordinary travellers', thought the diarist James Boswell, 'are a band of wild beasts.'

These wanderers in other worlds were fathers of the yarn and babbled their news of golden isles and holy peaks that rose beyond the amaranthine sea. They were progenitors of an immense literature of ship-talk and sea stories and allusions to the power and mysteries of the oceans. To spellbound listeners they insisted they had seen amazing creatures and the naked chiefs and nubile maids of paradise, swore they had glimpsed the horns of unicorns and fed from cornucopias in the orchards of play. They spoke of silver ceilings and sapphire floors and into European imaginations blew the scent of spices and the seeds of fantasy and desire.

Fairy tales and phantoms drew discoverers across the ageless sea. Legends, quests and crazes led them to disillusion and horrible deaths. Yet the pursuit of will-o'-the-wisps also paid dividends in steady accretions to the knowledge of geography and the oceans' ruling rhythms.

Bare-knuckle rope-and-canvas seafarers found the world and made it work with trade and deliveries of mail and migrants. 'As concerning ships,' wrote Robert Kayall in 1615, 'it is that which everyone knoweth and can say; they are our weapons, they are our ornaments, they are our strength, they are our pleasures, they are our defence, they are our profit; the subject by them is made rich, the Prince in them mighty; in a word, by them in a manner we live, the Kingdom is, the King reigneth.' It was late in the historical day that steamships displaced the art of masts. Clippers eked out their romance into the twentieth century, and Shackleton and Scott raised sail to conserve coal and oil in remote Antarctic waters. Workaday sailing ships all but vanished from the oceans ninety years ago, although there was the last hurrah of a grain race from Australia in 1939. Now museumized square-riggers seem like bottled ships upon the nation's mantelpiece. Towns which once did business daily with the sea lean upon their histories. Amphibious creatures like the British have come ashore and the sea has shrunk in the imagination.

In the North Atlantic, a thousand miles from land, our lonely ketch was a vestige of the centuries of sail, a wandering ghost beneath the armadas of clouds.

The sea, however, is only one part of my story of discovery. The narrative runs from the smooth chins of the Georgians through the bird's-nest beards of the Victorians to the assertive moustaches of the Edwardians. Essentially it spans the nineteenth-century epoch of discovery and industrial revolution. It is stretched at one end to embrace a prelude, the late eighteenth-century explorations, and particularly the voyages of Captain Cook, which might have seemed almost interplanetary to his contemporaries. It is extended at the other to mark the beginning of the age of flight.

It is not too fanciful to imagine a man born in 1801, the year in which a boat scattered ducks on a Scottish canal and ushered in the age of steam transport, living to see the Wright brothers, dressed in jackets and ties, pioneering the reality and science of aviation in 1903. Thus one long human life could have spanned the invention and proving of all of the commanding principles of powered travel on land, at sea and in the air. In that year of 1903 my centenarian would also have seen Britain's Parliament raise the speed limit for motor cars from fourteen to twenty miles an hour; and a record-breaking Great Western Railway locomotive thundering through Somerset at more than a hundred miles an hour.

Colonel James Fynmore of the Royal Marines saw many astonishments in his long life. He died in 1887, aged ninety-three, the last surviving officer of the Battle of Trafalgar in 1805. He fought as a midshipman aboard HMS *Africa*, part of the force of twenty-seven British ships and almost 18,000 seamen, which, in four and a half hours of brutal cannonade, smashed the thirty-three ships of the combined French and Spanish fleets and resolved the struggle for oceanic power between Britain and France. Within ten years the cannon of Waterloo completed the work of Trafalgar and raised Britain as the hammer of tyranny and saviour of Europe. The Royal Navy's iron guns were the bullion of the Pax Britannica.

For all the cost, distress and turmoil of twenty-two years of war with France, Britain in 1815 was free and secure, possessed abundant coal, iron, limestone and copper, and grew most of the food its 13 million people needed. It was the world's leading iron-maker and the king of steam, the advanced technology of the century. The themes of the time were unrestrained expansion, local initiative, self-reliance and ruthless money-making, a free-for-all in which the government hardly interfered. Barrels and bales of colonial sugar, spices, tea, furs, gold and silk and calico flowed into London, the mightiest city on earth. Loud with argument about the nature of society, politics, morality and justice, the British embraced their unruly age of tremendous possibility.

In the accelerando of technology there grew an obsession with speed, a belief that men could master time and distance. It seemed self-evident that velocity improved human existence. Thomas Macaulay, the historian and essayist, considered that with the exception of the printing press and the alphabet the inventions which abridged distance achieved most 'for the civilization of our species'. He himself experienced the slow voyage to India in 1834 and spent twelve weeks at sea as an intellectual python, digesting Gibbon's *Decline and Fall of the Roman Empire* and seventy volumes of Voltaire. On the return voyage he merely learnt German.

A tribe of new men, the mechanics, Vulcans who grew up with sparks and fire, led the escape from the oppression of distance. They made engines and machines, reshaped each other's work, and sought the mysteries and the future in their crucibles of blazing metal. Few of them sprang from the aristocracy and the educated elite. They were mostly humble and simply educated. Self-taught, observant, curious, persistent, occasionally visionary and heroic, they wrote the definitions of progress.

For most of the nineteenth century the British were better than anyone else at making iron, furnaces, steam engines, steamships and money. The spinnerets of industry cast nerves and sinews for the world, webs of girders, steel rods, cables, wires, rails and bridges. In time, the British, their European neighbours and the Americans saw their purposeful machines and tamed metal as instruments of a Western mission to bring progress to the world; and their domination of technology as the plain fact of superiority, and even divinity.

Victorians in their iron age lived with the idea of progress achieved and confidence in progress ahead. The navigators, engineers, soldiers, traders, planters and administrators formed a rolling wave of advance. Through commerce, opportunism and chance the British collected pieces of the world as if compiling a jigsaw. With their wealth and technological paramountcy so much was possible. The Americans, meanwhile, discovered, measured, annexed and bought their continent and bolted together a country as well as an ideal. The wonder of the iron age was the magnificence of human energy and endurance and the sheer scale of danger, disease, terror and failure.

Tennyson's phrase in 'Ulysses', 'To strive, to seek, to find, and not to yield', might have served as the watchword for the countless expeditions that ventured into the opaque. At the beginning of the nineteenth century vast areas of the world were still unrevealed and silent. Much of the Arctic could only be guessed at. No one had seen Antarctica. No European could know what lay in the heart of America, or beyond the fringes of Australia or where the Nile and the Niger began. Much of Africa and India were unmapped and unreported. What lay over the giant mountains was a mystery. The monsters and bunyips the map-makers drew in empty spaces might have existed.

The travellers who found the answers sometimes rode horses or camels, or paddled canoes; but mostly they walked. The wheel played little part. Even in Europe it was of limited use beyond the cities until the 1780s. Travel was almost everywhere laborious, and the age of the dashing stagecoach lay in the future. Coastal shipping was far more efficient at delivering people and goods. Only the rich could afford a carriage seat and a view over a horse's backside. At harvest time crowds of Irishmen sailed to Liverpool and walked across England to bring in the corn. Welsh drovers shod their cattle and herded them hundreds of miles to London to be made into the roast beef of old England. Every summer the 'garden girls' of Wales, barefoot to save their shoes, walked more than 200 miles to tend the fruit and flower gardens of London and Kent, and knitted as they went.

The armies of civil infantry vanished as railways, trams and buses ended the pedestrian centuries. The walking classes became virtually extinct and walking itself a matter of leisure and culture rather than work and travel. The brothers John and Robert Naylor made the first hike from John O'Groats to Land's End in 1871 and wrote a book about it. Eccentrically, they walked for enjoyment.

Yet progress was usually an inching forward, short steps rather than leaps. The nineteenth century was an age of hybrids, the old and new co-existing for many years. Steamship tonnage only outstripped that of sailing vessels in the 1880s. True, the railways quickly saw off the stagecoaches in Britain, but in the developing world of America, Australia and Africa the coaches worked into the first decades of the twentieth century. Horse traffic dominated British city streets into Edwardian times. In 1903 congested London had two motor taxis and 11,000 horse cabs, each animal dropping seven and a half tons of manure a year.

In the high latitudes in which we sailed, the Atlantic nights were short and darkness was diluted by the embers of sunset or moonlight's satin on the swells; and even when the wind was still, the sea whispered along the hull like hushed voices in a cloister. There were many times, however, when the seas rose up in surly force, like slate tips on the run. Alone on the midnight watch, when my three companions slept, I sometimes wondered what a nineteenth-century sailor would have made of us. Fools, I imagined. Seamen in the age of sail believed that anyone who went to sea for pleasure would take his holidays in hell.

I watched shearwaters swing in long parabolas, caressing the waves with their wings, and storm petrels and Arctic terns keeping faithful company. A thousand years ago the Vikings finding their way through these waters, from Norway to the Shetlands and on to Iceland and Greenland, observed the flight of such birds as part of their reading of the texts of sea and sky. They noted the direction of currents, the shape of waves, the clouds over islands, the changing colour of the sea, the green of the continental shelf giving way to the indigo of the abyssal ocean. And, of course, they looked for familiar friends among the stars.

Long before they knew the oceans and the earth, men knew the heavens. Before they had even an inkling of the existence of continents they charted the patterns of stars and wandering planets and the fiery streaks of the comets. Astronomy was the oldest science and navigation the first application of it. For hundreds of years science in general was muddled by alchemy, astrology and superstition; but navigation depended on observable certainties, the punctual phases of sun and moon and stars and the behaviour of the sea itself. It was the art of exactness even if in practice men were often lost on the ocean, hundreds of miles awry, because of the state of the sea and the weather and the huge errors they made in measuring time and angles with crude instruments on see-saw platforms. The improvement of navigation was the commanding scientific quest of the centuries.

In the history of seafaring the breakout into the Atlantic was the seminal adventure. The ocean that was for centuries forbidding and taboo, 'the green sea of darkness', became irresistible to the human mind and, in time, the chief stimulus to the construction of larger and better ships. Navigators learnt that the Atlantic's prevailing winds and currents formed an oceanic carousel turning clockwise. Steering south they found the trade winds to blow them to the winterless West Indies. Homeward bound to Europe, they harnessed themselves to the westerlies and the huge warm river of the Gulf Stream.

Like our predecessors in sail, we could not say, as we watched the land slip away and set us free, how long we would be on the ocean. Braced against the relentless rolling I read histories of Atlantic travel. Many of the people who sailed in the late sixteenth and early seventeenth centuries saw the ocean as a baptismal ordeal and their survival as the evidence of God's will. The queasy Pilgrim Fathers aboard the *Mayflower* in 1620 endured the jeering of a young sailor who promised they would die and be thrown into the grey ocean. It was he, however, who was the first to be heaved over, snuffed out by disease, and the pallid pilgrims observed the heavy splash and spoke of 'the just hand of God'. The bedraggled early migrants stumbling ashore in America looked repulsively pale, sick and hairy to the Indians who helped and fed them. The empire of British America made its tenuous handhold, dependent on Indian doles and kindness. For their part the settlers were dismayed that the Indians had nothing worth stealing.

From northern Scotland we sailed for Cape Cod and crossed the Grand Banks of Newfoundland, waters once rich and fished for centuries by Englishmen, Basques, Bretons and Spaniards. Grand Banks cod fed Europeans in their winters and was a sacred Lenten substitute for meat. It took fishermen five centuries to ransack the Grand Banks, and when the cod dwindled away in the 1990s the fishing was officially stopped. We sailed over a plundered and melancholy sea.

The Grand Banks province of the Atlantic is notorious for its brutal weather. The Gulf Stream and the chill Labrador Current conspire to brew dense billows of fog and, engulfed in an icy shroud, barely able to see the bow, we watched for the icebergs that drift down from Greenland. In the days of sail men's instincts were so sharpened that they could smell them. South of our track, on the calm night of 14 April 1912, the invincible *Titanic* raced almost at full speed towards a drifting cliff of ice; and speed was all.

We worked our way out of the comfortless fogs and down the long jagged coast of Newfoundland. Twenty-three days from Britain we made our landfall at Cape Cod, which gropes out from America like a claw. We crossed the old whalers' haven of Nantucket Sound to find an anchorage close to the pretty clapboard waterfront houses built by skippers who made their fortunes from whalebone and oil. After a week or so we returned to the ocean to find the Gulf Stream and the westerly wind. Soon we were wayfarers in wild waters, away with the dolphins and the whales, the bow a ploughshare turning the sea.

For most of history this is how it was.

1

The Copper King

'One cannot look at the sea without wishing for the wings of a swallow.'

Sir Richard Burton, 1863

Seafarers longed for speed. A slow vessel was a bane, a fast one adored. The drag created by the ribbons of weed and clusters of barnacles below the waterline made vessels slower and multiplied the perils and wretchedness of voyages. One remedy was careening, a Mediterranean practice adopted by British captains in the late sixteenth century. They sailed their ships on to suitable beaches and when the tide receded set their men to scraping off the barnacles.

As seamen ventured beyond inshore waters and into the oceans a more destructive and often deadly enemy was waiting, a worm that ate their ships. The warm-water shipworm, *Teredo navalis*, was not a worm at all, but a clam with a worm-like body. It entered ships' hulls in its brief larval stage by boring a microscopic hole and then fed itself on wood particles. Its two hard shells worked as a drill or rasp and it grew to a length of two feet or more, with a diameter of half an inch or larger, making a fatal honeycomb of ships' timbers. It also devastated the thick wooden piling of wharves and the supplies of masts and booms stored in dockyard ponds.

To fight the worm shipbuilders coated hulls below the waterline with anti-fouling mixtures such as pitch, lime, tallow and sulphur; or they nailed on laminates of canvas and thick brown paper sealed with lime. In 1553, for the first time, English shipwrights hammered wafers of lead to the keels of ships to protect them on long voyages, but captains found that these were easily torn off by the sea. Greek seamen had tried the same defence in the fourth century. More effectively, carpenters nailed a sacrificial shield of pine planks over the hull and renewed it when the teredo infested it. The *Endeavour* was protected in this fashion while being fitted out in London in 1768 for the first of James Cook's three world voyages. The planks were laid over a coating of lead and grease and secured with numerous nails, their heads the size of a large coin, beaten in close together as a form of armour.

The voracious teredo crippled the ships while scurvy and fevers crippled the men. It thrived in the Gulf of Mexico, the Caribbean and along the American coast. Captains calling at Chesapeake Bay to load Virginia tobacco learnt that they could mitigate the assaults of shipworm by anchoring where the tidal streams flowed most swiftly. Others moved out of teredo range by anchoring in the freshwater reaches of rivers. Smaller vessels were hauled ashore and out of danger during the six-week loading period.

For the Royal Navy's admirals and managers in the West Indies, whose task was to maintain a fleet strong enough to confront French and Spanish forces, the teredo was one of an array of problems. Their ships were, first of all, a long way from home, a two-month voyage from England, and beset by problems of repair and supply. It was difficult to find enough experienced sailors to make up a crew. The West Indies then had no reputation for delight. Sailors quite reasonably feared such islands as Jamaica, Antigua and Barbados as places of death. Many of their shipmates had ended up on muster lists as DD, discharged dead. Seamen were safer at sea. It was when they came ashore to refit or replenish their ships that they succumbed to the mosquito-borne diseases of malaria and yellow fever and the venereal poxes of the wharves; and there was little that eighteenth-century naval surgeons could do about them. The Admiralty in London found it hard or often impossible to meet requests from the West Indies for more sailors. Impressment, the seizing of landmen or merchant seamen, never produced the numbers or experience required and always rankled. Desertion was on a serious scale but, whatever the short-sighted disciplinarians thought, the penalties of hanging and flogging were not deterrents; and many captains took a common-sense view, preferring to have a man haul on a rope rather than hang from it. Indeed, crews were such a precious resource that officers were instructed to be reasonable and lenient with them.

As they struggled to fill the gaps left by disease and runaways, administrators received almost daily reports of ships battered and broken by bad weather and riddled with teredo. Inspections revealed timbers dangerously wormy after only a few months' immersion in West Indian waters. Many men-of-war would barely last three years. To beat the worm some of them were ordered back from the Caribbean to the cooler waters of Britain. Those that remained were careened to have their hulls scraped, repaired and pasted with a sulphur goo; but the process of hauling a ship over in a careenage was lengthy and often damaged the frames and planking.

The sheathing of hulls with thin copper sheets was first suggested to the Royal Navy in 1708 but rejected. The sheets were hand-rolled in workshops and so were expensive to manufacture. By the middle of the eighteenth century, however, the menace of the teredo had to be addressed whatever the cost. In 1761 the navy sheathed the thirty-two gun frigate HMS *Alarm* with copper sheets and dispatched her to the West Indies for two years to see what happened. On her return she docked at Woolwich and Lord Sandwich, the First Lord of the Admiralty, led the inspection of the hull and was impressed. He recommended the coppering of three more ships and King George III approved.

It became ever more clear that the growth of British colonial commerce, security and prosperity depended upon clean hulls. 'For God's sake and our country's,' pleaded a captain in a passionate letter to the Admiralty, 'send copper-bottomed ships to relieve the foul and crippled ones.' For coppered ships always outsailed their uncoppered sisters. They steered better, tacked more nimbly, turned in a third of the usual distance, were demonstrably quicker on a broad reach, the fastest point of sailing, and were livelier in light airs. A captain wrote to the Admiralty in 1780 that coppering 'exceeds the expectation of everyone. The advantage from the helm alone is immense.' French admirals watched in dismay, but digested the lesson, when they saw coppered British ships dance away from danger. Admiral Rodney was adamant that the greater speed and manoeuvrability of his coppered ships gave him victory over the French in the Battle of the Saints in 1782, and thereby saved Jamaica and its sugar and rum empire from French invasion. 'To bring the enemy to action,' he pronounced, 'copper-bottomed ships are absolutely necessary.' In other words a coppered vessel would always catch a fleeing uncoppered one.

It took dockyard workers about six weeks to sheathe a ship with copper sheets and the cost was around £600. This was expensive but the Admiralty's ever-assiduous accountants reckoned that a sheathed ship would remain at sea for long periods and spend much less time under repair in dock. Effectively, the fleet actually at sea would be a third larger than in the pre-copper era.

The American War of Independence increased the premium on the speed of ships. In 1778 the British Post Office was ordered to sheathe its Atlantic mail ships 'with all possible dispatch'. In that year the mail vessel *Hillsborough* reached nine knots, more than ten miles an hour, in a passage from Jamaica to Falmouth. Though 'she never was esteemed a remarkably fast vessel, the copper has qualified her to sail as fast as any ship'. The mail agent at Falmouth made the point that had more ships been coppered they would very likely have escaped the American and French privateers, the commerce raiders that prowled the Atlantic.

The teredo's ravages made the case for copper all the more pressing by accentuating Britain's shortage of ship timber. Ships were largely built of English oak and shipbuilders rated the trees of Sussex, Surrey, Hampshire and Kent as the finest. A single tree, a century old, yielded about one and a quarter tons of timber. Around 5,750 were felled to construct a first-rate ship of a hundred guns, the size of HMS *Victory*, Lord Nelson's flagship. Thus the forests were pillaged. Visiting the Forest of Dean in Gloucestershire in 1802, Nelson himself was appalled by what he saw and wrote to the Admiralty about the pigs which were eating all the acorns, gobbling the raw material of Britain's naval future.

Shipbuilders knew that the timbers cut from freshly felled trees quickly succumbed to dry rot and were therefore useless. Timber was best left to season in the air for some years. The oak frames and planks selected for the building of HMS *Victory* at Chatham in 1759 were seasoned for thirteen years. In the late eighteenth century the British East India Company, merchants trading in India and elsewhere in Asia, asked shipwrights in Bombay to build them ships of teak grown along the Malabar coast of western India. It was called 'Hindustan oak' and its natural oil defended the hull against worm. The Company established botanical gardens in Calcutta in 1787 specifically for growing teak for shipbuilding. Before coppering was introduced oak-built ships were usually good for only four voyages from Britain to the east, but Indian-built teak ships famously sailed for many years.

For the Romans the chief attraction of Wales was its gold, silver, lead and copper quarried by British labourers and slaves. In 1761, the year that HMS *Alarm* was sheathed, and in 1768, prospectors located copper among the old Roman diggings on Parys mountain in the heart of Anglesey. Here, as Erasmus Darwin, Charles's grandfather, ornately put it, were 'azure ores bosom'd in rock'.

Two landowners claimed rights to the mountain and both of them asked Thomas Williams, the local solicitor, to disentangle their claims. Williams was as diplomatic as he was shrewd. He became manager of both men's mines. In a few years he rose to the overlordship of what turned out to be a copper mountain at a time when pernicious worms, French hostility and British naval interests combined to make copper a most valuable strategic resource. For Thomas Williams it was better to be sitting on a hill of copper than a pile of gold.

Just as the ironmasters made their wealth from the army's need for guns and cannonballs, so Thomas Williams prospered from the Royal Navy's demand for copper. Although he had no experience of industry he soon revealed a trader's acumen and an ambition for commercial power. He rapidly became boss, or dictator as his enemies saw it, of much of the British copper trade and before long emerged as the greatest British industrial magnate of the late eighteenth century.

He was a pioneer businessman and showed how a proper tycoon should operate. He was hard-headed, made decisions quickly and did not go back on them. To build his empire he acquired mines, mills, smelters and factories in England, Scotland and Wales and travelled frequently to meet his managers and workers. He was a master of manufacture and a far-sighted organizer of transport and distribution. He sent his ore and copper ingots by sea and by the network of inland canals which transformed British transport between 1760 and 1800. Canals were far more efficient for bulk transport than roads. There was a simple formula. A large horse could carry on its back a burden of around 300 pounds. A load of more than twice that weight could be drawn if it were on a cart. If the cart were on rails this same horse could pull two tons. But a single horse could tow twenty tons of freight loaded into a canal barge.

In 1780 Williams built a copper works at Holywell in Flintshire which was 'stupendous in extent, expense and ingenuity'. The ruins are still there and still to be marvelled at. The copper rollers were turned by wheels powered by the copious waters of Holywell spring. This was for centuries, and remains today, a destination for Christian pilgrims, the place where Henry V gave thanks for his victory at Agincourt and the Gunpowder Plotters sought blessing.

Thomas Williams's smelting and manufacturing works hugely increased his power. As well as copper sheets it turned out huge quantities of brass and copper wire, pans for refining sugar and distilling rum in the West Indies and copper plates for etching and printing. Williams minted his own copper coinage to pay his men and himself amassed the then astonishing fortune of £1 million.

Contemporary industrialists regarded the 'Copper King', as he became known, with mixed admiration and awe. Matthew Boulton, the Birmingham manufacturer, wrote of him in 1785: 'He has certainly done more in the copper trade than all the other drones in it, he is taking gigantic strides.' But admiration turned to acid, especially in the years from 1787 to 1792 when Williams had the monopoly of British copper. His rivals complained of his hard-headedness. Boulton called him 'the despotic sovereign of copper'; and James Watt wrote of him as 'a perfect tyrant who will screw damned hard when he has got anybody in his vice'. In 1800 Williams's business enemies, supported by William Pitt, the Prime Minister, accused him of controlling and inflating the price of copper. Williams, by then the Member of Parliament for Marlow, Buckinghamshire, called for a parliamentary investigation. It cleared him.

Only two years after his Holywell mill started producing its shining rolls and ribbons of copper, Williams had faced disaster and the prospect of extinction when the Admiralty came close to abandoning the magic metal. Because copper nails were soft, shipwrights used hard nails and bolts of iron to fasten copper sheets to hulls. Seawater caused electrolytic rotting between the copper and the iron fastenings, and the holes exposed the timbers to the teredo. Inspectors found that an iron bolt an inch and a quarter in diameter had eroded over two years 'to the size of a quill'. Similarly, copper rudder fittings fastened with iron crumbled away. Many of these were substantial, the bolts twenty feet long and the rudder bands and braces more than a ton and a half in weight. In 1782 catastrophic corrosion was suspected in the loss of four ships and the deaths of hundreds of men off Newfoundland. It may also have been a factor in the loss in the same year of HMS *Royal George* which rolled over and sank at Spithead, the naval anchorage off Portsmouth, drowning 900 seamen and their female visitors. An examination of HMS *Victory* at that time disclosed widespread rot in the timbers beneath her corroded copper sheathing, while a naval officer inspecting the hulls of warships in 1783 found decaying planks and iron bolts reduced to 'rusty dust'. The wooden walls were crumbling. The Admiralty complained that the marriage of iron and copper had contributed to naval losses in the American war. Ships were being lost for the want of a nail.

Inevitably, a strong prejudice quickly grew up against copper, but steadier minds saw that the remedy for corrosion was to fix the sheathing with fastenings of the same metal. It was a matter of finding a way to make hard copper nails and bolts. Thomas Williams had scant technical skill himself but among his men he found two who did. In 1783 they successfully hardened copper by pressing it in cold rollers; and the manufacture began of the new nails, bolts and rudder bands, 'harder, stiffer than any iron bolts'. The relieved Admiralty ordered all ships to be fitted with them.

French naval experts visiting the Holywell works during the brief spells of peace with Britain found the nails better than anything made in France; and Williams, as hard as one of his own nails, employed travelling agents to supply his copper to the navies of France, Spain and Holland. He also pointed out, in his businesslike way, that coppered merchant ships kept their value and lasted longer. In 1799 he urged Liverpool owners to copper their ships, citing the example of 'a vessel belonging to Liverpool that was copper bolted and sheathed in 1785 ... and has sailed from thence on her sixteenth voyage to Africa, the West Indies and home'. Well into the nineteenth century shipowners made a point of emphasizing in their advertisements that not only were their vessels copper-bottomed, they were also copper-fastened.

Thomas Williams's hardened copper bolts and fittings, the nails as small as two and a half inches, were technological details that made a significant difference in the British struggle against the French threat. In the Napoleonic wars the Royal Navy's ships, with their vital skins of copper, sailed at their best possible speed, the fastest of them making around five or six knots. But, more importantly, ships were able to remain continuously at sea for many months and maintain the pressure of the blockade on French ports. This strategy prevented Napoleon's admirals gaining control of the English Channel and opening the way to invasion of Britain.

During the blockade thousands of sailors did not step ashore for almost two years between 1803 and 1805. Theirs was a unique achievement of endurance and seamanship, month upon month of tedious and exhausting patrol along notoriously dangerous coasts, from Brest to Toulon, waiting for the enemy to emerge. Nelson took pains to prevent disease and discontent among his men by seeking out traders in small Mediterranean ports who would sell him meat on the hoof, lemons and the sailors' favourite vegetable, onions. At last the exasperated Napoleon, who knew little of the sea, ordered his fleet to sail, come what may. After the arduous years of their blockade, the men of the storm-beaten ships of the Royal Navy were toughened, determined and ready to finish the matter; and this they did at Trafalgar.

Thomas Williams died in 1802, remembered with affection by the people of Anglesey, including those hundreds of men, women and children who bent to the gruelling labour of hammering his copper ore into walnut-sized pieces to prepare it for smelting. He was, for his time, a decently paternal employer and long after his death people referred to him as Twm Chwarae Teg, Tom Fair Play. He was thrice-buried, first in Bath, then in Llanidan in his native Anglesey, and finally, twenty years later, in Llandegfan church.

In the absence of his powerful personality his copper empire gradually broke up. In 1926 part of it formed the core of Imperial Chemical Industries. His money made some of his female descendants acceptable brides for dukes and earls. Sir Kyffin Williams, the artist, who has Thomas Williams on one branch of his family tree, describes entertainingly how Thomas Williams's decadent progeny frittered much of the fortune. In the new age of the railway timetable Thomas's eccentric grandson insisted on telling the time by the sun, thereby, and to his fury, missing many trains at Menai Bridge station. Between trains he fathered eight extravagant and scandalous children, two sons and six pretty daughters, one of whom, Edith, 'became shamefully involved' with the Prince of Wales and Lord Blandford. This affair led to the Prince challenging Lord Randolph Churchill to a duel. Lord Randolph was Blandford's brother and Winston Churchill's father. The whole seedy affair was firmly squashed by Queen Victoria who looked down her nose at this generation of Thomas Williams's descendants and called them a bad lot. Naughty Edith spent some of her share of the Copper King's money on a carriage drawn by four ostriches.

2

With dauntless feet

'And he said to me: "By the Arctic Sea there's a treasure to be won".'

Robert Service, 'The Ballad of the Northern Lights'

In log trading posts beside the rivers across the Canadian wilderness, where hunters and trappers gathered to deal in furs and buy supplies, Indians told tales of wealth and saw a gleam in the eyes of listening white men. In the distant north, they said, there was a river that no European had ever visited, and its rocks were rich in copper. For proof, they showed the copper knives and ornaments made by the Indians who travelled and traded beyond the Arctic rim. In 1767 two Chippewa Indians arrived in Prince of Wales fort, the Hudson's Bay Company base on the western shore of the great bay, and sought an interview with the commander. They had just walked and canoed from the far north and they unrolled a deerskin on which they had drawn a map showing the river where copper deposits lay. In the following year the commander sailed for London to meet the Company's directors. He showed them the Indians' skin chart. With its crude lines, its promise of treasure beyond the tundra and its authentic sniff of the north, it was evocative indeed.

The Hudson's Bay Company, more formally known as the Honourable Company of Adventurers of England Trading Into Hudson's Bay, was founded in 1670 to develop and administer the fur trade in Canada. King Charles II granted a charter to his cousin, Prince Rupert, chief of a coterie of eighteen directors who were declared the 'absolute lords and proprietors' of Rupert's Land, a vastness of forests, rivers and lakes. In time this barely explored northern territory grew by acquisition to cover two-fifths of modern Canada, a twelfth of the world's land. It stretched from Labrador to the Rocky Mountains and incorporated northern Quebec, most of Ontario, Manitoba, Saskatchewan, large tracts of Alberta and the North West Territory, and some of North Dakota and Minnesota.

Its bleakness commended it to some in Britain as an ideal penal colony: there were always those who sought a moon of banishment, and it was suggested that the more hardened criminals should toil as the slaves of Eskimos. The land was only sparsely peopled but had an abundance of valuable fur-bearing animals, bear, muskrat, lynx, wolf, mink, marten, otter, red, silver and white fox and, the monarch of them all, the personable and ingenious beaver. This builder of dams and river lodges was the heart of the early economy and history of Canada, and for that reason has a place in the country's coat of arms. Beaver pelts, called 'plews', were the finest possible material for hats. From the seventeenth century into the nineteenth a hat of beaver felt was a practical and fashionable necessity, and often expensive. The Morgan family tomb in Skenfrith church in Monmouthshire makes the point: on one side of it are carved relief images of four sons, wearing doublets and hose, and the picture is completed by separate carvings of beaver hats, status symbols showing that in life the boys had done well; a posthumous swank.

The Company formed its own empire. Its charter allowed it everything it could possibly desire, 'the whole trade of all those seas, straits, and bays, rivers, lakes, creeks and sounds . . . that lie within the entrance of Hudson's Straits'. It was run by a secretive and privileged circle, its shareholders anonymous, even its Governor's name discreetly concealed. It was almost religiously devoted to profit, ruled by the stern regulations of thrift and untainted by benevolence.

From its Canadian headquarters on the St Lawrence River near Montreal the Company cast a web of trading posts and forts across the north. It put much of the management of them into the hands of men from Orkney, as rugged as their native rock. This was a tradition and a convenience. The Company ships, sailing from London to Hudson Bay, always stopped for water at Stromness in the Orkneys, and while filling the barrels the captains hired likely young Orcadians happy to commit themselves to an existence of long exile in pitiless winters. Determinedly self-sufficient, these recruits found wives among the Indians and the people of mixed European and Indian blood with whom they traded. During the nineteenth century the Company was so rooted in Canada's history, so old indeed, that its initials HBC were said to stand for Here Before Christ. Just as the Romans counted time from the founding of their city-state, so the Company dated its calendar from the year of the royal charter: 1670 was Year One. Through its sheer reach and authority the Company shaped the politics, character and boundaries of Canada and built a bulwark against the probing and elbowing of American colonizers.

In May each year the Company's ships made a three-month voyage across the North Atlantic to Five Fathom Hole in Hudson Bay. Almost as large as the Mediterranean, its shoreline curving in a horseshoe of more than 7,500 miles, the bay was usually navigable for three or four months, from July to October. The ships unloaded guns, axes, knives, iron, liquor and trading goods and filled their holds with the pungent bales of pelts for the warehouses in London.

King Charles's charter obliged the Hudson's Bay Company not only to trade in furs but also to send men to find the North West Passage, the icy grail of northern discovery. The Passage existed in men's imaginations as a short cut from the Atlantic to the Pacific, an easy avenue to the wealth of India and the East Indies, and especially of China, poetically known as Cathay. The quest for the Passage had become pressing after the overland spice roads from the East to Europe fell into the hands of the Ottoman Turks who seized Constantinople in 1453 and threatened the supplies of pepper, cinnamon, nutmeg and cloves that preserved and flavoured the Europeans' winter meat and made life more enjoyable. The only ocean routes to the East, by way of the Cape of Good Hope and Cape Horn, were hazardous and extremely long; and in any case they were for years controlled by Spain and Portugal, for in 1494 the world had been divided by Pope Alexander VI into zones of Spanish and Portuguese influence.

Merchants and mariners believed that the North West Passage would make dukes of its discoverers. The challenge was to find the zigzag way through the monstrous jigsaw of islands along the northern rim of Canada; or to discover a river on which a ship could sail clear across North America to the far ocean. The possibility of sailing to the East across the top of the world grew into a peculiarly British passion and torment. In 1745 an Act of Parliament offered a reward of £20,000 to the explorer who could unravel the enigma. For more than three centuries it was a siren that drew men to the ice and then killed them, one of the grand delusions of exploration.

By 1769 critics of the Company of Adventurers were drumming their fingers and complaining that while the 'absolute lords' were profiting from the beaver they had done little to meet their charter obligations to find a passage through the Arctic labyrinth.

Enter Samuel Hearne, an ambitious sailor aged twenty-four. Born in London, he entered the Royal Navy as a captain's servant at the age of eleven and, suitably toughened and educated, joined the Hudson's Bay Company as mate of a whaling sloop; and there could be no harsher life than hunting whales in the Arctic. Anxious to make a reputation as an explorer, he petitioned his masters for an assignment. He was just the man they were looking for. They commanded him to walk north-west from Prince of Wales fort, the most northerly British outpost in Canada, cross the scrub, tundra and permafrost of the region which would later be called the Barrens, and strike out for the shores of the Arctic Ocean. He was to see if a North West Passage led westward out of Hudson Bay and to find and evaluate the copper-bearing rocks in the Coppermine River described by the Indians. He would have to make a round trip of more than 3,500 miles over country no European had ever travelled. He prepared for the journey by studying navigation at the fort with the amiable William Wales, an astronomer deeply involved in the problem of fixing longitude. Wales was there to observe the transit of Venus across the sun, a counterpart to the observation of the transit being made by Captain Cook in Tahiti.

On 6 November 1769 a seven-gun salute cracked in the freezing air as Hearne, with two white men and two Cree Indians, marched out of the fort. Three weeks and 200 miles later, Indians stole the party's muskets and food and ran off laughing. Hearne and his men survived the journey back by trapping rabbits when the going was good, and chewing their hide coats for nourishment when it was bad. He had seen famished Indians eat their clothing: 'sometimes a piece of an old half-rotten deerskin or a pair of old shoes were sacrificed to alleviate extreme hunger'.

Hearne rested at the fort for two months and set out once more, this time with an Indian guide who claimed to know the river where the copper lay. Six months later Indians robbed Hearne of his tent, coat and gun, and his guide fled. In a few days he began to suffer the agonies of hunger: 'For want of action the stomach so far loses its digestive powers that it resumes its office with pain and reluctance. Another disagreeable circumstance of long fasting is the extreme difficulty and pain attending the natural evacuations.'

Now, far from the fort, without shelter, alone and with little hope of finding food, he faced a freezing death. His rescue was a true story of the north, as utterly astonishing as it was melodramatic. A party of Chippewa, led by their chief, Matonabbee, found the desperate Hearne and gave him food and a suit of otter skins. They fashioned saplings into snowshoes for him and guided him back to Prince of Wales fort. Matonabbee, a striking man six feet tall, had been born in the fort and knew English. Indeed, he had been one of the Indians who had travelled to the distant north and had drawn the deerskin map that had so roused the excitement of the Hudson's Bay Company. As Hearne and Matonabbee snowshoed towards the fort, a rapport grew between them, a mutual affection that bridged their cultures and seemed to make a link between the world of the modern European and what white men saw as the realm of the primitive. To the grateful Hearne Matonabbee appeared a noble man, 'the most sociable, kind and sensible Indian' he had ever met, a paragon with elegant table manners and 'the vivacity of a Frenchman, the sincerity of an Englishman and the gravity and nobleness of a Turk'. He was also 'remarkably fond of Spanish wines, though he never drank to excess'.

Matonabbee told Hearne that a man had no hope of getting far if he made the fundamental mistake of journeying without women.

Success in northern travel, he pointed out, depended on them. Women were made for labour, were inured to it from childhood, and one woman could lift as much as two men. They carried tent poles, skins and other baggage, in summer shouldering loads of eight or ten stone, more than a hundred pounds, and in winter even heavier burdens. 'They pitch our tents,' Matonabbee said, 'make our clothing and keep us warm at nights. There is no such thing as travelling any distance without them.' Also, Matonabbee added, wives were cheap to maintain and in hard times needed little food, 'for as they always cook, the very licking of their fingers in scarce times is sufficient for their subsistence'.

Hearne had been more than eight fruitless months on his second thrust into the north but he recuperated in the fort for only twelve days before setting out once more, in December 1770. This time, as he headed into the spongy tundra and wind-scoured snowfields, he knew exactly what to do. He put himself into the hands of Indians led by his new friend Matonabbee, who had a platoon of wives. He committed himself to living off the land, depending for food on berries and roots and on the wandering herds of caribou which migrated northward every spring and summer, and were for native people the difference between life and death. Hearne took very little: a gun, a quadrant to fix his position, some tobacco, a coat, a spare pair of underdrawers and his journal.

Hearne's companions inspected him closely. To their eyes he looked a strange creature. His hair, they told him, was like a stained buffalo tail, his pale eyes resembled a gull's and his face seemed like meat soaked in water and drained of blood. 'I was viewed as so great a curiosity that whenever I combed my head some of them never failed to ask for the hairs that came off which they carefully wrapped.' In turn he looked at his companions and wrote a description of them: they were strong and robust, their skins smooth and polished and free of smell, the men mostly beardless and neither men nor women having much body hair.

On freezing nights in the whirling snow they all huddled together like puppies for warmth, on spruce-bough beds if they were lucky, the bitter wind tearing like a wild animal at their tents.

The party varied in numbers as Indian families joined and left, but sometimes it was sixty strong and required large supplies of food. The women butchered and cooked the animals the men hunted, but did not themselves eat until all the men were sated. 'In times of scarcity,' wrote Hearne, 'it is frequently their lot to be left without a single morsel.' From time to time, he observed, the men stuffed themselves until they fell writhing, clutching their distended stomachs in pain.

Hearne took to the hunter's diet; not that he had much choice. He enjoyed the raw brains of animals and the freshly plucked foetuses of deer and beaver, 'the greatest dainties that can be eaten'. The sex organs of male animals were a men-only dish but both men and women were 'remarkably fond of the wombs of buffalo and deer'. They all ate kidneys warm, straight from the animal, even if it were still alive, and sucked fresh blood from the hole made by a musket ball. The vitamins in raw food helped them to avoid scurvy. A favourite meal was the half-digested and fermented contents of a caribou stomach, particularly tasty if the animal had been feeding on lichens. Into the stomach the Indians poured blood and also threw lumps of fat which had first been chewed into tenderness by men and boys. 'To do justice to their cleanliness I must observe that they are very careful that neither old people with bad teeth nor young children have any hand in preparing this dish.' The stomach bag was smoked over a fire and when cooked gave off steam 'in the same manner as a joint of meat, which is as much as to say, Come, eat me now'. He thought the dish 'exceedingly good, with an agreeable acid taste, that were it not for prejudice, might be eaten by those who have the nicest palates'.

Hearne declared that 'whoever wishes to know what is good must live with the Indians'. He liked the strips of meat and fish set out on top of the women's bundles to dry in the sun on the march. He shrank, however, from joining his companions when they snacked on the flies and lice they picked from their clothes and heads. Matonabbee, he saw, was so 'fond of those little vermin he set five or six of his wives to work to louse their deerskin shirts, the produce of which he eagerly received with both hands and licked them in fast'. Clothing made of deerskin provided delicacies other than lice. It was also a source of worms and warble flies, the latter as large as the joint of the little finger. 'They are said to be as fine as gooseberries,' Hearne reported. 'When I acknowledge that the warbles and lice are the only things of which I could not partake I trust I shall not be reckoned over-delicate.' He himself once kept going through three foodless days on a pipeful of tobacco.

As they trekked north Matonabbee indulged a whim and bought another wife from some passing Indians. He already had seven, 'most of whom', in Hearne's opinion, 'would for size have made good grenadiers'. Matonabbee took pride in their strength and stamina. A woman who gave birth on the trail soon slung her baby on her back, took up her burden of skins and poles and kept marching. A sick woman, like an ailing man, would be left by the trail with a deerskin cloak and some food, and the rest of the group would 'take their leave and walk away crying'. The hard lives of Indian women made 'the most beautiful among them look old before thirty', thought Hearne, 'and several of the ordinary ones are perfect antidotes to love and gallantry'. They were, nevertheless, desirable for all sorts of reasons. An Indian would take another man's wife simply by wrestling for her in a stag-like brawl. On the other hand it was common for men 'to exchange a night's lodging with each other's wives . . . esteemed as one of the strongest ties of friendship between two families; and in the case of the death of either man the other considers himself bound to support the children of the deceased'.

One of the skills a man prized in a woman was her ability to cure and shape skins and make fur clothing. The Indians prepared for winter by killing deer with spears and bows and arrows in late summer before the hair grew so long that it would drop off. Hearne recorded that a warm winter suit for one person required eight to ten deerskins; and each individual also needed another ten skins to make leather stockings, shoes and lengths of rawhide, a total of about twenty skins for each person each year.

However hard the going over ice and swamp, and through stormy snowy days when there was nothing to eat, Hearne turned with a remarkable discipline to his journal. It was a companion to him; and we can imagine how curious he would have seemed to the Indians as he hunched himself over the pages in his squaw-made deerskin robe and scratched and scribbled away with half-frozen fingers. He was a born reporter and never failed to make detailed notes of everything he saw, be it vegetation, animals or birds. He did not, for example, omit to write a description of the dung of the musk-ox, that shaggy and bulky relic of the last Ice Age. He composed a minutely observed account of the intercourse of porcupines, recording that 'their mode of copulation is singular, for their quills will not permit them to perform that office in the usual mode. To remedy this inconvenience, the usual mode is for the male to lie on his back, and the female to walk over him beginning at the head till the parts of generation come into contact.'

He described how his companions fashioned canoes from birch bark and saplings, how they made sledges and how they foraged and hunted. He watched them kill moose when the animals were at their most vulnerable, crossing water, and saw two boys stalk one, killing it 'by forcing a stick up its fundament when they had neither gun nor bow with them'. Moose nose and tongue, he added, were 'most excellent' to eat. In the dark winter days when the pale sun barely rose halfway up the trees he observed how 'the brilliance of the aurora borealis and of the stars made some amends, for it was frequently so light all night that I could see to read very small print. In still nights I have frequently heard it make a rustling and cracking noise like a large flag in a fresh gale of wind.'

To his observant gaze and open mind the habits, spirituality, beliefs and techniques of the Indians were of constant interest. He looked on as medicine men treated the sick.

> They use no medicine in ordinary cases, sucking the part affected, blowing and singing to it, hawking, spitting and uttering a heap of unintelligible jargon. For some inward complaints, griping of the intestines, difficulty of making water, &c., it is very common to see those jugglers [by which he meant the medicine men] blowing into the anus, or the parts adjacent, till their eyes are almost starting out of their head: and this operation is performed indifferently on all, without regard to age or sex. The accumulation of so large a quantity of wind is at times apt to occasion some extraordinary emotions, which are not easily suppressed by a

> sick person; and as there is no vent for it but by the channel through which it was conveyed, it sometimes occasions an odd scene between the doctor and his patient . . . they frequently continue their windy process so long that I have seen the doctor quit his patient with his face in a very disagreeable condition. However laughable this may appear to a European, custom makes it very indecent to turn anything of the kind to ridicule.

Hearne and Matonabbee's people walked to the 300-mile-long Great Slave Lake and headed north to the Coppermine River. The Indians prepared enthusiastically for a clash with the Eskimos, their traditional enemies, and painted their faces red or black and cut their hair short. Hearne himself, 'fearing I might have to run', took off his cap and stockings and tied up his hair. He was a horrified witness when his companions fell upon an encampment of twenty Eskimos at a place known ever since by the name Hearne gave to it: Bloody Fall. The Indians speared the Eskimos one by one and Hearne possibly embellished the story for publication; certainly he spared no horrible detail of the skewering and squirming of the victims.

In July 1771 the party arrived at the mouth of the Coppermine and Hearne stood on the shore of the Arctic Ocean. He searched for a few hours in the river valley for the legendary deposits of copper ore, but found only 'a jumble of rocks and gravel', and some evidence of Indian mining. There were no large deposits of copper, no fortune for the Company. He also concluded, rightly, having crossed no salt-water stream on his journey, that no North West Passage issued from Hudson Bay. When the party turned and began the long trek back 'we travelled so hard that my feet and legs swelled, several toenails dropped off, the skin chafed off the tops of my feet . . . I left the print of blood almost at every step I took'.

On the return journey they came across a young Indian woman who had been kidnapped by another tribe, had escaped and had lived for seven months on her own by snaring squirrels and partridge. She was good-looking and the men in Matonabbee's party began to wrestle each other for the right to take her as a wife. Matonabbee thought he would wrestle, too, but one of his harem joked mischievously but unwisely that 'he had more wives than he could properly attend'. Hearne, who had written admiringly of the 'benevolence and universal humanity' of his noble friend, now saw his savagery. The wife's 'piece of satire proved fatal, for the great man Matonabbee, who would willingly have been thought equal to eight or ten men in every respect, took it as such an affront that he fell on her with both hands and feet and bruised her to such a degree that she died'.

It took eleven months for the party to walk, snowshoe, sledge and canoe back to Prince of Wales fort by way of the Great Slave Lake and the Slave River. They arrived on the last day of June 1772, after eighteen months and twenty-three days away. Hearne had won his place in history, the first European to travel overland to the Arctic Ocean. It was a northern epic. Much of the land he traversed would not be penetrated again until the 1890s.

Three years later, in 1775, when Hearne was thirty, the Hudson's Bay Company rewarded him with the governorship of Prince of Wales fort. He expanded the fur trade with Indian trappers, kept beavers as household pets and set up home with a gentle half-Indian girl of sixteen. This agreeable life was abruptly disturbed in 1782 when French warships sailed into Hudson Bay and a party of soldiers seized the fort, taking the furs and blowing up the ramparts. Hearne's charming young mistress fled into the wilds where she eventually died. His old friend Matonabbee, his morale broken by the destruction of the fort and the belief that the French would murder him, hanged himself. The French soldiers confiscated Hearne's journal, the account of his northern journeys, and gave it to their commander, Jean-François de Galoup, Comte de La Perouse, a navigator and admirer of James Cook. Within three years he would be leading a scientific expedition in the Pacific. He read Hearne's journal with approval and recognized it as a masterpiece of Arctic travel. He was struck by the author's humanity which, he may have considered, was much like his own: he would later be noted for his kindly dealings with Pacific islanders. The perfect French gentleman, he returned the journal to Hearne, urging him to publish it.

Hearne was forty-two when he sailed for England from Hudson Bay in 1787. He spent his last years in Red Lion Square in London and continued to enjoy his fish cooked in the way he had enjoyed in Canada, barely scorched and almost raw. He knew how to survive on the roughest of frontiers but in London he seemed to be out of his element. Perhaps because he felt a misfit he drank heavily. A publisher gave him a generous advance of £200 for his book but he never saw it in print. He had been dead three years when it was published in 1795.

In the nineteenth and twentieth centuries many European travellers in the world's wild places, certain of their higher cultural rank, despised indigenous skills and customs and died, hungry or thirsty, frozen or baked, while around them people survived. Hearne was broad-minded and shrewd enough to see that in crucial matters of diet, clothing and adaptation to severe conditions the people of the land knew how to live. Their native lakes and forests did not seem terrible to them. Hearne himself did not regard the land as a foe, nor did he consider his companions primitive and inferior. He demonstrated that one of the keys to survival was humility.

3

Elephants for want of towns

'Tales, marvellous tales
Of ships and stars and isles where good men rest . . .'
James Elroy Flecker, 'The Golden Journey to Samarkand'

As Samuel Hearne demonstrated, a journey needed a journal. An explorer without words was merely a man with scratches and sore feet. The tracks that travellers etched in distant snows, and the wakes inscribed by sailors on the sea, were only the overture to discovery. Those who vanished over the horizon and entered the voids coloured by legend and imagination had a duty to respond to the hunger for information. A raconteur was all very well; serious-minded scholars, navigators, merchants and publishers demanded something more satisfying: detailed notes and maps and drawings bearing the unmistakable watermark of truth.

'Take with you paper and ink and keep a continual journal of remembrance day by day.' Thomas Randolfe's instruction to men about to journey beyond the known world went to the point of exploration. As Queen Elizabeth's ambassador to the Russian court in 1588, he authorized three Englishmen to explore the seas north of Russia and ordered them to commit to memory everything they saw, 'not forgetting or omitting to write it'.

Note it, he added, by way of emphasis, 'that it may be showed and read at your return'.

Samuel Johnson underlined the desire for facts. 'I can only wish for information,' he said in a letter to Warren Hastings, the Governor of Bengal, in 1774. He urged him to encourage an era of research and scholarship in India, 'to inquire into many subjects of which the European world thinks not at all, or thinks with deficient intelligence and uncertain conjecture . . . our books are filled, I fear, with conjecture about things which an Indian peasant knows by his senses'.

Similarly, in 1803, Thomas Jefferson, President of the United States, instructed Meriwether Lewis and William Clark, on the eve of their departure for the American West, lands unpenetrated by any white man, to make careful records of all they experienced. He wanted their evidence undecorated and warned them not to allow their imaginations to fill the mysterious spaces.

The first great European travel journal and the influential impulse of the European centuries of exploration was Marco Polo's account of his journeys in the East. It burst like sudden sunlight in the minds of those who read it. 'I do give you all to understand', the author wrote by way of modest introduction, 'that since the birth of our Saviour and Lord Jesus Christ there has been no man, Christian or heathen, that has come to the knowledge and sight of so many diverse, marvellous and strange things as I have seen and heard . . .'

It was no doubt true. In the quarter of a century between 1271 and 1295, in the company of his father, Nicolo, and his uncle Maffeo, Marco Polo saw a remarkable number of cities and countries. Nicolo and Maffeo had already made a nine-year trading journey to Cathay and had been welcomed by the emperor, Kublai Khan, grandson of Genghis Khan, the Mongol conqueror of much of Asia. Marco Polo was seventeen when he set off with Nicolo and Maffeo. Travelling east from their native Venice they crossed Persia, Armenia, Afghanistan, the Pamir mountains and Kashmir and walked and rode through the Gobi desert into China to reach the court of Kublai Khan in Peking, modern Beijing. The emperor took to Marco Polo and gave him the roving commission that enabled him to travel widely and to regale Kublai Khan with stories of the amazing things he had seen. The three Venetians stayed in China for twenty years before making their way home through Java, Sumatra, Ceylon, the Andaman Islands and along the Indian coast. Marco Polo had seen more of the world than any other man and his reports were so fantastic and ranged across such distances that people called him 'Marco of the millions'. He dictated his recollections to a writer of romances in Pisa; and these accounts circulated in Latin, Italian and French manuscripts and were read throughout Europe. They were first published as a book in 1477, in German. In 1579 John Frampton translated them from Spanish and published them as *The Most Noble and Famous Travels of Marco Polo*, 'most necessary for all sorts of persons, and especially for travellers'.

Its readers fell into a wonderland: of amazing cities, diamond mines, pearl fisheries, harems and beautiful women, rhinoceroses, polar bears, spice gardens, sparkling islands and mountains of lapis lazuli. Marco Polo investigated the story of the dried pygmies exported from India as a delicacy and found it 'an idle tale', for they were monkeys preserved in camphor. It was sometimes hard to tell exactly where he had travelled. The stories had gaps and confusing passages and contained much hearsay (indeed some scholars doubt that he went to China at all). But all of his narrative was electric, compelling and kaleidoscopic. It gave men dreams to dream.

The book of Marco Polo's travels contained no map, but maps in the fifteenth century were in any case few and consisted of limited and speculative drawings of perhaps a quarter of the world, all that was known. No Roman map survives. The Greek geographer Strabo envisaged the inhabited world as a northern hemisphere island. In the second century AD the Graeco-Egyptian astronomer Claudius Ptolemy, a citizen of Alexandria, a renowned centre of scientific study, compiled a world gazetteer and a collection of maps. His ideas were exciting because he allowed the possibility of inhabited lands as yet unexplored. He pioneered the mapping grid and confronted the problem of how to render the spherical earth on flat paper. Trawling widely for information he incorporated the reports of sailors and traders as well as his own observations. He set down the names and geographical coordinates of 8,000 places and included the Mediterranean and parts of India, China, Africa and south-east Asia. He also marked the British islands which apparently had been circumnavigated by the Greek mariner Pytheas who sailed from Marseilles in about 320 BC.

Pytheas may also have visited Shetland and Norway. He saw the midnight sun, the darkless days of the far north's high summer, and the pack ice he picturesquely called the curdled sea. His accounts introduced into European imagination the white north, the Arctic, which took its name from the nymph changed into a bear and placed in the sky as Arctos, the constellation of the Great Bear.

Ptolemy's works vanished from Western view for more than a thousand years. Christian mobs burnt Alexandria's library in the fourth century, but Ptolemy's outstanding book *Geography* was saved. It was carried to Damascus and Baghdad and taken through the Islamic world to Moorish Spain and thence to Florence early in the fifteenth century where it was translated into Latin. Scholars reconstructed Ptolemy's maps from his writings and published them in Italy and Germany in 1477, the year that Marco Polo's journeys were first available in book form. Although the maps were a view from the Mediterranean of classical times, the vision of a scholar who had been dead for 1,300 years, they were nevertheless the most influential idea of what the world looked like, the beginning of modern cartography.

Map-makers used information supplied by seafarers and overland travellers. They often drew on Marco Polo's luminous reports of the splendours of China and its neighbours. To fill some of the gaps they borrowed from the legend of Prester John, the mighty Christian king who supposedly ruled a twelfth-century Asian realm rich in sapphires and free of snakes, scorpions and noisy frogs. And they dipped their buckets into the entertaining fictional well of the travels of 'Sir John Mandeville', first written in French in the 1360s. From these sources and their imaginations they coloured their empty spaces with pictures of kings and palaces, prancing birds, demons, dragons and monsters.

In the fifteenth century China and the East Indies seemed to Europeans not only magical but almost unattainable. Marco Polo had journeyed overland to the East; but the growth of the Ottoman empire's power in Turkey and the Middle East raised a barrier to easy European access by land. The question of whether the spices and treasure of the East could be reached by sea became a pressing one. John Frampton had expressed the hope that his translation of Marco Polo 'might give great light to our seamen if ever this nation chanced to find a passage out of the frozen zone to the South Seas'.

Ptolemy's authority as an astronomer ensured the survival into the sixteenth century of his belief in the idea that the sun revolves around the earth. His opinion that the earth was rather smaller than it really is, about three-quarters of its true size, also complicated the endeavours of the first Atlantic navigators. Certainly Christopher Columbus, of Genoa, was influenced by Ptolemy's theory. He reasoned that the Atlantic was a fairly narrow sea that could be crossed in a few days. He proposed to make a landfall in the Indies and find his way to the gate of the earthly paradise which, many believed, lay just beyond Asia.

Columbus took years to win backing for this idea but at last he persuaded the King and Queen of Spain to sponsor him. After thirty-four days' sailing he reached the Bahamas on 12 October 1492 and then found Cuba and Santo Domingo which he thought were stops on the way to paradise. In his cabin he had a Latin edition of Marco Polo which he frequently consulted and annotated. Sailing through the Caribbean, he believed he was in the very Indies the book described, although he could not find the promised wealth. Making landfall in Cuba he was convinced he had reached Cipangu, or Japan. He made three more Atlantic voyages in the service of Spain, in 1493, 1498 and 1502, and believed to the end of his life that the places he visited, in the Caribbean, Venezuela and Central America, were the spice islands of the East. He thought he had sailed close to the Ganges and had no conception of an American continent. The significance of his voyages was that for all his mistaken geography he joined the Old World irrevocably to the New. He sailed into the unknown convinced he would find something of importance; and after his visits the New World was permanently on the map. John Cabot, an Italian living in England, was also ignorant of the existence of an American continent. He sailed from Bristol in 1497, saw Newfoundland and landed, perhaps, in Nova Scotia, thinking he had reached Asia.

The year 1498 was an extraordinary one for exploration. The Portuguese Vasco da Gama found the sea route to India. He rounded the Cape of Good Hope, which was given that name because it promised to mark the way to the treasures and bounty of India, and rode with the monsoon to Kerala, on India's southern coast. There he bought a cargo of the pepper, cloves and ginger that Europe craved. As a banner-bearer for Christianity, however, he was dismayed to find that St Thomas, Doubting Thomas, had beaten him to it by fourteen centuries and that there already existed a Christian community which wanted nothing to do with the Pope. Columbus made his third Atlantic voyage in 1498 and John Cabot set out from Bristol on his second and was never seen again.

Amerigo Vespucci, a Florentine banker, crossed the Atlantic in 1499 and deduced that Columbus was mistaken in his belief that the Caribbean was the island-strewn edge of the Indies. Vespucci thought it was the gateway to a New World. He went to Brazil in 1501, the year after it was discovered by a Portuguese fleet, and described its people as innocent, naked, clean and beautiful, free of greed and property, living in a fertile and flowery land, and enjoying life to the age of 130. So much did the women enjoy making love, he told his readers, that they applied biting insects to their men's penises to make them swell.

In 1507 Martin Waldseemuller, a German priest and geographer, drew a map eight and a half feet wide on which he inscribed the word America, as a salute to Amerigo Vespucci whose book about Brazil he knew well. Richard Ameryk has a few champions in the eponym argument: he was King Henry VII's customs officer in Bristol and a backer of John Cabot's first voyage to Newfoundland in 1497. But Waldseemuller was clear that 'Americus Vespucci' was the man for posterity and he feminized Americus to America for consistency's sake because Asia and Europa were goddesses. Waldseemuller was the first to use the term America and his map, which bore a small portrait of Vespucci, has been picturesquely called America's birth certificate. Curiously, the map indicated an ocean west of the American continent, although at that time two key events concerning the Pacific lay in the future: Vasco Nunez de Balboa's march over the isthmus of Panama and his sighting of the Pacific in 1513; and Ferdinand Magellan's pioneering Pacific voyage in 1520.

A Portuguese in the service of Spain, Magellan believed that a passage could be found around the tip of South America that would open a route from the Atlantic to Asia, and especially to the 'nutmeg rocks and isles of cloves'. He had no way of knowing the breadth of the ocean to which he gave the name Pacific. It was, as one of his biographers wrote, 'a sea so vast that the human mind can scarcely grasp it'. Its immensity was an important finding of Magellan's voyage of 1519–22; and although it was by then well known that the earth was a globe, the circumnavigation settled any doubts. Magellan's ships, with one Englishman, Master Andrew from Bristol, aboard, took nearly four months to traverse what seemed an ocean empty of land. In that time their crews sighted just two uninhabited islands which they called the Unfortunate Isles. 'Had not God given us so good weather,' wrote one of the men, 'we would have all died of hunger in the exceeding vast sea.' Magellan found the Philippines and was killed there by natives. Few of the 280 men who set out with him reached Spain: some accounts say thirty-five survived, some eighteen. 'I believe', concluded one of them, 'that no such voyage will ever be made again.'

In 1569 Gerhard Kremer, born in Antwerp of German parents, who had followed the habit of Renaissance scholars and latinized his name to Gerardus Mercator, published the map that changed the way the world was perceived. Hitherto, map-makers had approached the difficulty of representing the world's spherical shape by drawing it as two circles or as an oval or a heart. Mercator's projection, familiar to us today, flattened the sphere by stretching the polar regions and exaggerating the area of land masses in the north and south, so much so that Greenland looks the same size as India when, in reality, it is half its area. The map's significance for navigators lay in Mercator's drawing of the parallels of latitude and longitude as straight lines crossing at right angles. This was a gift for the ocean sailor. A line drawn on the Mercator chart is a line of constant compass bearing, much easier for the mariner to follow, but a concept too radical for Mercator's contemporaries. Its scientific approach was not appreciated for more than a century and the French were the first to overcome a sailorly conservatism. 'I wish all seamen', said Admiral Sir John

Narborough in the 1680s, 'would sail by Mercator's chart, which is according to the truth of navigation.'

Mercator illustrated the title page of his volume of maps with a picture of the giant Atlas, one of the Titans of Greek mythology, supporting the heavens on his shoulders, as he was condemned to do by Zeus. Thereafter any book of maps was an atlas. Scientific as he was, Mercator could not resist dipping his pen into Marco Polo's coloured inkpots when he came to fill the blank spaces. Indeed the way in which map-makers continued to tantalize by embellishing their work with imaginary islands and fanciful pictures led Jonathan Swift to comment tartly in 1733:

> So Geographers in Afric-Maps
> With Savage-Pictures fill their Gaps;
> And o'er unhabitable Downs
> Place Elephants for want of Towns.

Publishers did not eradicate monsters and fairy-story kingdoms from the map of the world until the late eighteenth century, and it was well into the nineteenth before explorers laid to rest the riddles surrounding the sources of great rivers and the existence of inland seas. The fact that Greenland is an island was not proved until 1892. No man set foot on the Antarctic continent until 1895. It was not until the twentieth century that expeditions reached such romantic targets as the North and South Poles. Conquest of the highest mountains, like Everest, which for many years seemed unassailable, was not achieved until the second half of the twentieth century. Until the middle of the 1990s more than a million square miles of Antarctica were a cartographic blank, and satellite mapping of some of the interior and the coast was not completed until early in the twenty-first century. Mysteries, fortunately, remain, and, like the yeti, stimulate and entertain. In a nod to the romantic beasts drawn by the old map-makers, whimsical tourist maps still show Loch Ness inhabited by a monster.

4

The appointment with Venus

'I am tormented with an everlasting itch for things remote.
I love to sail forbidden seas and land on barbarous coasts.'
Herman Melville, *Moby Dick*

More than two centuries after Magellan put the Pacific into the realm of knowledge, it remained largely unknown. A few British captains sailed there to rob Spanish imperial outposts of their gold and silver. In his circumnavigation between 1577 and 1580, Francis Drake seized Spanish treasure on the Pacific coast of South America and paid his London backers 4,700 per cent profit. George Anson sailed for the Pacific in 1740 as an obscure naval captain and returned a celebrity, knee-deep in treasure. He paraded his sensational loot through the streets of London in thirty-two squeaking wagons heaped with silver and pieces of eight, preceded by a band of drums and trumpets; and a newspaper crowed that Britain was mightier than ancient Rome.

Such voyages provided no answers to the pressing intellectual questions. In the middle of the eighteenth century the Pacific, covering a third of the earth's surface and speckled with 20,000 islands, was still a blank. With America being steadily explored, European attention turned to another new world, far away in the South Seas.

To the discoursing classes interested in science and geography the Pacific became the quest and question of the time. How large was it, what sort of people lived there and, in particular, where exactly was Terra Australis Incognita, the great land that supposedly lay in the south of the ocean? From ancient times cosmographers insisted that since the world was spherical the continents of the northern hemisphere had necessarily to be balanced by a southern counterweight. Mercator, among others, showed a southern continent spread over the bottom of the world and extending almost to Africa and South America. It nagged at the imagination, like the puzzle of the lost civilization of Atlantis in the Atlantic Ocean. It was said to be 5,000 miles broad and the home of millions of tall and handsome people. Some speculated that parts of Australia and New Zealand sighted by early navigators were, possibly, northerly capes of Terra Australis Incognita.

After the crucial Seven Years' War of 1756–63, in which Britain and France fought each other in India, Canada and the West Indies, Britain emerged as the leading world power. But rivalry persisted. Among other things it sharpened both British and French appetites for influence and knowledge in the Pacific. The impulse was inquiring, strategic and predatory. There was always the belief that newly found lands would yield such commercial prizes as cotton, spices, precious stones and timber. But whatever lay at the far side of the world, Britain had to respond to French pressure and assert its own power and reach at sea. To that end it needed the best navigational information.

In 1768 the Royal Society persuaded the Admiralty and the young King George III, its patron, of the importance of sending observers to the Pacific to record the transit of Venus across the sun and thereby calculate the distance between the earth and the sun, knowledge that would benefit astronomy and navigation. It would also help to find an answer to the mariner's vexing problem of determining his longitude. An attempt to observe the transit in 1761 had failed and the next one would be in June 1769. After that there would be no transit until 1874 and 1882 and none at all in the twentieth century.

The Royal Society had been founded in 1660 as the Royal Society of London for Improving Natural Knowledge, 'to study nature rather than books', and it also had the admirable desire to make English simpler and clearer, the better to explain science. From its foundation it was strongly journal-minded and instructed 'seamen bound for far voyages . . . to keep an exact diary'. Soon, on the recommendation of Christopher Wren and of Samuel Pepys, who was secretary to the Admiralty in the 1670s and 1680s, a drawing school was founded at Christ's Hospital in London. Boys intent on a naval career were taught to sketch coasts, headlands and islands as aids to navigation; and to attain high standards of topographical accuracy.

For the appointment with Venus in the Pacific the Admiralty bought a three-year-old Whitby collier and named it *Endeavour*. It was fitted out in London for one of the most significant expeditions in the history of exploration and King George gave £4,000 towards the costs.

Astronomy and navigation had their place, but the King, the Admiralty and the science-minded expected much more from the voyage than dull calculations. They wanted the curtain lifted on the Pacific itself, the discovery of new lands and encounters with creatures and human beings never seen before. So it was that aboard the rugged little *Endeavour* room was made for the 'ingenious gentlemen'. These were naturalists and artists who longed to be the first to tread the sands and forests of the islands. They would collect the Pacific, parcel it up and return it to their own world. They were led by Joseph Banks, twenty-five years old, personable, handsome and energetic, an ardent botanist and well-connected member of the Royal Society. He floated buoyantly on a handsome income of £6,000 a year from his estates and 200 farms at Revesby Abbey in Lincolnshire. Not for him the usual Grand Tour of the young English dilettante, the classical stones and other pleasures of France and Italy. 'Every blockhead does that,' he said breezily. Sailing around the world he would have the grandest tour of all.

His entourage, which he paid for, included his Swedish friend and fellow botanist Dr Daniel Carl Solander, whose presence underlined the team's professional credentials. The others were Herman Sporing, a Swedish naturalist who acted as Banks's secretary; two professional artists, Sydney Parkinson and Alexander Buchan; and two black servants and two white servants from Revesby Abbey. Banks also took two greyhounds. The artists were to make a comprehensive record of everything they saw in the South Seas;

and Banks looked forward to returning in triumph with the first pictures of the Pacific world.

Endeavour departed Plymouth in August 1768. Ninety-four were aboard, a fair crowd in a ship of 370 tons, 107 feet in length and just over twenty-nine feet in the beam. The Admiralty insisted on discharging its duty to disabled servicemen by sending John Thompson, who lacked his right hand, to serve as the ship's cook. It had been an Admiralty rule since 1704 to give precedence to such 'cripples and maimed persons' in the galley. The oldest of the crew was a sailmaker aged sixty-seven but most were in their twenties. The youngest was a boy of twelve, in keeping with the naval custom of taking children to sea to train them for the seafaring life, literally to learn the ropes, as servants, seamen and pupils. Nelson entered the service at eleven. Some lads went to sea at six. In manning the upper masts and rigging boys had the catlike agility that older men had lost. A ship's company's mixture of social types, experience and tender youth mirrored a society ashore in which children went to work at an early age. All this was a good hundred years before the condition called adolescence was identified and long before teenagers were invented.

Endeavour's commander, James Cook, a Yorkshireman of Scottish descent, was thirty-nine. If not exactly a unicorn he was rare. The names of the ships he led to the Pacific, *Endeavour*, *Resolution*, *Adventure*, *Discovery*, serve as a motto for his life. Born in 1728, he became a grocer's apprentice and then embraced the sea, learning seamanship in tough vessels like *Endeavour*, carrying coal from Whitby down the North Sea to London. When he was twenty-seven, in 1755, he entered the Royal Navy as an able seaman and rose by merit, not through family connection or influence. Shrewd officers noted his gifts and helped him. He saw action in a fight with a French ship. At twenty-nine he qualified as a ship's master and entered a professional non-commissioned elite. He wore no braid or uniform but was a senior seaman, the navigator and, in many ways, the man with heavy responsibility at sea. In nine years of meticulous service he surveyed and charted the St Lawrence River and the difficult fissured coasts of Newfoundland and Nova Scotia. Meanwhile his account of an observation of an eclipse of the sun brought him to the attention of the Royal Society. The Society had in fact wanted its own man to lead the Pacific expedition: Alexander Dalrymple, a navigator and believer in Terra Australis Incognita, who would have a master to do the sailing. The Admiralty wisely rejected a divided command.

When it looked around for a navy man there was Cook in his prime: a six-footer of considerable presence, seasoned, square, serious and authoritative, also genial and passionate. He was modest but admitted an ambition to sail 'as far as I think it possible for man to go'. Appointed to *Endeavour* he was promoted Lieutenant. He was on the threshold of the years in which he established himself as the greatest maritime explorer, who charted more than any individual before him and drew much of the Pacific and Australia and New Zealand into the world's understanding.

Cook, the professional, the simply educated son of a farmhand, shared his cabin space with Banks, the amateur, the wealthy landowner schooled at Eton and Oxford. It was not always easy for captains to share confined quarters with civilian scientists. The French Pacific explorer Captain Nicolas Baudin wrote: 'I must say that those captains who have scientists aboard their ships must, upon departure, take a good supply of patience. I admit that though I have no lack of it the scientists have frequently driven me to the end of my tether.' Cook and Banks, however, got on well and founded a friendship that endured for the rest of Cook's life, each recognizing that he had much to learn from the other. Banks had a flair for getting on with people and liked and respected Cook; and Cook was fifteen years his senior, with all the authority of a captain. They bridged their social differences. It was an age in which class and caste mattered but could be accommodated in subtle ways. In British society at that time, more than in any other European one, there was room to rise.

It was difficult, in any case, to maintain strict class distinctions in a crowded ship. Seagoing was dangerous and everyone aboard had a shared interest in survival and self-discipline. Cook's striving for harmony was, not least, a matter of practical seamanship.

The question of a suitable base in the Pacific for an observation of the transit of Venus had been conveniently resolved in May 1768 when Captain Samuel Wallis arrived home in HMS *Dolphin* with news of an exquisite island he had discovered during his recent circumnavigation. It was called Tahiti. Its position in the middle of the Pacific, as well as its animals, fruit, water and agreeable people, commended it as the perfect place for astronomical observation as well as for rest and refitting. Captain Wallis's goat, which had supplied milk for the officers' coffee during his voyage, was now performing the same function aboard *Endeavour*.

Before turning into the Pacific Cook stopped in Tierra del Fuego at the southern tip of South America. Banks, as if unleashed like one of his own greyhounds, led a party ashore. He and Dr Solander gathered plants while Buchan sketched a group of the native people. A sudden snowstorm trapped the party and forced them to spend the night in the open. Banks's two black servants froze to death, Buchan had an epileptic seizure and Solander fell dangerously ill.

Cook rounded Cape Horn and on 13 April 1769, eight months from Plymouth, *Endeavour* anchored in the entrancing setting of Matavai Bay in Tahiti. Her crew beheld a crescent of beach and coconut groves beneath high and densely forested mountains. They watched with delight as scores of laughing, chattering, flower-decked islanders paddled out to them in a flotilla of canoes. The islanders made everyone welcome and offered feasts of roast pig, fish, lobster, roast dog as sweet as lamb, tender rats, breadfruit and coconut. Soon many of the crew had the company of girlfriends; and sex could be bought for an iron nail. Sydney Parkinson, the artist, disapproved of such easy-going relationships, 'as if a change of place altered the moral turpitude of fornication'. Cook looked tolerantly on the liaisons, as he had permitted a measure of rum-soaked Christmas partying at sea. But, for all his appreciation of Tahitian charm, he himself kept the women distant and retained his authority, taking in good part the teasing that his age placed him beyond pleasure.

Enchanted early visitors to Tahiti trawled their lexicons. The French navigator Louis de Bougainville called the island a Pacific version of Cythera, the island of Aphrodite, goddess of love; and, for good measure, the Garden of Eden. He conjured the ineradicable image of South Seas charm in his description of the tawny Tahitian girl who glided over his quarterdeck in 1768 and 'carelessly dropped a cloth which covered her'. This vision 'appeared to the eyes of all beholders such as Venus showed herself to the Phrygian shepherd, having, indeed, the celestial form of that goddess'. Another stunned witness of this tableau murmured that the islanders 'know no other God but love'. Images of nymphs in palmy islands scampered through European minds; and the word Tahiti itself took root as a synonym for allure.

Young and easy, Banks submitted to delight, thinking himself wafted to a classic Golden Age of Greece, to 'an Arcadia of which we are going to be Kings', and decided that the Tahitian 'gentlemen, like Homer of old, must be poets as well as musicians'. He began to learn Tahitian and fell dewily in love with the island, its languorous life and its lithe, sensuous people. A bewitching girl 'with a fire in her eyes' ignited him. In common with many of *Endeavour*'s men he bared one of his arms to a Tahitian tattooist and so became a founder of the tattoo tradition among British sailors. He gave classical nicknames to some of the Tahitian men, Ajax and Hercules, for example, and likened the women to the graceful statue of the Venus of Medici in Florence, writing of their elegance and innocent and mischievous ways. All this, and even bread itself grew on trees, in the form of breadfruit.

Cook, however, did not view the Tahitians as idealized noble savages. He mingled with the people for three months, compared with Bougainville's thirteen days, enjoyed their warmth and spontaneity and was not shocked by their sexual straightforwardness. He was sympathetic but saw them as all too human and from time to time as perfectly pesky. Their notions of property differed from his. They stole and picked pockets and saw no harm in it. But dark shadows fell. One day an islander snatched a musket and as he ran off with a crowd of companions a midshipman ordered marines to fire into the thick of them. They did so, said the artist Parkinson, 'with the greatest glee, as if they had been shooting at wild ducks', and killed one. 'What a pity', Parkinson said, 'that such brutality should be exercised by civilized people upon unarmed ignorant

Indians.' The bloodshed angered Cook, too. He did not believe that killing taught lessons. 'Contrary to the opinion of everybody,' he wrote, 'I would not suffer them to be fired upon.'

Cook's instructions from the Admiralty for dealing with people in the Pacific were plain. He was to observe their 'genius, temper, disposition and number . . . to cultivate a friendship and alliance, making presents of such trifles as they may value, inviting them to traffic, and showing them every kind of civility and regard'. He agreed with the moral view of James Douglas, the Earl of Morton and President of the Royal Society, who had written to him before he sailed saying that the killing of native people was 'a crime of the highest order' and urging him to check the wanton use of firearms and the petulance of seamen. The natives, Lord Morton said, 'are the natural and legal possessors of the regions they inhabit'. But with the shooting of the islander, blood was drawn and the stain was there. Cook brooded on the violence of European intrusion, just as he brooded on the venereal disease that Europeans inevitably transmitted.

Parkinson's sadness over the violence deepened when Buchan, his fellow painter, died of epilepsy. Buchan's passing was a serious blow to Banks who valued his artistic team highly and looked forward to showing their works in London. Journals were important, but to his mind 'no account of the figures and dresses can be satisfactory unless illustrated'. Parkinson now took on the bulk of the recording work and laboured prodigiously to draw and paint hundreds of plants and birds and other specimens collected by the botanists. Tahiti was a paradise for Banks but an ordeal for the working artist: fly traps of molasses were ineffective and swarms of flies ate the colours from Parkinson's paper as fast as he could lay them on.

Cook and the astronomer Charles Green established their observatory and set up their telescope and clock to record the moment that the shadow of Venus crossed the sun on 3 June. The following month *Endeavour* sailed from Tahiti and Cook steered south in accordance with the Admiralty's secret instructions. 'There is reason to imagine', they said, 'that a continent or land of great extent' lay in the South Pacific. The instructions reminded him that the discovery of new countries would 'redound greatly to the honour of this nation as a maritime power', and should he find the mysterious continent he was 'with the consent of the natives to take possession of convenient situations in the country in the name of the King of Great Britain'.

Banks was one of those who believed in the existence of a southern land. Cook was not. *Endeavour* sailed to latitude forty degrees south. Finding nothing of the 'aerial fabric called continent', Cook followed the Admiralty's orders and turned west towards New Zealand, which the Dutchman Abel Tasman had found and named in 1642. Cook had Tasman's narrative in his cabin. In October *Endeavour* reached Poverty Bay on the eastern coast of North Island, and, as they did elsewhere in New Zealand, her crew made a cautious and uneasy contact with warlike Maoris who thought the white men might be supernatural creatures. Indeed, eighty years later, a Maori chief who had been a small boy when Cook arrived recalled that his people first thought these strangers were goblins. The old man remembered Cook's air of authority and his 'perfect gentlemanly and noble demeanour'. Cook gently touched his head and presented him with a nail. Some years ago, an elderly chief in the highlands of Papua New Guinea told me of his first sight of a white man in the 1930s, the pilot of an aircraft; and he mimed for me the way he and his fellow tribesmen crept forward as the man, assumed to be a god, emerged from the cockpit.

In six months Cook circumnavigated New Zealand and demonstrated that it was not part of Terra Australis Incognita. He surveyed 2,400 miles of its coastline and drew charts so accurate that they were in use for two centuries. A French naval officer later wrote of their 'astonishing thoroughness'. In April 1770 Cook crossed the Tasman Sea, not knowing what he would find, and after sixteen days saw the coast of what would later be known as Australia.

He presently sailed into a bay edged by low wooded hills and anchored in clear green water. He went ashore with a small party and some Aborigines threw spears at them and were scattered by musket fire. But the people were essentially indifferent or timid rather than hostile. After observing them for a while Cook concluded that although they appeared wretched 'they are far happier than Europeans. They live in tranquillity which is not disturbed by the inequality of condition.'

No one, I imagine, forgets his first encounter with the landscape of Australia, the scent of the air and the quality of the light, the heat, the evocative smell of the dry land, the distinctive perfume of the eucalypt's long leaves, the stillness broken by the notes of bright birds, the austere beauty; it is like nothing else on earth. So it must have been for Joseph Banks and Daniel Solander as they explored; and also for Sydney Parkinson who carefully drew the plants while they were still fresh. The explorers wandered the shore and foraged for shells and shrubs and flowers never seen before, adding hundreds of specimens to their collection. They caught giant stingrays, which to Parkinson tasted like stewed turtle, and shot gorgeous cockatoos, preserving the skins and feathers and eating the flesh in parrot pies. The *Endeavour* was by now a museum afloat, a garden on the sea. Cook, in his good-natured way, considered all the seeds, shells, leaves and drawings crammed into the cabins, noted the joyful faces of the naturalists and named the anchorage Botany Bay. For good measure he called its portal headlands Cape Banks and Point Solander.

Three miles north of the bay Cook sailed past the spectacular sixteen-mile-long haven he named Port Jackson. In time it would be the site of Sydney. He headed north to chart the unexplored eastern coast, picking his way through the shoals. He knew nothing of the deadly Great Barrier Reef, a submarine mountain chain 1,200 miles long. Threading through its islands one June night *Endeavour* was pierced by a coral spear. Cook and his crew saved her by the coolest seamanship, throwing six half-ton cannon overboard and pumping furiously. Everyone noticed the curious fact that in this crisis the sailors did not swear. Banks and the rest of the 'ingenious gentlemen' took their turn to sweat at the pumps. At last the ship was floated off the reef and beached in Endeavour River for seven weeks of repair. It was fortunate that the coral had jammed in the hole and partly plugged it or the ship would have foundered. While the carpenter and his mates worked on the repair, Banks and Solander botanized insatiably. The strange, elusive and definitively Australian kangaroo was seen, shot, sketched and eaten, and its skin scraped ready for a taxidermist in London.

Before leaving Australian waters in August 1770, Cook landed on an island and hoisted the British flag. In King George's name he claimed the 2,000-mile coastline he had just travelled and surveyed. Marines fired a volley. Cook called the territory New Wales, a name confirmed by Banks in his journal. It was not that anyone saw a resemblance between the coasts of Wales and Australia, simply that a New England and a New Scotland, Nova Scotia, already existed. The term New South Wales appeared in a copy of Cook's journal sent to the Admiralty; and New South Wales it became (though in 1840 the northern section of the territory was renamed Queensland).

Cook sailed to Java to refit his ship in the port of Batavia, modern Jakarta. At this stage in the voyage he had lost eight men: two drowned, one dead of drink, another of tuberculosis, two of hypothermia, one of epilepsy, one by suicide. The remarkable fact was that not a single man had fallen to the scurvy that commonly wiped out half a ship's company. Batavia, however, was a swampy harbour and Cook could do nothing about fever. Seven men died of malaria, including the surgeon. Cook sailed for Cape Town and during the eleven-week voyage twenty-four more died of malaria and dysentery. Green the astronomer was one, the secretary Sporing another.

The artist, Sydney Parkinson, also died. His legacy of 1,300 sketches and paintings is in the National Maritime Museum in Greenwich and is his monument, the fresh gaze of a gifted young man on the people and landscapes of a new world, the first pictures from paradise.

Everyone who went to sea accepted the risks of sudden death, of being washed overboard or falling from the rigging. Many of the men who had been in action had seen the dismemberment and evisceration caused by shot and flying splinters. The 'butcher's bill' of battle, however, was relatively small. In the Napoleonic wars around a twelfth of the deaths at sea were attributable to combat. At the Battle of Trafalgar 449 of the 17,772 British sailors were killed, compared with 4,408 of the 30,000 French and Spaniards. The principal killers of seamen were the frightening diseases whose causes were unknown and difficult to prevent or cure. Many died of dysenteric fever, the 'bloody flux', others of the yellow fever and malaria that made sailors dread service in the West Indies. They all knew the grim story of Admiral Hosier's Caribbean expedition of 1726 in which 4,000 of the 4,750 sailors had died of fever.

Seamen also died of smallpox and typhus or succumbed to calenture, in whose fevered and rhapsodic throes they imagined the sea was a green meadow and jumped over the ship's side to walk in its lush grass. But the most serious was the nutritional deficiency written variously as scorbutus, scorbuto, scarby, scorby and most often as scurvy, the occupational disease and scourge of the long-distance voyager.

As soon as ocean travel became commonplace, in the sixteenth century and after, scurvy's horrible symptoms became well known. By the time ships had been at sea for about six weeks their crews had consumed all the fruit, vegetables and fresh meat aboard and had turned to a diet of salted and pickled meat, dried peas, bread, biscuit and cheese. All this was laved down with large quantities of beer and watered spirits, especially rum. The deficiency of ascorbic acid, Vitamin C, in this monotonous menu eventually revealed itself in the most distressing way. Swollen gums grew spongy, teeth worked loose, joints ached, limbs ballooned, old wounds reopened, breathing became laboured and sufferers fell into a fatal lethargy. The foul breath of scorbutic men made it difficult to stay near them: ashore, a sufferer could clear a room with an exhalation. As the victims deteriorated they were seized by pain and died, and their bodies were pushed over the side. Aboard a ship sailing down the west coast of Africa the weary remnants of a crew who had disposed of the bodies of more than sixty of their fellows caught a large shark on a hook. Desperate for fresh food they cut it open and discovered the body of one of the shipmates they had lately thrown overboard. Hungry as they were, they could not bring themselves to eat.

Many ships sailing to and from India were wrecked because crews disabled by scurvy were too weak to handle the sails or take their turn with the steering. Vessels arrived in Cape Town from the East Indies and reported scores of their men dead from scurvy. Most of the hundreds of deaths during George Anson's voyage in the 1740s were caused by the disease. The Admiralty had eased the problem of overcrowding at Chelsea Hospital by sending Anson 500 pensioners, some aged more than seventy. Those who had the limbs and strength to do so deserted from Portsmouth, but 259 invalids sailed away to die at sea, as the Admiralty knew they would. Survivors remembered the rotting of their shipmates' flesh, the reappearance of long-healed injuries and bodies rolling about the decks because the men were too weak to throw them overboard.

Sir Richard Hawkins, veteran of the action against the Spanish Armada, who commanded a privateering voyage to South America in 1593, gave an account of scurvy and added: 'I wish that some learned man would write of it, for it is the plague of the sea and the spoil of mariners. Doubtless it would be meritorious work with God and man, and most beneficial to our country, for in twenty years I account of 10,000 men consumed with this disease.' Like other captains, including Vasco da Gama, Hawkins had seen that citrus juice mended men sick with scurvy. 'The most fruitful for this sickness is sour oranges and lemons . . . a wonderful secret of the power and wisdom of God that has hidden so great and unknown virtue in this fruit, to be a certain remedy for this infirmity.' Anson, too, saw sick men recover if they lived long enough to reach islands where oranges and lemons grew. No one, however regarded citrus fruit as a preventive; and it was, in any case, difficult to buy and store in Europe.

From the sixteenth century into the nineteenth the Admiralty and naval officers gave serious thought to scurvy because it undermined the navy's fighting efficiency. Officers obviously had a professional interest in keeping their ships and crews as healthy as possible. It was not surprising, given the extent of medical knowledge, that many of them should have accepted the opinion that scurvy was caused by cold and damp air, by a foul atmosphere between decks, by dirty living quarters and clothing or by laziness and lack of exercise. It became the custom to wash decks with water and vinegar, to ignite gunpowder as a fumigant and to ensure that men had clean clothes. These measures did not prevent scurvy but they made life at sea more tolerable.

James Lind, a Scottish naval surgeon, investigated scurvy after learning of the deaths on Anson's voyage. In 1747 he conducted experiments with a group of twelve sailors, all with scurvy, and while ensuring that they all had the same three meals a day, divided them into six pairs and for two weeks gave each pair a different and popular scurvy treatment of the day. Two, for example, were given cider, two had vinegar, two sulphuric acid and two had sea water. The two dosed with oranges and lemons made the most rapid recovery and Lind concluded that citrus juice was the remedy for scurvy; and this, as we have seen, was already known to many captains. Lind, who would be hailed as the father of nautical medicine, published his findings in his *Treatise of the Scurvy* in 1753, and the Admiralty was later criticized for failing to adopt the lemon juice cure immediately.

The problem was that Lind was by no means clear himself. He persisted in his belief that scurvy was caused by blocked pores and by cold, damp air. He denied that the absence of vegetables and fruit in the diet was the only cause of scurvy and noted that sailors had lived for months on salt meat without getting the disease, although he recommended the use of vegetables to 'correct' the quality of hard, dry food. In 1758 he became the first physician in charge of the new Royal Navy hospital at Haslar, Portsmouth, and saw many hundreds of scurvy cases in the twenty-five years he was there. He made his name as a pioneer researcher but he was disappointed, in spite of all the work he did, that he did not find a cure for the disease. He never specifically called on the authorities to dose seamen with lemon juice.

The Admiralty was beset by competing theories about scurvy and possible cures. Vitamins would not be discovered until 1912. Fresh meat and vegetables were evidently good for the health and morale of sailors. In the 1750s warships in the Channel and Western Approaches which had to remain on station for months on end were supplied with live bullocks and other victuals hoisted aboard in slings. It was expensive and difficult work but paid dividends in health on board. For distant-water duty salted or pickled meat and hard cheese remained the notorious staples; but the Admiralty's Sick and Hurt Board tried out all manner of foods and substances which were thought to prevent or alleviate scurvy. Portable soup was one of them, made from cattle livers and kidneys boiled down into what we would recognize as stock cubes. Other treatments included powdered orchid roots, dried malt to make beer, and sauerkraut or pickled cabbage.

A voyage like Cook's circumnavigation was an ideal opportunity for the Sick and Hurt Board and the Victualling Board to test a range of preventives and remedies. Aboard the *Endeavour* Cook loaded an antiscorbutic armoury of a thousand pounds of portable soup, a year's supply of sauerkraut at two pounds a week for each man, malt, mustard, vinegar, molasses, sugar and a supply of lemon and orange juice in bottles. This latter was intended as a treatment, not a prophylactic. Cook could not know, however, that because the juice had been boiled to make a concentrate it had lost much of its Vitamin C and that it would further deteriorate in storage.

Cook regarded the prevention of scurvy as a matter of proper seamanlike care, a practical humanity. He could not sail his ship and navigate if men were sick and dying. He was furious with other commanders who neglected simple remedies and allowed their people to fall ill: to his mind they had failed in their duty. He knew at first hand how scurvy could chop down a crew. In HMS *Pembroke* in 1758 he crossed the Atlantic for the first time and saw twenty-six men die during the voyage. Later, during survey patrols along the Canadian coast, he wrote that 'the scurvy never fails to pull us down in great numbers'. We do not know if he read James Lind's work. Probably he did not. But at the time he sailed for the Pacific in *Endeavour* he was well up with the debate on scurvy, its cause and cure. Richard Hakluyt, the naval chronicler of the late Elizabethan era, had written that sailors passed their years in 'great and continual hazard and so few grow to grey hairs'. Cook's ambition was to show his men how they could travel to the earth's ends and live to grow grey, that whatever the other risks they ran they could avoid dying of a preventable disease.

Unlike most naval officers Cook had started his service in the foul conditions of the forecastle and knew ships and sailors inside out. Other captains may have held their noses in respect of the seamen's quarters and let their men live in a midden, but Cook believed passionately that a clean ship was a happy and healthy ship, and he was a most determined enforcer of the Admiralty's regulations regarding cleanliness. To these he added more of his own. Hence his men regularly swabbed out their accommodation with solutions of vinegar and gunpowder, rinsed the bilges, brought out their bedding for airing and washed themselves and their clothes frequently. Cook headed his own health police and examined his men's hands: he cut the rum ration of any man with dirty fingers. He reacted quickly in the Cape Verde islands when his crew came aboard, after a run ashore, with pet monkeys on their shoulders. The beasts excreted everywhere. Cook ordered them thrown overboard at once.

He saw to it that in cold weather the men were kitted with thick trousers and warm wool jackets, fearnoughts as they were called. He wrote in his journal that in sleet and snow, 'with all the rigging covered with ice and the air excessive cold, the crew stand it tolerable well, each being clothed with a fearnought jacket, a pair of trousers and a large cap made of canvas and baize, these together with an additional glass of brandy every morning enables them to bear the cold without flinching'.

His dietary rules followed naturally from his views on clothing and cleanliness. Experience and instinct led him to insist that when they went ashore the men should gather and eat fresh food. Wherever he anchored he set the crew to picking greens such as nettle tops, wild celery and herbs, fruit, berries and roots, and the shoreline herbs known as scurvy grass. He ordered them to brew a piquant beer from spruce leaves, which had variable antiscorbutic qualities. The sailors also caught fish and hunted fowl. As long as meat was fresh, whether it was stewed penguin, boiled walrus or a roasted Pacific island dog, Cook told his men to eat it: it was better than salt pork.

By example and persuasion, by cajoling and insistence, he ensured that food like sauerkraut went down the throats of sailors. They were as reluctant to eat it as is a cat to swallow a worming pill. Cook was the master of making grumbling sailors consume what was good for them, although they complained behind his back. 'It was no uncommon thing', an officer recorded, 'for the men when swallowing to curse him heartily and wish for God's sake that he might be obliged to eat such damned stuff mixed with his broth as long as he lived.'

Cook had a particularly canny way of dealing with the men's stubbornness. Seeing that at first they would not eat sauerkraut he had it served to the officers every day. 'Such are the tempers of seamen', he observed, 'that whatever you give them, although it be ever so much for their good it will not go down with them and you will hear nothing but murmurings against the man that first invented it; but the moment they see their superiors set a value upon it, it becomes the finest stuff in the world and the inventor a damned honest fellow.'

In 1776 Cook's work in the war on scurvy was saluted by the Royal Society with the award of its Copley Medal. He completed his second voyage the year before, having lost not a man to the disease. The medal citation read:

> If Rome decreed the civic crown to him who saved the life of a single citizen, what wreaths are due to that man, who, having himself saved many, perpetuates the means by which Britain may now, on the most distant voyages, preserve numbers of her intrepid sons, her mariners; who, braving every danger, have so liberally contributed to the fame, to the opulence, and to the maritime empire, of their country!

Such florid praise reflected a contemporary view that Cook's practices aboard ship inaugurated a new era in healthy voyaging. He did not, however, invent a wonder cure for scurvy. There was no connection at all between his regime of cleanliness and the defeat of scurvy. The disease was not contagious. He succeeded because he employed a range of measures and let fly at scurvy with everything he had. On his second voyage he added carrot marmalade and soda water to his repertoire. But he had no way of telling which of the preventives were efficacious. He was enthusiastic about fermented malt, although it seemed in practice to be a better laxative than an antiscorbutic. Nevertheless, a scrubbed ship was far better than a dirty one and Cook's determined multiple attack on scurvy certainly helped to give most of his men the opportunity to achieve grey hair. He deserved the medal.

In the 1780s the Scottish naval surgeon Gilbert Blane pressed the navy to make a regular issue of lemon juice to seamen. A determined man, he relentlessly and aggressively badgered senior officers and Admiralty civil servants. He was perfectly clear that scurvy could be prevented by vegetables and fruit, particularly oranges, lemons or limes. Fifty oranges or lemons, he said, could be considered the life of one man. In 1796 the Admiralty ordered a daily issue of three-quarters of an ounce of lemon juice to each man aboard warships. The order was timely, for the lemon juice ration contributed significantly to the well-being of British seamen in the years of the naval blockade of Napoleon's navy, the prelude to Trafalgar. Nelson ordered large quantities of the fruit for the Mediterranean fleet: he believed that 'the great thing in all military service is health'. Sips of lemon largely eradicated scurvy and, like copper nails, played their part in victory at sea.

5

Simple fornication

'We knew the merry world was round
and we might sail for evermore.'
Alfred, Lord Tennyson, 'The Voyage'

On *Endeavour*'s return to London in July 1771, the self-effacing James Cook wrote formally to the Admiralty expressing the hope that his journals, charts and drawings would be 'found sufficient to convey a tolerable knowledge of the places they are intended to illustrate'. The self-effacing ship's goat survived her second circumnavigation, and Dr Johnson wrote a Latin couplet saluting her for 'her never-failing milk'. Joseph Banks, who was not self-effacing and was triumphant and bubbling with tales, was granted immediate celebrity. The burbling newspapers were full of 'Mr Banks's voyage'. In all the excitement Cook was second fiddle, except in the sober judgement of the seadogs at the Admiralty who commended his captaincy and understood perfectly well the magnitude of his achievement.

Everyone agreed that it had been a great voyage. Cook and Banks had brought a hemisphere home: the Pacific Ocean, New Zealand and Australia. They carried a hoard of sensations of science and geography, 30,000 specimens of plants, birds, fishes, reptiles, insects and mammals, covering 3,000 species of which more than 1,500 had hitherto been unknown. Banks posed for his portrait in a Maori cloak and submitted happily to the adulation of London society.

It was not long, though, before he had to bear the sniggering. People were anxious to hear about the maidens of Tahiti, 'naked and smiling . . . Like Eve unapron'd ere she robbed the tree'. Banks believed he had sauntered in the islands of Eden but some in

London were sceptical and resisted his notion of the noble savage. There was much tittering of the tabloid sort, and sharp pens sought to prick this particular South Sea bubble:

> With us this makes no Variation;
> Still is it simple fornication.

Increasingly the Pacific became a moral question. How was it that savages without the benefit of Christianity could live so simply and contentedly in blessed sunshine without labouring for their daily bread, while church-garrisoned Europeans were saddled by sin, pain and hard work, and tormented by all the corruption of the world? A dissenting view emerged, that far from being an existence of classical grace, the South Sea island life was one of anarchy, ignorance, idolatry and brutality. People began to say that what Eden needed was God and missionaries.

Banks, meanwhile, pressed for a second world voyage, with his friend Cook as commander. The Admiralty agreed and secured two Whitby colliers, *Resolution* and *Adventure*. In his hubris Banks insisted on spacious quarters for himself and a party of fifteen artists and scientists as well as servants and two musicians. To accommodate this miniature university an extra deck was fitted to *Resolution*, but it was soon found that the extension rendered the ship unseaworthy and the Admiralty ordered it removed. Banks threw a tantrum. He would not go to the Pacific. He chartered a ship and sailed off to explore Iceland instead. On the return voyage he visited Staffa in the Hebrides and saw Fingal's Cave whose geological wonders he introduced to a wider public.

Resolution and *Adventure* departed in July 1772. There were still those who believed in the existence of a southern continent and one purpose of the expedition was to settle the matter for good. Cook would sail into the deep south, beyond sixty degrees latitude, and circumnavigate the globe. There was also the pressing question of longitude, of finding a ship's position east or west of a prime meridian. Latitude was the easier matter and could be determined from the position of the sun or the north star. Without knowing his longitude a captain could not say where he was, could not mark an island or a cape on his chart with any accuracy and, having made a landfall, could not return to it with certainty. Navigation was often, at best, educated guesswork. In 1714 Parliament passed the Longitude Act and offered a prize of £20,000 to the individual who resolved the problem. The Act followed a navigational disaster in 1707, the wreck of Sir Cloudesley Shovell's ships and the drowning of 2,000 sailors in the Isles of Scilly. As was often the case, the navigators had become hopelessly lost.

Cook found his longitude by the standard method of his time, taking sights of the moon by sextant, noting the angle of its elevation and then working through complex calculations which took half an hour or more. A clear view of the moon was not always possible. A better system for determining longitude was to have precision clocks on board. A navigator could fix his longitude if he knew the time at the prime meridian of Greenwich as well as his ship's time. When it is noon at Greenwich it is eleven a.m. at fifteen degrees west, ten a.m. at thirty degrees west. At sea, however, a clock's accuracy was always disturbed by the ceaseless rolling and pitching, by variations in temperature and barometric pressure, and by problems with lubrication and friction. One of Cook's tasks on his second voyage was to evaluate a chronometer, a watch constructed with balances and compensatory devices to enable it to keep accurate time. It was made by Larcum Kendall and was a copy of the chronometer built by John Harrison, a Yorkshire carpenter and self-taught clockmaker, who devoted most of his life to perfecting it. Four of Harrison's chronometers remain in working order at the Royal Observatory.

Aboard *Endeavour* Joseph Banks's naturalists and artists had brilliantly established the value of teams of scientists and observers. For the second voyage the Admiralty appointed the astronomers William Bayly and William Wales. Wales was a Yorkshireman, the brother-in-law of Charles Green who had died in the *Endeavour*. He later taught mathematics to Charles Lamb, Leigh Hunt and Samuel Taylor Coleridge at Christ's Hospital. The naturalists were Johann Forster and his son Georg. The artist was William Hodges, aged twenty-eight, a blacksmith's son who had started his working life as an errand boy and whose talent was discovered by the Welsh painter Richard Wilson. The young George Vancouver was a fourteen-year-old midshipman, learning surveying under the eye of the headmasterly Cook while Wales taught him navigation. In this way he was trained for the famous survey of the Pacific coast of North America that he would make in the 1790s.

Cook sailed towards the Antarctic, steering as far south as possible, beyond sixty-seven degrees, in his search for Terra Australis Incognita. He was the first explorer to cross the Antarctic Circle and to record the aurora australis. As Cook worked his way through floating icy boulders along the edge of Antarctica, Hodges sketched and painted experimentally with pen and pencil and wash, capturing the vistas of sea, ice and clouds, and the play of Antarctic light. It was pioneering art, for he had no examples to show him the way. There were no classical landscapes here; only the desolate spectacles of sapphire mountains afloat in brooding inky seas. Hodges developed a seamanlike eye in his observations of the skies and the weather during long passages in the ocean and this influenced his painting deeply. He was helped by his friendship with Wales, who was a meteorologist as well as an astronomer. Later, during *Resolution*'s wide swing through the Pacific, Hodges confronted the difficulties of rendering tropical light in the dramatic landscapes he painted in Tahiti and in his remarkable pictures of Dusky Bay in New Zealand which *Resolution* reached in March 1773.

In June, still in New Zealand, Cook reflected that the coming of civilization was not necessarily a benefit to the Maoris. During his previous voyage in *Endeavour* sex between sailors and Maori women had been conducted 'in a private manner'; but now the Maori men had become

> the chief promoters of this vice, and for a spikenail or any other thing they value will oblige their wives and daughters to prostitute themselves and not with the privacy decency seems to require, such are the consequences of a commerce with Europeans and, to our shame, civilized Christians. We debauch their morals already too prone to vice and introduce wants and perhaps diseases which they have never known before and serve only to disturb that happy tranquillity they and their forefathers enjoyed. If anyone denies the truth of this

assertion let him tell me what the natives of the Americas have gained by the commerce they have had with Europeans.

Cook sailed once more to the Antarctic Circle and reached a latitude of more than seventy-one degrees, farther south than anyone had penetrated. Then he turned north and anchored at Easter Island. Hodges painted the enigmatic statues. *Resolution* crossed the Pacific again, and called at Tahiti, the Fiji islands, the New Hebrides and New Zealand. Then she sailed for home, traversing the Pacific once more, rounding Cape Horn, calling at South Georgia and replenishing at Cape Town.

Cook anchored off Plymouth in July 1775, completing one of the most magnificent journeys of history, three years and eighteen days of exploration 'in which time I lost but four men and only one of them by sickness'. He had charted numerous islands and added much to the knowledge of geography and the natural world. He was delighted with William Hodges's innovative and entrancing pictorial record of the voyage; 'masterly', he called it. Hodges avoided sentiment in his studies of Pacific people. No noble savages or ethnic stereotypes entered his portfolio. He sought, as he said, 'the real and natural character'.

The chronometer earned Cook's praise, too: it was a significant step in the struggle for exactness in determining longitude. The problem of a timepiece's accuracy at sea, however, would not be completely resolved until the coming of radio. Cook's patient search, Antarctic penetration and circumnavigation demolished the idea that a fruitful southern continent inhabited by handsome people lay somewhere in the deep south of the Pacific. He showed that New Zealand and Australia were not connected to any other land mass; and that any southern land must lie in the inhospitable polar region, too cold to support human life.

In July 1776, only a year after returning from his second voyage, Cook embarked upon a third. He need not have gone. No one expected him to volunteer. He was forty-seven and his fame and reputation were secure. He was not as fit as he had been and the years of responsibility at sea had worn him down. He had suffered bouts of painful illness during the second expedition, and modern analysis of his symptoms suggests that he had a severe intestinal infection which undermined his concentration and temper. His reward after his two circumnavigations was promotion to post-captain and an undemanding job at Greenwich Hospital; but, as he wrote to an old friend, 'My fate drives me from one extreme to another; a few months ago the whole southern hemisphere was hardly big enough for me and now I am going to be confined within the limits of Greenwich Hospital, which are far too small for an active mind like mine . . . whether I can bring myself to like ease and retirement, time will show.' He volunteered to command a third voyage in HMS *Resolution*, in company with HMS *Discovery*, and wrote to his friend: 'I have quitted an easy retirement for an active, perhaps dangerous voyage.'

The main purpose of this third expedition was to search for the North West Passage from the Pacific side, to sail through the Bering Strait and probe eastward. Cook had the talented William Wales with him again, as astronomer, and his sailing master was a twenty-one-year-old Cornishman, William Bligh. His artist was John Webber, who had trained in Berne and Paris, and although he was hired to make records of plants and drawings of coastal scenery, he was instructed to go further, to be journalistic in his work, illustrating and bringing alive the reports of the voyage. Such illustrations, said Cook, would 'serve to make the result of our voyage entertaining to the generality of readers'. Webber seized his opportunity and worked prolifically. He drew and painted the landscapes and peoples of Siberia, southern Africa, New Zealand, Hawaii, the Pacific islands, China and south-east Asia. No other voyage was ever so abundantly illustrated by one man.

Discovery joined *Resolution* at Cape Town. In company they sailed to Kerguelen Island in the southern Indian Ocean, called at Tasmania and went on to New Zealand, Tonga and Tahiti, where Cook met old friends. From Tahiti the ships sailed north and in January 1778 sighted the Hawaiian islands which Cook named after Lord Sandwich, the rakish First Lord of the Admiralty and inventor of the convenience food that bears his name. Cook and his men were the first Europeans the Hawaiians had seen and were warmly embraced. Cook himself was treated as a lord. He was astonished to discover that he was among Polynesians and that those of his men who spoke some Tahitian could communicate with the Hawaiians.

To begin his search for the North West Passage he then crossed the North Pacific and sailed up the coast of America and Canada into the Gulf of Alaska and the Arctic Ocean. He was the first navigator to cross both polar circles. In August, at more than seventy degrees north, he was defeated by pack ice near Icy Cape and sailed to the Russian coast of the Arctic Ocean. It was clear that if the North West Passage did exist the ice and bitter weather would have made it very difficult to negotiate. Cook turned southward to rest his men in Hawaii and anchored in Kealakekua Bay in January 1779. The ships left on 4 February but it was Cook's ill fortune that shortly afterwards *Resolution*'s foremast was damaged in a storm and he had to put back to Hawaii for repairs.

This time the welcome was not enthusiastic. Cook, who had originally been received cordially, was now apparently seen as an intruder in Hawaii. The atmosphere was further soured by the white men's angry reaction to the islanders' thieving. Cook was not the tolerant and genial man he had once been. In failing health, he was fatigued and irritable and ordered punishments aboard his ship that were uncharacteristically harsh. Even the navigation that was the core of his talent was occasionally careless.

Here in Hawaii, then, he was not his usual humane and patient self. A conflict broke out with the Hawaiians which rapidly became a dangerous tumble of misunderstandings and provocation. Cook and his men thought that the Hawaiians would be cowed by musket and pistol fire, and in this they were mistaken. Cook had himself remarked during his second voyage five years earlier that white men should not believe that guns made them invincible in their dealings with island people. On 14 February, undeterred by gunfire, Hawaiians speared and bludgeoned Cook on the beach at Kealakekua. They divided his skull and bones among four of their chiefs. Some of the fragments were recovered and a positive identification was made of a severed hand bearing the distinctive scar made by a powderhorn explosion in the years when Cook was surveying the coast of Newfoundland. Cook's sad companions buried his remains at sea and began the long voyage home.

The contributions that Cook's expeditions made to navigation, anthropology, medicine, botany and cartography broadened the world's mind. More than two centuries on, the Pacific scenes painted by Parkinson, Buchan, Hodges and Webber remain movingly evocative of discovery and adventure. Cook's techniques set the pattern for scientific inquiry in the years that followed. The French and Americans awarded him a status beyond enmity, regarding his work as so valuable, so sacred even, that when they were at war with Britain they undertook not to molest him. Cook's genius included generosity and humanity. He pondered on the impact of exploration and the effects of European intrusion. His name endures in islands, straits, capes, mountains and towns. John Cawte Beaglehole, his biographer, proposed that Cook's memorials were Geography and Navigation. 'If we wish for more,' he added, 'an ocean is enough.'

A remarkable company of men traced their professional lineage to their voyages in Cook's ships and formed a family tree of sorts. Banks, Solander, Wales, Vancouver, Bligh, Parkinson, Hodges and others looked back on *Endeavour*, *Resolution* and *Discovery* as old schools, reckoned Cook a god of the quarterdeck and were wise enough to know that their experience with him was the time of their lives. For midshipmen in their early teens there could hardly have been a more thorough and marvellous education. Vancouver went on to continue Cook's work. Edward Riou, a midshipman on the third voyage, became one of Nelson's captains and was killed at the Battle of Copenhagen in 1801. Isaac Smith, a midshipman in *Endeavour* and later an admiral, was in the boat rowed to the bone-white beach at Botany Bay where Cook told him: 'Isaac, you shall land first,' and invited him to plant the inaugural European footprint in Australia.

Some in their youthful pride and pleasure thought themselves a special band of brothers. David Samwell, the Denbighshire doctor and bard, who served as a surgeon during the third voyage, wrote after a reunion that 'it is an article of faith with every one of us that there never was such a collection of fine lads'.

6

The flowerpot men

'... come to these new
Bluenesses, bizarre to us strangers;
Elegiac, pale and matte, the swags
Of gum leaves bunched where there are bell-birds ...'

Ted Walker, 'Australia Blue'

If a pie existed, the finger of Joseph Banks was in it. In the half-century following his return from the Pacific the reputation of 'the celebrated Mr Banks', Banks of the South Seas, grew like a tropical plant.

An impresario of exploration, Banks never lost his boyhood passion for botany and saw to it that natural history was recognized as a serious branch of science. He was thirty-six when he was elected President of the Royal Society in 1778 and he held the post for forty-two years, giving full freedom to his large and enthusiastic personality. The position made him the government's principal adviser on science and he was phenomenally active. On the one hand he was the formidable, ubiquitous, bossy and interfering despot; on the other, the urbane, benign, inquisitive and accessible uncle, always ready to encourage and fund scientific enterprise. His letters flowed in a copious stream; he was a champion attender of dinners even when hobbled by gout; his home at 32 Soho Square in London was a permanent open-door symposium; and his Thursday breakfast parties were famous. When he was seventy he advised a dithering young botanist to take the risks of travel, just as he himself had, telling him that 'if I had listened to the voices raised to dissuade me ... I should probably have attained no higher rank than that of a country Justice of the Peace. Let me hear from you how you feel inclined to prefer ease and indulgence to hardship and activity.'

Through botany Banks formed a long and amiable association with King George III, a keen gardener and farmer. They were both young when they first met in 1772, Banks twenty-nine and the King thirty-four. The King consulted him over the future of Kew House and its ornamental gardens which had first been laid out by his mother Augusta, the Dowager Princess of Wales, ten miles from London beside the Thames. Out of the discussions evolved the Royal Botanic Gardens which matured into the commanding international institution of plant collection and research; and Banks became its scientific adviser in 1773. The King adorned him with a baronetcy in 1781.

Banks set the pattern of his operations by dispatching the gardener Francis Masson to South Africa on a pioneering plant-gathering expedition. He was so delighted with the haul of hundreds of specimens that he next sent him to the Canary Islands and the Caribbean, and again back to South Africa.

Then he spread his net wider and persuaded travellers who knew something of botany to send him seeds, plants and bulbs, anything of interest. James Bruce, a Scot who was an early explorer in Africa, parcelled up seeds from Egypt, the Nubian desert and Abyssinia; and Banks gave the name of *Brucea* to one of the plants he sent. Like Banks, Bruce was dynamic, rich and used to having his own way. He was also a tower of a man, six foot four with a booming voice, a hard rider and a prolific lover. In 1770 he reached the source of the Blue Nile in the northern mountains of Ethiopia, 2,750 miles from the Mediterranean, and followed it down to Khartoum. He thought he had found the headwaters of the Nile itself, though another century would pass before the true source of the parent river, the White Nile, a thousand miles away, would be identified.

Banks's own collecting foray in Botany Bay in 1770 played its part in the British settlement of Australia. The American War of Independence ended the transporting of criminals to the American colonies. Banks advised a Parliamentary committee in 1779 that Botany Bay would be the ideal destination for the convict overflow. Nothing happened until 1783 when James Matra, who had been a midshipman on *Endeavour*, renewed the idea that a colony should be founded at Botany Bay. With his eye on possible future profit, he suggested that it would yield two raw materials vital to the Royal Navy: flax for sail canvas and pine trees for masts. Both, to Britain's strategic disadvantage, came from Russia by way of the Baltic. Naval timber was at a premium and always on the Admiralty's mind. Masts, booms and yards, worn out or smashed by heavy weather and enemy action, had constantly to be replaced.

The notion of a rich source of flax and timber in a Pacific outpost interested some, but the real pressure on William Pitt, the Prime Minister, came from the swarming jails, the squalid prison hulks, rioting convicts and the indignation of Members of Parliament who demanded the dumping of criminal rubbish somewhere far away. A committee considered and rejected the founding of penal settlements in Africa and Madagascar. It consulted Banks, the Australia expert, who again listed the virtues of Botany Bay; and that was the committee's recommendation. In 1786 the government ordered the establishment of a colony of chains on the far side of the world, thinking that at the very least Britain would thereby have a foothold in that distant orbit.

The 'First Fleet' of eleven crowded ships sailed from Portsmouth in May 1787 carrying more than a thousand people, 736 of them skinny convicts, mostly London thieves in their twenties, although the youngest was a boy of nine and the oldest a woman of eighty-two. We can only guess at their fear. They had no idea of the size of the oceans, or of where they were going, and they lived in the darkness and stench below. In January 1788, 252 days out of Portsmouth, the fleet arrived in Botany Bay, the largest party of people ever to make a long-distance voyage and the forerunners of more than 160,000 prisoners transported to the Australian settlements. Forty-eight died on the voyage. Captain Arthur Phillip, the fleet commander, saw immediately that Botany Bay was a dry bone with little grassland and not enough water to support a colony. He sailed six miles north and turned through the sandstone headlands into the blue harbour that Cook had noted but not entered. This was the place: there was plenty of good soil and water; and here he settled and founded modern Australia and called it Sydney, after the Home and Colonial Secretary Lord Sydney.

Many of the seeds and plants discovered by Banks's botanical contacts prospered at Kew; and Banks developed a two-way traffic, dispatching seeds to British colonies. 'As Banks planned it,' said Dr H.C. Cameron, one of his biographers, 'Kew was to be more than a garden where exotic plants might grow. It was to be a comprehensive exchange house of the Empire, where the possibilities of acclimatizing plants from one part of the globe to another might be tested . . . It was to serve as an advisory centre for all practical activities in botany while it controlled the development of botanical exploration and experiment.'

Around 7,000 plants were introduced into Britain through the Banks network and many were sent abroad. In 1778 Banks recommended that tea bushes should be transplanted from China, where almost all tea was grown, to India, where it was not known. Tea had been drunk in Britain from early in the seventeenth century and by the 1770s was a huge business. Banks said tea was 'of the greatest national importance' to Britain, but it was not cultivated in India in his lifetime. Wild tea was found in Assam in 1820 but the first cultivated Assam leaf only reached London eighteen years later.

Plantation owners in the West Indies, looking for cheap food for the increasing numbers of their slaves, asked Banks whether the breadfruit he had seen in Tahiti could be transplanted to the Caribbean. Banks thought it could. The King approved an expedition and the Admiralty found a small ship, fitted and coppered her, renamed her HMS *Bounty* and ordered thirty-three-year-old Lieutenant William Bligh to sail her to Tahiti, load her with breadfruit seedlings and take them to the West Indies. David Nelson, a young Kew gardener who had sailed with Bligh on Cook's third voyage, was engaged by Banks to harvest the Tahitian breadfruit. To protect the seedlings the crew of forty-five were forbidden to bring dogs, cats, goats or monkeys aboard. The ship's rats, they were told, would be exterminated by poison and no complaint should be made about the reek of their corpses.

Bounty called at the Cape of Good Hope where, at Banks's request, Bligh collected apple trees which he planted in Tasmania on the way to the Pacific. He anchored at Tahiti at the end of

October 1788 and remained there for twenty-three weeks while his men loaded breadfruit plants. A number of the crew, bewitched in the usual fashion by the island and its coconut-scented women, incubated the emotions that erupted in mutiny the following April. Bligh kept himself aloof from the girls. The Manxman Fletcher Christian, his twenty-three-year-old sailing master, took a chief's daughter for a mistress. He also earned Bligh's disapproval by having himself tattooed. Later Bligh would say that he could recognize each mutineer by his tattoos.

For all the rigour of naval discipline, wise captains knew that no ship's company worked well under the lash and that ships were most safely sailed by men treated fairly and working together. William Bligh was a brave seaman and an excellent navigator, but he lacked the good captain's ability to mould his people into a team. They did not know where they stood with him. He blustered and threatened and insulted officers and men alike. During the *Bounty*'s stay in Tahiti discipline frayed and Bligh had bad-tempered confrontations with a number of the men. He and Christian fell out and when they resumed the voyage Christian sulked. In front of the men Bligh foolishly accused Christian of stealing coconuts from the *Bounty*'s stores. That was the breaking point.

On the morning of 28 April 1789, three weeks after *Bounty* had sailed from Tahiti, her gaping crew watched the showdown on the quarterdeck. The sweating, bandy-legged and overwrought Christian was gripping a cutlass and shouting: 'I am in hell!' Bligh was his prisoner, dressed only in his nightshirt, pinioned, undignified and angry. Bligh reminded Christian that they had been friends, that Christian had dandled Bligh's children on his knee. And that was the heart of it. This was a quarrel between men who were socially equal. Christian was an officer, too, the second in command. Bligh was disliked by many of the crew, and for his acrid temper they called him 'Bligh, the *Bounty* bastard'. But this was not a lower-deck rebellion: such events were rare. Without Christian's leadership there would have been no mutiny and no seizure of the ship. The descendants of Bligh and Christian, as I have found in discussion with them, take sides to this day, loyally defending their forebears.

The mutineers tossed the 1,115 breadfruit pots into the ocean and cast Bligh and eighteen men into the *Bounty*'s twenty-three-foot launch, knowing that their chances of survival were small. Bligh landed briefly in Tofua but fled when natives attacked his party on the beach and killed a man. He did not stop again. In six weeks he sailed 4,000 miles west to Timor, an astonishing feat of seamanship and will. After Tofua not a man was lost, although David Nelson, Banks's gardener, died later in Batavia. Back in Britain in 1790 Bligh claimed that one of the causes of the mutiny was the sensual nature of Tahiti itself, and its 'handsome and cheerful' women.

The following year Bligh sailed HMS *Providence* to Tahiti to fulfil Banks's breadfruit mission. This time there was no trouble and the seedlings grew rapidly in Jamaica and St Vincent. A tree was pointed out to me in St Vincent as a Bligh original. The slaves, however, disliked the taste of breadfruit and would not eat it.

One of Bligh's midshipmen aboard *Providence* was Matthew Flinders; and just as Cook had taught Bligh to navigate, so Bligh taught Flinders. The young man was a born explorer. With his friend George Bass, a fellow Lincolnshireman, he sailed a small boat around Tasmania in 1799 and showed that it was an island; and Bass gave his name to the strait between Tasmania and the mainland.

At that time most of Australia was not known. It was shapeless. No one knew how large it was, and it was not yet named Australia. Banks had always longed to see it more thoroughly explored and when, in 1800, he received a letter from twenty-six-year-old Lieutenant Flinders, proposing an expedition of discovery, it struck the right note. Flinders became Banks's protégé and was given command of a 334-ton sloop, HMS *Investigator*, and a commission to chart the coast. Banks paid for the scientific equipment. He also appointed Robert Brown, a Scot aged twenty-seven, to be the expedition botanist: he would become one of the outstanding naturalists of the nineteenth century. In addition Banks sent two artists, Ferdinand Bauer and William Westall, and a Kew gardener, Peter Good. One of Flinders's midshipmen was his nephew, John Franklin.

In 1801 Flinders sailed from England to Cape Leeuwin on

the south-west corner of Australia and turned east to chart the southern coast where the wind blows powerfully and huge seas hurl themselves against the cliffs. He sailed into the Spencer Gulf and trekked inland to climb in the mountains named after him, the Flinders Ranges. After refitting in Sydney he sailed up the coast of New South Wales and through the Great Barrier Reef, completing Cook's survey, and into the Gulf of Carpentaria and along the north coast. Although the *Investigator* was by now half rotted away and the crew were suffering from scurvy, Flinders completed the circumnavigation in 1803. His 'scientific gentlemen', Brown and Bauer, had amassed a valuable collection of plants, minerals, insects, bird skins and drawings while he himself had charted the harbours that would become Hobart, Brisbane, Perth, Adelaide and Melbourne, and outlined the shape of a continent almost as large as the modern United States. He gave it the name Australia.

In August 1803 Flinders left Sydney for Britain in HMS *Porpoise.* The ship was wrecked on a small island on the Barrier Reef. Flinders and thirteen men retrieved a small boat from the wreckage and sailed it 700 miles to Sydney to raise help for the castaways who remained on the island. Determined to get home he took the only vessel available, a twenty-nine-ton schooner, and sailed it across the Indian Ocean to the French colony of Mauritius. It was his bad luck that, unbeknown to him, war with France had been renewed. The graceless bureaucrat of a governor imprisoned him for more than six years, ignoring release orders from Paris. By the time he reached home, Flinders was in poor health and he devoted his remaining strength to writing the magnificent account of his Australian voyage. On the day it was published, in July 1814, the first copy was hurried to his home in London and put into his hands, but he was already unconscious and within hours he was dead.

On Banks's recommendation the government sent William Bligh to govern New South Wales. By now he was fifty-two. The eighteen-year-old colony of 7,500 people was raw and turbulent and run by crooked army officers who used the convicts as slaves, owned most of the livestock and controlled the trade in rum, the chief local currency. This scoundrelly 'Rum Corps' tolerated not a mutter of criticism. Bligh arrived in 1807 to restore order but his aggressive and vengeful nature hardly suited him to the job and quite soon his enemies labelled him 'Caligula'. Within a year the officers mutinied, arrested Bligh and packed him off home.

As far as we know Bligh gathered no specimens for his constant supporter, Sir Joseph Banks, but by now Banks was receiving seeds and plants from a network of amateur botanists in many parts of the world. After Captain Cook's death he did all he could to use and extend Cook's discoveries and actively supported voyages to the north-west coast of America. He was impressed by Archibald Menzies, a navy surgeon who had worked at the botanic garden in Edinburgh and who sent Banks plants from America, the Caribbean, China and Japan. Banks secured him a job in HMS *Discovery* which in 1790 departed under the command of George Vancouver on a long survey voyage to South America and the Pacific coast of North America. Vancouver was an outstanding navigator but, like Bligh, a prickly pear. He was irritated by Menzies's constant plant-hunting, begrudged the space for his specimens, and once, in a rage, shut him in his tiny cabin. Menzies, however, botanized cheerfully on. He brought home from Chile the seeds of the monkey-puzzle tree which became a Victorian status symbol.

Another of Banks's fruitful contacts was Sir William Hamilton, the British ambassador in Naples, whose wife, Emma, was Nelson's mistress. Hamilton was an authority on volcanoes and sent Banks descriptions of the eruptions of Vesuvius.

It was typical of Banks that he listened sympathetically to John Ledyard who sought him out in London. Ledyard was a Connecticut man who had served as a Royal Marines corporal aboard *Resolution* during Cook's third voyage and witnessed Cook's death in Hawaii. Anxious to become a fur trader in Canada he decided to walk there from London and, aiming for the Bering Strait, trekked through Sweden and Finland to St Petersburg and then on for more than 3,500 miles to Yakutsk in Siberia, one of the coldest places on earth. Today it takes six hours or so to fly to Yakutsk from Moscow: I went there once and felt my nostrils crackle and freeze in the temperature of minus forty centigrade. In

Yakutsk Ledyard found Joseph Billings, an officer in the Russian navy, who had also served with Cook. The suspicious Russian authorities ended Ledyard's eastward march and ordered him back across Russia to Poland. Such determination commended him to Banks as leader of an expedition into Africa. Banks gave him money and sent him to Cairo in 1788. Everything was ready and Ledyard was about to head south with his men when he fell ill and died.

Mungo Park, a Scottish doctor and naturalist, was another of Banks's protégés. Banks found him a job as assistant surgeon aboard an East Indiaman bound for Sumatra; and to Banks's delight he returned with a collection of fish and plants. In 1795 Banks sent the twenty-four-year-old Park to West Africa with instructions to travel inland and find the source of the Niger. After a terrible journey, almost a year of exhaustion, fever, hunger and attacks by bandits, he found the Niger River in July 1796, 'glittering in the morning sun, as broad as the Thames at Westminster'. He set a standard for sangfroid and understatement. After he was robbed and stripped in the bush he wrote in his journal: 'I saw myself in the midst of a vast wilderness, in the depth of the rainy season; naked and alone; surrounded by savage animals, and men still more savage. I was 500 miles from the nearest European settlement . . . I confess that my spirits began to fail me.'

It took Park eleven months to make his way back to his starting point. As he travelled he saw columns of slaves filing slowly down to the coast. At his journey's end he had no choice but to take passage in the only vessel available, a slave ship bound for the West Indies where he boarded a packet to Falmouth.

Banks also encouraged Park's second journey into the African interior in 1805. It was a disaster. The expedition was well organized and accompanied by thirty-six soldiers; but fever soon struck and the troops sweated towards their deaths in their thick red uniforms. Men who lacked the strength to ride or walk begged Park to leave them to die beside the trail. At night some of the party awoke to find jackals nipping and gnawing at their feet. All but four of the forty-four white men in the expedition had already died when spear-wielding warriors attacked the survivors as they tried to escape along a river. Park and his three companions jumped from their boat and drowned. Some years later Park's son went in search of his father and died of fever.

Banks's mark was also on the historic British expedition to China in 1793. That country, too, was another of the voids in European minds, a puzzle, closed to all but a few travellers. The curious longed to learn more. The Chinese, for their part, knew little of the British, and some of them believed stories that the British enjoyed eating Chinese people, uncooked. Hoping to improve trade with China, and especially an increase in the supply of tea to which the British were growing addicted, King George III sent the Irishman Lord Macartney to Peking as the first British ambassador. Macartney sailed with a team of ninety-five including Sir George Staunton and his son George Thomas Staunton, a prodigy of twelve. The boy's former mathematics teacher, John Barrow, was recruited as scientific administrator. There was also an artist, William Alexander, and two botanists, advised by Banks, who were to search for plants and new varieties of the tea shrub.

The expedition had numerous cultural and diplomatic collisions. Imperial courtiers in Peking were pleased with gifts of Wedgwood ceramic ware but horrified by the presentation of three exquisite London carriages. It was impossible, they pointed out, for the Emperor to sit in a coach driven by a coachman perched on a higher seat and with his back insultingly turned to the Emperor. Macartney met the Emperor, even though he had angered officials by making it clear he would not kowtow, or prostrate himself; but he deigned to descend to one knee to present the King's letter to the Emperor. The warmest part of the proceedings was the performance of young George Staunton, a clever linguist, who spoke to the Emperor in Chinese. Afterwards there was a chilly banquet and the Emperor inspected the other British gifts without enthusiasm. The mandarins coldly told Macartney to leave Peking. It was a defeat but not a failure. Everything that Macartney's people had seen and experienced added much to British knowledge of the Chinese way of life, politics, science and medicine, and Alexander made more than a thousand paintings and sketches which went some way to satisfying British curiosity about the country.

Later George Thomas Staunton became a friend of Banks, worked for the British East India Company in Canton and translated the basic laws of China, the first Chinese book rendered into English and published in 1810 when he was twenty-nine. It was an invaluable guide for British traders and scholars. Among many other things it provided details of the postal regulations. Government postmen covered the 1,200 miles between Peking and Canton in twelve days, and such was the insistence on punctuality that messengers who were late were punished by twenty blows with a bamboo rod. Postmasters who failed to keep post stations properly staffed were sentenced to fifty blows. More seriously, the punishment for high treason was 'a slow and painful death' by a thousand cuts as well as the beheading of all the guilty man's male relatives over the age of sixteen. Yet any judge who unjustly ordered an execution was himself put to death in like manner. British merchants proposing to trade in China learnt that the punishment for a master who killed his servant was strangulation.

'If English visitors at the court of Peking had been permitted to remain,' Staunton wrote of his time in China, 'they might have determined that a considerable proportion of the opinions entertained by Chinese and Europeans of each other was to be imputed either to prejudice or misinformation . . . as men of science we have much to learn respecting the arts successfully cultivated by an eminently industrious and ingenious people.'

John Barrow returned from the Macartney mission as an author and authority on China; and, following four years in the Cape Colony, an expert and writer on southern Africa, too. In 1804, aged forty, he was appointed second secretary at the Admiralty, the equivalent of a permanent secretary today, a considerable achievement for the son of a humble Lancashire farmer. Barrow was one of the officials who saw Nelson briefly before the admiral hurried off to board his ship for Trafalgar. As a member of the Royal Society Barrow got to know Sir Joseph Banks well, attended events at Banks's house in Soho Square and listened to his expansive views on exploration. Banks died in 1820 and to the end of his life remained fascinated by the mysteries of geography, curious about the natural world and ever enthusiastic in his support for exploration. He had the sort of eminence Barrow wished for himself.

Barrow, by contrast, was seized by the romance of imperial discovery, by a passion to see the British flag planted in the unmapped places, and between 1816 and his retirement in 1845 he sent many expeditions to Africa and the Arctic, and promoted Sir John Franklin's ill-fated expedition. No doubt Barrow admired both Joseph Banks's grandeur and his longevity, his forty-two years as monarch of the Royal Society, but Barrow himself did not do badly as an inspirer and orchestrator of expeditions by gallant men. He reigned in his Admiralty post for nearly forty years.

In the graveyard of old St Mary's church by the River Thames in Lambeth, London, lie two men whose lives were shaped by botany. The Tradescants, John the Elder and his son John the Younger, pioneered English horticulture; and the disused church was restored in the 1970s as a museum of gardening honouring their genius. Britain has been immeasurably enhanced by its immigrant plants: the Romans introduced the walnut tree and grew vines, the Norway spruce and the pomegranate came in the fourteenth century, the hibiscus from Africa and the yucca from America in the sixteenth. But botany truly flowered in the seventeenth century: the first botanic garden opened in 1621 at Oxford, the cedar of Lebanon made its debut and the Tradescants filled gardens with novel colours, shapes and perfume. Tradescant senior worked successively for the Earl of Salisbury, Lord Wotton and the Duke of Buckingham and was Keeper of His Majesty's Gardens, Vines and Silkworms. He planted gardens with the hundreds of shrubs, flowers, vines and trees that he collected during forays in France and the Low Countries in 1609 and 1611. From a journey to Russia in 1618 he brought back the Arctic rose and the Siberian larch. In 1620 he voyaged to Tangier aboard a ship sent to smash Barbary Coast pirates and returned with clover, turpentine trees, roses and much else. He grew the first English horse chestnut and raised dozens of varieties of tulips, apples and plums. His son voyaged three times to the New World and returned with the swamp cypress, the tulip tree and the acacia; and he was appointed his father's successor as Keeper of His Majesty's Gardens.

The Tradescants collected more than plants. Their Ark of Curiosities was the founding nucleus of the Ashmolean Museum in Oxford, being intended, as John the Younger said, to be 'a benefit to further enquirers into the various modes of Nature's admirable works'. It included Russian snowshoes and the deerskin cloak of Powhatan, chief of the first American Indians the English settlers met in Virginia in 1607.

The Tradescants made the gathering of seeds a British enthusiasm. Joseph Banks's energy made plant-hunting an adventure in an age when sailing ships could range almost anywhere and so much was new. A whole tribe of botanists and nurserymen travelled to the earth's ends to find the shy and the beautiful. Often exhausted and ill, they limped back from mountains and forests with ever more gorgeous and subtle plants and revolutionized the look of the British landscape and British gardens. At Banks's direction they made botany a servant of empire, dispatching seeds to the colonies to create plantations and new economies of rubber, sugar, tea and coffee, all requiring gigantic armies of slaves and labourers to work them.

Close to where the Tradescants lie in St Mary's churchyard is the tomb of Vice-Admiral William Bligh whose life was also shaped by botany; and, in particular, by his botanic majesty, Sir Joseph Banks.

7

The African meteor

'Let us be satisfied with our gains and, being rich, let us try to become righteous.'

Henry Brougham, 1803

Thomas Williams, the Welsh copper magnate, pointed out to shipping merchants that a slave ship with a coppered hull sailed more swiftly across the Atlantic than an uncoppered one. He had no need to remind them that a good bow-wave was good business. Profits in the markets of the Caribbean and South and North America depended on the condition of the human merchandise. Slaves too sick and weak to fetch any price at auction were dumped on the beach and left to die. Whether Williams's observation had an element of the humane as well as the practical we do not know; but of course he was right. Coppering and advances in rigging and navigation improved the speeds of slaving vessels and increased the survival rates of the captives cooped and fettered beneath the hatches.

Slaves resilient enough to begin work as soon as they were ashore sold for high prices. A typical vessel landing 268 slaves in Jamaica in the 1760s cleared a profit of more than £8,000. In the ten years to 1793 British ships carried about 400,000 slaves from Africa to the West Indies and America and made £15 million in profits for their backers.

These were phenomenal sums; and the chance of making big money was not limited to the big slave-trade players, the wealthy merchants of London, Bristol and Liverpool. Small-time businessmen, too, bought eagerly into the commerce. The slave cake was so rich that even a thin slice of it delivered handsome dividends.

'Almost every man in Liverpool is a merchant,' it was reported in 1795. 'The attractive African meteor has so dazzled their ideas that almost every order of people is interested in a Guinea cargo. Many of the small vessels that import about a hundred slaves are fitted out by attorneys, drapers, ropers, grocers, tallowchandlers, barbers, tailors . . .' With luck a single voyage could yield a man enough profit to found a family fortune, build a mansion, patronize the arts and transform his daughters into brides for aristocrats.

Sugar was the fuel. The European taste for it was an expensive addiction. Honey was the traditional sweetener but everyone who could afford to do so poured sugar into the new drinks of tea, coffee and chocolate, and used it to make puddings and jam. Around 1700 Britain's annual sugar consumption was about five pounds a head; and within seventy years it was three times as much. Today the average Briton eats a pound a week. For three centuries the islands of the West Indies were Europe's bottomless bowl of sugar, the cause of jealousy and war between Britain, France, Spain and the Netherlands. These four countries, with the United States, were the major slave traders and the architects of massive servitude. In the eighteenth century Britain developed the largest share of the Atlantic slave business. The routes to Africa and thence to the Americas and back to Britain were the roads to riches.

Profit from the Caribbean sugar and slave colonies financed the large industries of shipbuilding, ship repair and supply, sailmaking, chandlery and rope manufacture. It also supported the export of wool, cotton and firearms. Plantation owners swelled with money and influence and formed a political lobby, the West Indies interest, with forty Members of Parliament speaking for it. Since the slave trade earned a large part of Britain's overseas income, perhaps four-fifths of it at one time, the owners raised a powerful voice.

Portuguese sailors were the first European slave dealers. Working their way south along the coast of West Africa in the middle of the fifteenth century they found slave markets that had flourished for more than a thousand years. Here the rulers and merchants of African city states sold captives to Arab traders who crossed the Sahara desert from the Mediterranean and returned with their purchases to sell them on as household servants, agricultural labourers, harem eunuchs and soldiers. For Africans and Arabs alike the business was as natural and traditional as it had been in the civilizations of ancient Egypt, Greece and Rome; and the commerce in slaves, black and white, had continued in Europe into the Middle Ages. African kings and traders rarely sold their own clans into slavery. To get their supplies they usually raided their neighbours and the forest tribes far inland.

The Portuguese were the dominant Europeans in the West African slave trade for two centuries. Britain became a participant through the Royal African Company chartered by King Charles II in 1672 to replace the failed Royal Adventurers into Africa. Two years earlier the King had licensed the Hudson's Bay Company to trade in Canada: animal skins and slaves were mercantile commodities. The increasing participation of Europeans in the slave trade made it a wholesale business and multiplied its horrors. African kings, no more or less greedy than white merchants, conspired in the trade without qualm and profited as much as any Liverpool or London tycoon. The slave-catchers tore their victims from their homes and families, yoked them neck to neck and drove them in stumbling coffles towards the coastal markets, forcing them sometimes to tread the crumbling bones of their predecessors who had died on the march. After weeks of travel the slaves saw, for the first time in their lives, the shining ocean; and many of them shrank from it in terror, their fears intensified by the sight of ships which they took to be huge birds with white wings. Here by the sea they had their first encounters with the strange long-haired white men who to their eyes seemed so alien.

The traders corralled them in stone compounds for weeks or months to await the arrival of the slave transports. Men and women were brought naked to a ship's surgeon who cursorily checked them over, examined their teeth and genitals and made them jump into the air as a crude test of their fitness. When the deals were struck in the market place, overseers crammed the quivering slaves into the ships for the voyage across the Atlantic to the West Indies, the Middle Passage as it was called.

Depending on its size, a ship carried between 200 and 600 people, sometimes more, and each individual's living space aboard was five or six square feet with headroom as little as four feet two inches. Captains chained the slaves together, not least to prevent them jumping overboard to seek death by drowning or in the jaws of sharks. Those who refused food in the hope of starving to death were flogged to encourage them to eat. 'They have a more dreadful apprehension of Barbados', remarked one captain, 'than we have of hell.' The voyage usually took from thirty to fifty days. Between a tenth and a fifth of the slaves died at sea of dysentery and smallpox, dehydration and suffocation. When disease struck, captains tried to limit its spread by throwing the seriously ill overboard. Some slaves died following the exemplary whippings and tortures ordered by ships' officers in the aftermath of desperate uprisings. For all their civilized graces at home, merchants and slave ship captains were all parties to cruelty and murder, as they well knew. Some captains were more humane than others, but no nation was better than another in its treatment of slaves.

During four centuries of slave-dealing, European merchants carried more than 11 million Africans across the Atlantic and transplanted them in Brazil, the Caribbean, Mexico and the United States. It was believed that only black slaves could endure the harsh conditions of tropical plantation work. Slave muscle and slave sinew cleared the forests and hillsides of the Americas. They sowed, harvested and processed sugar cane, coffee, cotton and rice; and they worked the silver mines of Mexico. Colonists could not get enough slaves and for many years insisted that they could not possibly survive and flourish without them. Until the 1820s, for every white settler who went to make a new life in the Americas five black men and women were transported there, most of them to endure lives of suffering.

Slaves formed the core of the Atlantic trade but not the whole of it. On the first leg of the triangular voyage ships left Europe for ports in Gambia, Ghana, Benin, Dahomey, Sierra Leone and elsewhere along the west coast of Africa. British ships took a wide range of trade goods to Africa: wool from England, Wales and Scotland, cotton checks from Manchester, muskets from Birmingham, blankets, iron, copper sheets, brass bowls, gunpowder, glass, salt, trinkets, beads, hats, horses, knives, pots and pans, tobacco, brandy, rum and gin. They also carried pretty cowrie shells from the Indian Ocean. African merchants valued them as a useful currency. Cowries could not be counterfeited.

The inhabitants of Barbados and other islands were informed that a slave ship was in the offing by its foul smell upon the breeze. Those captives who had survived the weeks in the dark holds were auctioned for cash or for sugar. Purchasers often turned the sales into rowdy scrums, pinching the flesh of the frightened slaves and shouting out their offers to the auctioneer.

In the museum housed in the old British military prison on the edge of Bridgetown in Barbados, I saw what at first looked like a piece of silver jewellery. It bore the initials GHC. In the years around 1800 it belonged to George Hyde Clarke, a magistrate and sugar planter. It was his personal brand. Heated in a fire it seared his mark of ownership on the upper arms, breasts and shoulders of the young men and women he had just bought in the sales. What made it more shocking to modern eyes was that an instrument of pain and subjugation had been fashioned into a decorative piece. Slaves were almost always branded. The Society for the Propagation of the Christian Gospel scorched the word 'Society' on the chests of those who worked its Barbados estate in the eighteenth century.

'You cannot hide the past,' a Barbadian professor of history remarked. 'Some of my students tell me that when they visit the old sugar plantations they can hear the groans of their ancestors.'

Plantation bosses awarded names to their new slaves, Hannibal, Caesar, Jack, Eliza and Rachel, for example, and put them to work at once. Men were reckoned to be good for about ten years of labour, and their life expectancy in Barbados was twenty-nine. It was easier and cheaper to work them to death and replace them rather than feed and treat them well. Slaves who tried to blunt their hunger by chewing sugar cane were flogged. In the opinion of a British army officer, who had doubtless witnessed considerable harshness in his military career, Barbadian slaves were 'treated with the greatest cruelty'. He described a man he had seen, flogged so that he could not sit on lacerated buttocks and locked into an iron collar with long spikes so that he could not lie down and rest.

After they had sold their cargoes slave ship captains loaded rum, sugar and tobacco and sailed for home, the third leg of the iniquitous triangular trade. The protectionist Navigation Acts of 1647, 1650 and 1651 ensured that profits from sugar and slaves stayed in British hands. Cargoes bound to and from the British colonies were carried exclusively in British ships commanded by British captains and manned by crews who, under the law, had to be three-quarters British.

Of the 11 million slaves transported from Africa around 4 million were shipped to the Caribbean. Sugar and slavery shaped every West Indian society. The story of Barbados, for example, was replicated in most of the islands. A British party of sixty settlers and six Africans landed there in 1627, built a church on the west coast and cleared the thick forests. For a while they eked out a living from cotton and tobacco but by the 1640s some of them were planting sugar cane from Brazil and showing by their profits where the future lay. In 1642 the island counted 11,000 settlers, English, Irish, Scots, Welsh and Dutch, a varied lot of fortune-seekers, hard-up younger sons, exiles and runaways and a large body of luckless white indentured labourers. In the 1650s Barbados exported all it could produce of sugar, molasses and rum. Only twenty-one miles by fourteen it was the most important and wealthiest English colony in the New World, the sweet heart of Britain's empire of the Americas. By 1660 all the forests were gone and plantations dominated the landscape. The census of 1680 recorded 23,000 whites and 40,000 slaves working on 700 plantations. The estate owners, said a governor of Barbados, enjoyed 'sumptuous houses, clothes and liberal entertainment'.

By the eighteenth century the island had become a well-tilled garden planted mostly with sugar cane and also with millet, yams, potatoes and vegetables. In the hectic cane-cutting season between January and July, men chopped the ten-foot stalks and women tied them into bundles and loaded them on carts to be hauled to a wind-powered mill. Here in stifling heat and dust slaves fed the cane through crushers and rollers and boiled and rendered the juice into molasses and rough sugar. The cane was milled as soon as it was cut because bacteria rapidly reduced its sugar content. One of the sugar magnates, James Drax, built the island's first windmill, to a Dutch design, with rollers large enough to crush eight tons of cane a day. In the sugary heyday of the plantations the sails of more than 500 mills were a distinctive feature of the Barbados landscape, but only one survives today, the Morgan Lewis mill, built early in the eighteenth century on land owned by a Welsh planter. It worked for two and a half centuries and people remember that its canvas sails were still turning in 1945.

Like the Morgan Lewis mill, the old semaphore signal tower at Gun Hill, with its commanding view of the island and the sea, is a remnant of the sugar age and tells of the anxiety that haunted the sucrocrats and disturbed their rum-soaked ease. Following a rebellion on the plantations in 1816 the tower was one of a chain of six laid across the island by the military two years later to warn of any trouble. In Jamaica, which overtook Barbados in sugar production in the eighteenth century and became the Caribbean's chief producer, the fear of black revolt was increased by each of a dozen uprisings. Armed bands of escaped men took to the hills and attacked plantations. By 1778 there were eleven blacks to each white in Jamaica, but if the heads of planters often lay uneasily on their pillows up at the Great House their fears of insurrection did not lead them to treat their slaves better or to reduce the whippings they inflicted. Rather, they enacted severe laws to deal with what to their minds was insolence and insubordination. In the meantime, British ships continued to bring out iron manacles and chains manufactured by British workshops. There was always a steady demand.

In the latter part of the eighteenth century, a time when British ships were carrying more than 40,000 slaves a year from Africa, the opponents of slavery in Britain slowly gathered strength. Theirs was a movement of writers, clergymen and politicians, and although they commanded the moral argument they found the journey to abolition of the trade a very long one. A parliamentary petition failed in 1783 after the Home Secretary stated what was widely believed to be common sense and the reality of economics, that the slave trade was necessary to the financial well-being of every European country. Indeed, it was self-evident to businessmen that the prosperity of many cities and industries depended on the labour of colonial slaves.

The city of Bristol had a tradition of aggressive merchant

venturing. It accumulated wealth from sugar and slaving in the late seventeenth century and by the early years of the eighteenth was the country's chief sugar port. Into the nineteenth century the bulk of its wealth continued to be the West Indies commerce.

By then, however, it had been overtaken by Liverpool whose even more ambitious merchants gloried in the Caribbean trade, built larger ships, had better docks, paid their crews lower wages and cheerfully and illegally traded slaves to Spain. They secretly unloaded their rum and sugar cargoes from the Caribbean in the Isle of Man to avoid paying duty in Liverpool. These hardheaded men became renowned civic figures, mayors and Members of Parliament, the founders of dynasties whose sugar money funded banks and factories. The father of William Gladstone, the great Victorian Liberal leader, was a Liverpool merchant enriched by Jamaican estates. As the champions of the trade pointed out, it was not only London, Bristol and Liverpool which prospered from slavery. Manufacturers in Manchester and Birmingham agreed that much of their business depended on it; and other ports, such as Plymouth, Falmouth, Poole, Deal, Glasgow and Dublin also had a stake. In France the 'ebony merchants' made Nantes the chief slave-trade city. Its counterpart in Spain was Seville and in Portugal the prosperous port of Lisbon.

The slaving interests ranged a powerful economic case against the abolitionists and asked them whether they wished to be the authors of ruin. As for the moral question they suggested that, surely, the 'savages' had been removed from a benighted existence and introduced to a happier one; that it was better that they were slaves in the Americas than butchered corpses in Africa. They sought to show that the slave business was compatible with a Christian outlook, arguing that black people were inferior and intended by God to work for whites.

The abolitionists meanwhile compiled dossiers of horrors, collecting accounts of the realities of slave voyages, the stories of the captains who threw slaves overboard when water ran short, cut off their hands, flogged them and rubbed hot pepper into their wounds. The hero of this investigative work was Thomas Clarkson, a Cambridge graduate in his twenties, a brave and dedicated inquirer who travelled to Bristol and Liverpool to amass eyewitness accounts of atrocities. In the end he wore himself out, but in 1787 the compelling nature of his evidence brought about the launching of the Committee for Effecting the Abolition of the Slave Trade. William Wilberforce, the young Member of Parliament for Hull, joined the campaign and became its moving and determined spokesman. He had powerful supporters in Parliament: William Pitt, Charles James Fox and Edmund Burke. But these and other well-known figures were frustrated by the West Indian lobby which defeated them in Parliament time and again, countering the arguments against the 'greatest evil which has ever afflicted the human race' by warning that the end of the slave trade would be the end of British supremacy in commerce and at sea.

At the turn of the century, thirteen years after their campaigning started, the abolitionists observed only the slave trade's seemingly unstoppable success, record profits and the largest harvests of West Indian sugar. British ships were carrying more slaves than ever, running a regular supply service of 50,000 captives a year to the West Indies and America. In 1798 Liverpool sent three ships a week to Africa. Nelson, a seasoned West Indies hand who often visited Barbados, firmly supported the plantation owners, declaring that he had been 'bred in the good old school, and taught to appreciate the value of our West Indian possessions'. He promised to raise his voice against the 'damnable doctrine of Wilberforce and his hypocritical allies'.

The Royal Navy of Nelson's time counted many black seamen among the crews of its men-of-war. One of the numerous paintings of the Battle of Trafalgar shows a black sailor among the men around the fallen Nelson. The navy had relatively liberal views on recruiting black men and, with good hands always scarce, ability counted more than colour. The professional skills of a Jamaican mulatto seaman in the navy took him up the ladder to captain's rank in 1800.

Thousands of black sailors, servants and freed slaves lived in Bristol and Liverpool at that time and London, ever the city of opportunity and diversity, had a black population of around 20,000. The first African slaves had arrived in the capital in 1555, a source of curiosity and anxiety, and Queen Elizabeth I was moved in 1596 to tell the city authorities that there were too many 'diverse blackamoors brought into these realms'. But in the eighteenth century London, Bristol and Liverpool were trading cities of the world and it was natural enough for black seamen to circulate and settle. Slaves were imported to work in the British mansions of plantation owners, and black servants were fashionable in the households of merchants and aristocrats, and were given as gifts among acquaintances. Black soldiers who had fought for Britain in the American War of Independence also arrived to settle and joined London's poor. Notices in the newspapers advertised black servants for sale and also reported runaways, describing the absconders' appearance and noting the brand mark burned into the skin.

For all the blustering of the majority who shared Nelson's opinions about the value of the sugar islands, the moral mood in Britain began to change. An evangelical spirit and ideas of humanity and guardianship grew stronger; and the issue of the slave trade became insistent. Henry Brougham made the point that the British had been 'ringleaders in the crime', that 'the fruit of our iniquity has been a rich empire', and that it was time to be 'satisfied with our gains'. Warning of trouble ahead, he asked: 'When a fire is raging windward, is it the proper time to be stirring up everything that is combustible in your warehouse?'

Wilberforce battled on and reaped the reward for his persistence as well as his eloquence. Parliament outlawed the slave trade in 1807. Certainly the people's conscience had been pricked but by then slavery's importance to the British economy was waning and there was a growing acceptance of the argument that free men were actually more productive than slaves. The islands were no longer the places where money sprang from the warm earth. But above all, in a civilized country, an opinion had grown that the trade in slaves was not right. As new opponents of slavery the British could begin to see themselves as morally better than the awful slave-owning French and the hypocritical proprietors of the American 'land of the free'.

Nevertheless the finance houses of London continued to back slaving voyages and Liverpool still built slave ships for foreign customers. The navy, meanwhile, symbolized Britain's new moral stance by operating anti-slavery patrols off the African coast. The ending of the transportation of slaves was not by any means the end of the institution of slavery itself. It was not until the 1830s that slaves in the British Caribbean colonies were finally freed. Slavery persisted in the United States, Spain and Portugal into the 1860s and in Brazil into the 1880s. In Barbados, in 1838, 70,000 slaves composed a folk song on their emancipation, giving thanks not to those determined reformers who had worked patiently for years against the slavery lobby, but to the young Queen Victoria who had nothing to do with it. 'Hurrah for Jin-Jin,' they sang, Jin-Jin being their name for the Queen. In mid-century a British Prime Minister, Lord Russell, proclaimed Britain's moral integrity in its fight against 'the curse and crime of slavery' and saluted 'the high, moral and Christian character of this nation'.

8

The painted East

> 'The dignity of the artist lies in his duty of keeping awake the sense of wonder in the world.'
>
> G.K. Chesterton, 1928

By the second half of the eighteenth century the British were strongly drawn to the vividness of India. They knew something of the beauty of carpets and silks, muslins and chintzes and brocades. They had sniffed the suggestive smells of spices and perfumes. More than a few had tasted curry. In London they were startled by the gaudy Indian coats sported by swells on home leave from Calcutta, the capital of British commerce in Bengal. Indeed, William Hickey, a lawyer of the East India Company who kept a famous diary in his years in Calcutta, wore coats so garish when he first went back to London that women laughed behind their fans and men leered, obliging him to discard his oriental colours and wear clothes of duller stuff.

People knew something, too, of the extraordinary lives of the British nabobs, whose nickname derived from nawab, a Muslim nobleman's honorific. London newspapers derided these fortune-makers as 'banditti' and 'plunderers of the East' for the immense wealth they accumulated through their trade in spices and textiles and precious stones. A poem of 1773, 'The Nabob, or Asiatic Plunderers', satirized their excesses, as did Samuel Foote's play *The Nabob.* It was difficult for any writer to exaggerate the swashbuckling of the gentlemen of the East India Company, and there was much truth in the caricatured luxury and lavishness of their existence, the stupendous squandering, drinking and feasting. The nabobs' retinues were famously large, swarms of footmen, messengers, valets, cooks, hookah-men, hairdressers, stewards, palanquin-bearers, gardeners and grooms. William Hickey kept sixty-three, a modest household by comparison with many.

Company men went to India not to govern but to prosper. They were middle class, determinedly on the make and energetically 'shook the pagoda tree', the pagoda being a unit of gold currency. They grew gloriously rich and filled the holds of home-going ships with fat cargoes of pepper, cotton, silk, saltpetre, indigo dye, ivory and silver. There was something about them of the frantic modern money-makers in the City and on Wall Street. In their carousing and high jinks and frolicking with mistresses they acquired, as the historian Philip Mason put it, 'the vices of the aristocracy'. But there was a price. The nabobs gambled with their lives, knew the constant fear of sudden death and heard the funeral bell toll frequently for friends and colleagues extinguished by malaria, cholera and dysentery. Of twenty-eight young men who went out to work for the Company in 1762 seventeen were dead within ten years. If not disease there was always something else: in 1800 poor Rose Aylmer, aged twenty, died in Calcutta of 'a most severe bowel complaint, indulging too much with that mischievous and dangerous fruit, the pineapple'. Meanwhile, a joker composed an undertaker's song:

> This is the job
> That fills the fob,
> O, the burying a nabob for me.

But while the nabobs had their health they gulped their claret and danced until daybreak, usually ignoring the health warnings in the Calcutta newspapers about drinking during the monsoon. The survivors voyaged home with fortunes to buy country mansions and shares in banks and new industrial enterprises. Most of the Company's officers and gentlemen were English, but a significant proportion were Scots. A few were Welsh, among them five boys from Brecon who returned wealthy after their years in India and, variously, went into banking, canal construction and public life. David Jones was one of them. He arrived home after twenty-nine years' soldiering, minus a leg lost in action but rich in booty, and although he was always 'the jolly old major, drunk as usual' at social occasions, he was, when sober, a leading oriental scholar and authority on Persian. In this he was the other side of the coin, in the tradition of men who were serious students of Indian religion and literature in spite of the heat and sickness and numerous distractions.

All these old India hands could enthral their families and friends with their stories. The tales and treasures they brought back spread an idea of the excitement, strangeness and drama of India; but there were few pictures available to illuminate the words. Many of the drawings and paintings of landscapes, towns, villages, palaces and temples were made in hazy recollection, after travellers had returned. Books were illustrated by engravers who had never seen India and who embellished the sketches brought back by amateurs. India was a vision conjured by storytellers.

The expansion of British possessions and influence in India was matched by a growing curiosity concerning Indian life and a hunger for pictures that were true and not speculative and fantastic. There was a need for the corrective eye of the professional.

A number of portrait painters had worked in India but the first trained landscape artist to do so was William Hodges. In 1780, five years after he had made a reputation as the artist on Cook's second voyage, he sailed from London for Madras. The landing there was famously risky, and if you walk along the beach today, in the hot whipping wind, it is not difficult to imagine the drama and the fun when the East India Company ships arrived. They anchored offshore and awaited the messengers who paddled out with letters carried in their pointed waterproof hats. Passengers were unloaded into flimsy boats and ferried through the thundering surf to the beach. Here gallant young officers, manly, drenched and eager, waited with open arms to carry pretty and squeaking young women to safety.

The startlingly clear light and striking colours of the coast thrilled Hodges. It was, he said, all 'totally new to the eye of an Englishman, just arrived from London, who, accustomed to the rolling masses of clouds in a damp atmosphere, cannot but contemplate the difference with delight'. Under India's blue cloudless sky 'the mind assumes a gay and tranquil habit'.

From the start he was determined to keep as sharp and dramatic the surprise of the new. 'Of the face of the country, of its arts and natural productions, little has yet been said. Gentlemen who have resided long in India lose the idea of the first impression which that country makes upon a stranger: the novelty is soon effaced . . .'

Artists like Hodges painted at a time when art and the definition of beauty were evolving and being argued over by poets, philosophers and painters. The Reverend William Gilpin, a Cumbria man and headmaster at Cheam in Surrey, boated down the River Wye in Herefordshire in 1770 and was so struck by its loveliness that he published a book setting out laws on how landscape should be observed. Thus he invented the picturesque, the word, the cult and the cliché. Some of the devotees of his rules of observation stood with their backs to the landscape holding up Claude glasses, oval mirrors, framing a perfect picture. The code of the picturesque encouraged artists to improve on nature while remaining true to its spirit.

Hodges was captivated by the picturesque qualities and possibilities of Indian scenes, by the temples, the costumes and headgear of men and women, the pools, the banyan trees putting down roots from their branches. There was the classic among Indian pictures, the bathing scene: 'To a painter's mind a beautiful female form ascending steps from the river, with wet drapery, which perfectly displays the whole person'. Indian women were an irresistible subject for British artists and many painted them as classical beauties, romantic, charming, coy, erotic, innocent and always graceful.

Hodges offered a simple definition of his artistic purpose: 'To give dignity to landscape'. He moved beyond the established formula and sought to show not mere beauty but something of the character and the history of the scene under his gaze, the truth rather than 'fanciful representations'. In spite of his poor health, he travelled widely in India for three years, partly under the patronage of Warren Hastings, the Governor of Bengal. After visiting Calcutta he sailed up the Ganges and drew scenes at Benares, the holiest city of the Hindus and a place of pilgrimage for 3,000 years, whose riverside architecture he deeply admired. Here he witnessed a sati, a wife mounting her husband's funeral pyre and perishing in the flames, a ritual that horrified Europeans. Later he travelled to Agra and saw the Taj Mahal – 'a most perfect pearl on an azure ground' – and to the hilltop fortress of Gwalior and the gracious and cultured city of Lucknow. He returned to London in June 1784 and in the following years exhibited at the Royal Academy and published two volumes of engravings. In 1795 he fell foul of the Duke of York who, smarting from defeat in Flanders by a French army, disliked the moral tone in Hodges's paintings of war and peace. It was not the artist's job, the Duke said, to paint pictures which 'might impress the mind of the inferior classes with sentiments not suited to the public tranquillity'. Hodges was too political, and the Duke closed his 1795 exhibition. Two years later Hodges died penniless, perhaps of an overdose of laudanum.

Even so, his lively impressionistic pictures of temples and countryside, executed with vigorous brush strokes, helped to change British perceptions of India. 'Many other tours in that interesting country', he wrote, 'might be undertaken by the enterprising artist.'

Thomas Daniell, a hard-up Surrey landscape artist, who had started out as a bricklayer's mate and entered the art world by painting scenes on the doors of stagecoaches, was encouraged by William Hodges's example to seek his own fortune in India. By then he was a practised landscape painter but badly in need of work, and he saw hope in the new appetite for oriental pictures. Like Hodges, he was trained in the classical tradition, had also imbibed the ideals of the picturesque and was well aware of the public taste for it. He applied to the East India Company for permission to travel to India. The Company carried about a thousand passengers a year to China and India and strictly vetted and limited the number of artists because it did not want responsibility for lazy daubers. But Thomas was approved and, in April 1785, he and his nephew and apprentice William packed paints and pencils and boarded the Company vessel *Atlas* at Gravesend. It was bound for China and they would have to take a ship from there to India. Like all the lordly East Indiamen it was a ship well armed against attack by pirates and the French, and designed to carry large cargoes. Thomas was thirty-six and William fifteen: on the death of the boy's father Thomas had become his guardian.

The *Atlas* reached China after four and a half months, and it was another three months before the Daniells arrived in Calcutta. They set up at once as painters and picture restorers. Thomas started by producing twelve views of Calcutta which were immediately successful. William Hickey bought a set, being 'as great an encourager of merits as my humble means will allow'. The Daniells were also engravers, making aquatints which yielded 200 or so prints from a single copper plate. At first they were by no means expert at this new technology and took two or three weeks to produce a single plate. Thomas sweated over the process. 'It has almost worn me out,' he told a friend, 'a devilish undertaking.'

Industrious and determined, the Daniells were fortunate, too, that they kept their health at a time when many of their countrymen died suddenly of fevers, when it was a commonplace for a man to attend the funeral of a friend with whom he had dined the night before. By 1788 they had earned enough to finance a nine-month journey in the northern plains and hills, 'up the country', as people said. Since travellers risked ambush by bandits they married their art to their boldness. They sailed up the River Ganges, travelling light with only seven servants in two boats. At Kanpur, Cawnpore to the British, they joined a military party for the journey to Agra, with sepoys to guard the column of elephants, camels and horses. At that time forces of the Maratha kingdom were hostile in parts of northern India. In January 1789 the expedition camped by the River Jumna at Agra and the Daniells sketched the Taj Mahal, 'a spectacle of the highest celebrity', as they described it.

Uncle and nephew worked at folding tables, one drawing the basic lines, the other the details, the sketches passing to and fro between them. These, with notes on colours, formed the foundation of the elaborate finished pictures to which both Daniells contributed. For the sake of accuracy and speed they often used a camera obscura. A forerunner of the photographic camera, it comprised a large box in which a scene was focused through a lens on to paper and traced with a pencil.

The Marathas had kept William Hodges out of Delhi, but their tide had receded by the time of the Daniells' journey and the city gates were open to them. They had a scoop, the first English artists to draw Delhi's magnificent forts and monuments. The grand courtyard of the Jama Masjid, the largest mosque in India, remains exactly as they painted it; but the eastern gate, the noble subject of one of their pictures, is today obscured and fenced by tall iron railings against which scores of people have built rough shelters. The Daniells worked in light of exquisite clarity. In the past dozen years an obscuring cloud of pollution from industrial chimneys and vehicle exhausts has thickened over Delhi, although there are hopes that it may not be permanent. The Daniells painted the red sandstone fortress, the Purana Qila, and showed a lone camel passing its western gate, a tranquil scene. Today, although traffic thunders along the main road, there is a view of the gate roughly as the Daniells saw it. You may see a camel, too, but it will be frowning in a rolling cloud of exhaust.

The curves and triangles of the Jantar Mantar observatory built in 1724 in the heart of Delhi remain much as they were when the Daniells drew them, setting up their tables on what is now the palm-shaded lawn of the Imperial Hotel. Today every one of the hotel's four floors exhibits the works of the Daniells, Hodges and other early India artists.

After two and a half weeks in Delhi the Daniells turned north to the Himalayas. Again they were pioneers, travelling with four officers and fifty soldiers, exploring country barely known to the British. 'The traveller encounters no villages,' they warned. 'He must carry with him the means of subsistence, or perish.' They were the first Europeans to visit the Ganges town of Srinagar, in what is now northern Uttar Pradesh, and were awed by the panoramas of distant snowy mountains. The beauty of the Himalayas, they felt, was like the 'visionary effect of twilight' and the 'magical radiance of fireflies', beyond any artist's ability to capture.

They travelled from Lucknow to Benares where they painted one of the bathing ghats, the riverside steps crowded with pilgrims immersing themselves. Many of the buildings have changed since the Daniells were there, but not the timeless scene along the Ganges. Before dawn, worshippers flock through the alleys towards the ghats, feet pattering and clothes rustling silkily, to greet the sunrise in an ancient and moving spectacle. In the theatrical milky light, as the first red sliver of the sun appears, men and women dunk themselves in the broad river and raise their arms in salute.

The northern trip was successful, and on their return to Calcutta the Daniells sold enough of their work to fund a tour in the south. The British had only scanty knowledge of this region and it was only sketchily mapped, so that the Daniells had the excitement of journeying in what was, for Europeans, country largely untrodden. Fortunately for them, the East India Company's forces were completing their successful war against Tipu Sultan, the ruler of Mysore. Arthur Wellesley, who later became the Duke of Wellington, distinguished himself in this campaign. He had arrived in India weighed down by debt and returned after nine years with a fortune of £43,000. Again the Daniells had a scoop. They were the first artists to paint many of the battlefields and forts, and there was a large demand for such scenes.

From Madras they travelled on foot and on horseback and also in swaying palanquins shouldered by eleven men. With their grooms, baggage cart drivers and bearers, they had a team of forty-eight servants. Sketching busily, they went south to Trichinopoly and on to the stupendous temple at Madurai. No building in India is more gorgeous. Sculptors and painters have adorned its façades with thousands of figures of gods and sprites, saints and demons, elephants and cobras in a riotous celebration of Hindu art, all entwined, crowned and jewelled with breasts and bellies and cheeks swelling like ripe fruit.

The Daniells reached Cape Comorin, the southern tip of India and the meeting place of three seas, the Indian Ocean, the Arabian Sea and the Bay of Bengal, a holy place always crowded with pilgrims who come to worship in the temples and bathe in the pools among the brown rocks.

Thomas and William worked their way back to Madras by way of Rameswaram and Tanjore. Their southern trip was an artistic and commercial triumph. In February 1793, after eleven months in the south, they sailed from Madras to Bombay to paint in western India. Here they became friends of the artist James Wales who showed them, among other sites, the Elephanta Caves off Bombay. They left for Muscat in May, bound for London, but returned to Bombay on hearing that Britain and France were again at war. They continued to travel and work until at last they could make the long journey home with their immense collection of drawings, watercolours and field notes. This was their treasure chest, their rich store of work and inspiration, and for the rest of their lives they drew upon it to paint pictures of India.

In Fitzroy Square, London, the Daniells began the huge task of producing *Oriental Scenery*, 144 aquatints of landscapes, temples, mosques, forts and ruins, with details of people, boats, camels and elephants for scale and entertainment. These were their masterpieces, the fulfilment of their nine and a half years in India. Thomas Daniell acknowledged that men went to the East for money but he argued that 'it was an honourable feature in the late century that the passion for discovery, originally kindled by the thirst for gold, was exalted to higher and nobler aims than commercial speculation. Since this new era of civilization a liberal spirit of curiosity has prompted undertakings to which avarice lent no incentive.' There were naturalists, philosophers and artists who went for reasons of humanity as well as profit. The pencil, he said, was narrative to the eye.

Thomas and William Daniell collected, celebrated and brought back to Britain the wonder of India, reconciling the picturesque and the classical. Although they foreshortened perspective for effect, they were famously clear, accurate and meticulous in their detail. To the British public their pictures were true and compelling, their *Oriental Scenery* the outstanding collection of views. These works were displayed in numerous homes, and reproduced on wallpaper and on blue-and-white Staffordshire pottery. The Daniells shaped the way that India was visualized and contributed in no small way to British pride in possession. Their paintings influenced British architecture as Hodges's work had done. Most famously Thomas Daniell's work played its part in the design of Indianized buildings like the Brighton Pavilion and the onion domes and minarets of Sezincote, the Gloucestershire mansion of Sir Charles Cockerell, former Paymaster-General of Bengal.

Of all Britain's imperial involvements the affair with India was by far the grandest. For countless British men and women who lived and served in India, and for those at home who could only imagine it, the Daniells' distillation of India's magnificence and romance was the India they always wished to contemplate.

9

Pompey's Pillar

'"What do you see when you get there?"
"Creation!" said Natty.'
James Fenimore Cooper, *Leatherstocking Tales*

In 1535 the Breton navigator Jacques Cartier sailed from the North Atlantic into the St Lawrence River and began to believe that a waterway so promisingly broad might carry him clear across America to China. He followed the setting sun for more than 500 miles until he was stopped by the rapids near the village that would grow into the city of Montreal, headquarters of the Canadian fur-trading empire. Possibly it was he who wishfully gave the seething waters the name Lachine, meaning China. Farther south, and a century later, Jean Nicolet crossed Lake Michigan in a canoe in the belief that the western shore was the coast of China. Before disembarking he prepared for an audience with the Emperor by putting on an embroidered damask robe. Stepping ashore in what is now Wisconsin he was met by Indians of the Winnebago tribe who greeted him as 'the wonderful man' and entertained him to feasts of roast beaver.

The idea that a river might run across the continent to the Pacific, as an alternative to a passage through the northern ice, intrigued explorers for three centuries. Peter Pond was one of them, an American fur trader who journeyed westward for 500 miles from Hudson Bay to Lake Athabaska which straddles modern Saskatchewan and Alberta. He returned in 1778 in a canoe heavy with furs and his mind buzzing with the notion, formed from his own conjectures and the stories that Indians had told him, that a river flowed like a highway from the Great Slave Lake to the Pacific.

There was no clear idea at that time of the breadth of North America. No one, certainly no European, had crossed the continent to the west coast. Pond speculated that the distance from the Great Slave Lake to the ocean was short, a few hundred miles. He could not know that between the lake and the Pacific rose the Rocky Mountains. In 1787 he set down his gleanings and guesses on a map and showed it to his assistant, Alexander Mackenzie, a migrant from Stornoway in the Hebrides.

Mackenzie was sufficiently inspired to get together a canoe expedition to search for Pond's route. In June 1789, when he was twenty-five, he led a party of Canadians, with some Indian guides and their wives, from Lake Athabaska to the Great Slave Lake, and from there they followed a river that seemed to fit Pond's description. They paddled for forty days and found that the river debouched not into the Pacific but into the Arctic, a thousand miles or so east of the Bering Strait. It was later found to be Canada's longest river and was given Mackenzie's name; but at the time Mackenzie felt he had failed. Dismayed by the grossness of his navigational errors he retreated and battled back against the flow of the river. He realized he would never be an explorer if he could not find his longitude, and he sailed from Hudson Bay to London to spend a year studying navigation.

Properly schooled, he launched a new expedition in May 1793 and led six French trappers and two Indians westward along the Peace River, out of what is now Alberta. They travelled in a birch-bark canoe, twenty-five feet long and crammed with clothing, firearms and a ton and a half of food. They frequently unloaded it all to negotiate waterfalls and rapids, backpacking it in sacks to the next calm water while two men portaged the canoe. They entered the Rocky Mountains, the first whites to do so from the east, fought their way through the rocks and torrents of Peace River Canyon and hacked a trail in thick forest up and down the mountains. After days of pulling and pushing they abandoned the canoe early in July and, with ninety-pound loads on their backs, struggled on, the men sometimes growling and mutinous, close to giving it all up. At last they borrowed canoes from friendly Indians and paddled to the Indian settlement on the Bella Coola River north of Vancouver Island on the Pacific shore. There they learnt from

Indians that Captain George Vancouver, surveying the coast in HMS *Discovery*, had called at this very place the previous month.

'I mixed up some vermilion in grease', Mackenzie recorded, 'and inscribed in large characters on the rock: Alexander Mackenzie, from Canada, by land, the twenty-second of July, one thousand seven hundred and ninety-three.'

The expedition returned without loss to Peace River, taking thirty-three days, at thirty-six miles a day, a round trip of three and a half months. The journey across the continental divide revealed valuable information about the geography of the region, the forbidding nature of the northern Rockies and the distance between Athabaska and the Pacific coast; but the dreamed-of waterway did not exist, only the fury of dangerous rivers that tore boats to splinters. Mackenzie's path was too long and difficult to be a practical trading route.

The canoe was the essential transport of the Canadian wilderness. It became emblematic of Canada itself, a craft of the imagination and of mythology as well as a most practical boat. Given the trackless nature of much of Canada, long-distance travel by foot was arduous and in many places barely possible. The only practical highways were the canoe routes used by Indians, trappers and traders, the *voyageurs* or *coureurs du bois*. These men followed the St Lawrence River and the Great Lakes into the heart of the continent and explored the streams of the Canadian Shield, the country's foundation stone, 2 million square miles of territory rich in timber, furs and minerals.

From the earliest years of the fur trade every traveller mastered canoes and knew how to make and mend them with materials that grew beside the rivers. In *The Song of Hiawatha* Henry Longfellow described how the canoe-builder shaped cedar branches into a framework, sewed the birch-bark skin to the frame with larch-root sutures and waterproofed the joints and crevices with fir-tree resin so that the canoe floated

> Like a yellow leaf in Autumn
> Like a yellow water-lily.

A loaded twenty-five-foot canoe drew about eighteen inches of water and was propelled by six or eight four-foot cedar paddles at forty strokes a minute. Ten men paddled the forty-foot canoe, the *canot du maître*, that carried four tons of freight. When working against the river current the crews used long poles to push themselves forward. Negotiating a portage around rapids they loaded themselves like mules, the toughest commonly shouldering a sackload as heavy as themselves, sometimes more if they were strong enough and had not bust a gut, for hernias were frequent among *voyageurs*, just as drowning was a commonplace death on swirling rivers.

The grandest canoe of all was built for Sir George Simpson, the tyrannical Governor-in-Chief of the Hudson's Bay Company in the 1840s. He sat like a potentate while a team of Indians paddled rapidly and a bagpiper or bugler stood to announce his arrival at the trading posts. Once ashore he examined the books and searched for any evidence of luxurious living, scolding men over the amount of Company mustard they put on their meat.

The canoe is everywhere in the history of Canada, part of the national romance. It seemed fitting that the first time I entered the country I did so in a canoe, paddling up from the United States on clear lakes and rivers and camping on pine-perfumed islands. A Canadian, it is said, may be defined as one who can make love in a canoe. Pierre Trudeau, twice Prime Minister of Canada and a devoted canoeist, believed that the paddling Canadian imbibed the essence of the wild pure places and the 'values necessary to spiritual development'. Steering through the tricky rapids, modern Canadians may look like suburban excursionists, but in their imagination they are free *voyageurs* in buckskin.

In 1801 Alexander Mackenzie published the story of his exploits, *Voyages from Montreal through the Continent of North America to the Frozen and Pacific Oceans*. Thomas Jefferson, who had been elected the third President of the United States the year before, read it with profound interest. An author of the Declaration of Independence in 1776 he believed passionately that Americans should push westward and occupy the continent from sea to sea. One of his heroes, Daniel

Boone, personified the westering urge and in 1775 led pioneers through the Appalachian mountains to settle Kentucky.

The European relationship with Indians was evolving from one of more-or-less respect, land-sharing and accommodation to one of a large-scale and relentless white demand for the lands where Indians lived. Jefferson envisaged the occupation of America by a race of virtuous yeoman farmers and traders who would create a property-owning democracy. He set down the principle that the wilderness was owned by the United States government. A geographic survey of the country had started at 'the point of beginning' on the Ohio in 1785 and was steadily working through the task of dividing the country into mile squares which settlers could buy and to which they had rights. The land of opportunity was providing the opportunity to own land.

Mackenzie's expedition threw a shadow over Jefferson's plans, suggesting that the British might advance further, claim vast territories in the north-west and prevent the creation of his coast-to-coast United States. For some years Jefferson had been thinking of sending an overland expedition of discovery into the far west. Like others he believed there might be a river artery to the Pacific. In any case the expedition he had in mind would widen the fur trade and strengthen the claim of the United States to the ownerless region beyond the Rocky Mountains.

At that time Jefferson's United States was a republic of 5.3 million people, and it lay chiefly between the Atlantic and the Appalachians. Not much was known of the lands west of the Mississippi. Certainly the United States had no sovereignty over them. The broad middle of America was encompassed by the Louisiana Territory, owned by Spain, and included, wholly or in part, the thirteen modern states of Louisiana, Arkansas, Oklahoma, Kansas, Colorado, Missouri, Iowa, Minnesota, Nebraska, Wyoming, South Dakota, North Dakota and Montana. It extended, in other words, from the Gulf of Mexico to Canada and from the Mississippi River to the Rocky Mountains.

In 1801 Spain ceded the territory to Napoleon Bonaparte who dreamed of a French colonial empire in America. But the turn of international events brought Jefferson a great prize. Napoleon feared that in his imminent war with Britain the British might seize the Louisiana Territory from France; and he wanted no distraction from his ambition to emerge as dictator of Europe. He surely knew that, given the American hunger for land, the westward movement of Americans could never be contained and that in the long run the Louisiana Territory would be settled by them. In the spring of 1803 he sold it to the United States for $15 million and remarked that 'the sale assures forever the power of the United States, and I have given England a rival who, sooner or later, will humble her pride'.

Jefferson had no clear idea of the northern, western and southern boundaries of the land he had just bought. He knew only that his country had purchased an immensity. The 827,000 square miles of the Louisiana Territory doubled the area of the United States, and from that moment the American frontier moved rapidly and decisively westward.

By the time of the purchase Jefferson had already commissioned his western expedition and had appointed two seasoned frontiersmen to lead it. Meriwether Lewis, aged twenty-nine, was his secretary and protégé: William Clark, aged thirty-three, was Lewis's friend. They were to lead their party up the Missouri and strike out for the Pacific, charting, drawing and noting everything they saw, the landscape, plants, animals and birds; as well as the nature, languages and customs of Indian peoples. They were to find 'the most direct and practicable water communication across the continent, for the purposes of commerce'.

Jefferson sent Mackenzie's book to Lewis and Clark. He also sent them a map and some notes compiled by an explorer called John Evans. 'In my letter of the 13th inst,' Jefferson wrote to Lewis, 'I enclosed you a map of a Mr Evans, a Welshman . . . whose original object I believe had been to go in search of the Welsh Indians said to be up the Missouri.'

The contribution of John Thomas Evans to the exploration of the American frontier is as poignant as it is remarkable. This pious son of a Methodist preacher grew up in Waunfawr, a village on a mountainside above Caernarfon in north Wales. From an early age he was enchanted by the legend of Madoc, discoverer of the New World. According to the story, Madoc, the son of the Welsh prince

Owain Gwynedd, sailed from Wales for America in 1170 with 300 men. There he was the progenitor of a tribe of pale-skinned people, the Welsh Indians. The origin of the tale remains a mystery. Owain Gwynedd had no son called Madoc. The story, said to be derived from an 'ancient Welsh chronicle', first appeared in print in a pamphlet of 1583 which was intended to prove that Queen Elizabeth, not the King of Spain, had title to the New World. A few years earlier Doctor John Dee, a latter-day Merlin to Queen Elizabeth, conceived of a 'British Empire', the first time the term was used. He cited the Madoc story and suggested that since the Snowdonian prince had landed in America long before Columbus, there was a proper 'British' claim to the American territory.

Stories of Welsh-speaking Indians persisted down the years, one of the myriad legends of lost tribes and cities that punctuate history. The 'evidence' of the existence of Madoc's descendants, in reports brought back by travellers from America, aroused particular excitement and speculation among the literati and historians of the Welsh community living in London who were as susceptible as any other people to stories of stranded civilizations. The arrival in London in 1791 of William Bowles added to the buzz. Bowles was an Irishman who had lived among the Cherokee Indians, had taken an Indian wife and claimed to be a Cherokee chief himself. He therefore seemed an authoritative figure, and he informed the Welshmen of London that the descendants of Madoc were alive and thriving far beyond the Missouri.

Obsessed by the legend, John Evans resolved to find the Welsh Indians himself. It was noted that he possessed some of the 'giddiness' of the Methodists, and certainly he seemed to believe that he had an assignment from God. He sailed for Baltimore and arrived there in October 1792 to begin his search. He was twenty-two years old, courageous and purposeful. In the luggage he packed for his long journey into the interior he included a Welsh Bible so that he could bless the Indians in their own ancestral language.

Opinion suggested that the Indians of the Mandan tribe in North Dakota were the likeliest descendants of Madoc. They were reputedly pale and spoke what sounded like Welsh. Moreover, travellers reported that Mandan girls were pretty and chattered a lot, even while they made love, qualities considered the determining evidence of their Welshness. Descriptions of the Mandan showed that their rituals were significantly different from those of other Indians and that they depended on agriculture rather than hunting, differentia that aroused speculation about their origins.

In March 1793 John Evans started for the frontier, journeying from Philadelphia through the Allegheny mountains and taking boats down the Ohio and up the Mississippi to St Louis, a town of 1,300 people at the junction of the Missouri. At that time, being in the Louisiana Territory, St Louis was under Spanish rule; and for some reason, perhaps because they took Evans for a spy, the Spanish authorities locked him up. They freed him after two years so that he could join John Mackay, a Scot, on an expedition under the Spanish flag to ascend the Missouri and travel by way of the Mandan settlements to find a passage through the Rocky Mountains to the Pacific. Thirty men and four boats set out in August 1795. No one could be sure of what was out there – Indians, no doubt, but also, it was suggested, roaming mammoths and mountains of salt. The Mandan territory lay 1,800 miles from St Louis, and the expedition had covered half the distance when it was frozen in for the winter. In February 1796 Mackay ordered Evans to lead a small exploration party towards the Rockies; but after 300 miles Evans and his companions abruptly retreated, fleeing from howling war bands of Sioux.

Mackay sent Evans out again in June, and this time the Welshman travelled up the Missouri and through the Badlands of South Dakota to the Mandan settlement near present-day Bismarck in North Dakota. It was an astonishing journey, a brave and dogged Methodist with his hopes and his Welsh Bible in the service of Roman Catholic Spain. How high those hopes must have been when at last he walked into the Mandan villages. The Indians were always hospitable and received him genially. He lodged with them for seven months in a beehive-shaped house, fifty feet in diameter, through the most bitter of winters. He had plenty of time to talk with two chiefs, Big White and Black Cat. One of the conclusions he drew from his observations was that the Indians who had little contact with white men, like the Mandan, were of better character than those who were familiar with them. But as the winter weeks turned into months John Evans felt his hopes grow cold. He realized that his amiable hosts were not the lost tribe of his dreams.

Evans returned to St Louis in July 1797 after an absence of two years in which he had heard no Welsh, except in his own prayers. We can only imagine his crushed spirits as he wrote the verdict on his strange quest: 'Having explored and charted the Missouri and by my communications with the Indians, I am able to inform you that there is no such people as the Welsh Indians.' He died in St Louis two years later, aged twenty-eight, his end hastened by drink and perhaps by disappointment. His experiences and conclusions, however, did little to shake the belief among Welshmen who longed for the legend to be true, that somewhere in the American hinterland lived the progeny of Prince Madoc of Snowdonia.

The chart made of the territory John Evans travelled was carefully drawn, either by himself or someone using his sketches and notes, and is kept in the Library of Congress. It made a significant contribution to the cartography of the West. This was the map Jefferson sent to Lewis and Clark; and as they travelled they made their own notes and corrections on it. Their expedition left St Louis in May 1804 in three boats and 'proceeded under a gentle breeze up the Missouri', as Clark noted in his journal. Their Corps of Discovery included twenty-seven young single soldiers, some boatmen, a half-Shawnee, half-Canadian hunter, and York, a slave whom Clark had inherited from his father. Lewis took Seaman, his Newfoundland dog. Among their papers they had a letter of credit, signed by Jefferson, one of the 50,000 letters he wrote in his lifetime, that promised payment to anyone who helped them. 'Your party being small,' he wrote, 'it is to be expected that you will encounter considerable danger from the Indian inhabitants. Should you escape those dangers and reach the Pacific ocean you may find it imprudent to hazard a return the same way . . .'

The expedition covered about fifteen miles a day in a fifty-five-foot boat equipped with oars and a sail and accompanied by two dugout canoes. In October, after twenty-three weeks of rowing against the flow of the river, the party reached the

Mandan settlement in North Dakota, pretty well the edge of the known world, and spent the long winter in log cabins there. The Mandan took them out hunting, and the expedition could hardly have survived the temperatures of forty below freezing without the Indians' meat and furs. Lewis made notes of the Mandan language to see 'whether they sprang from the Welsh or not'. Two soldiers in the party decided that 'these savages have the strangest language but they are the honestest savages we have yet seen . . . we take them to be the Welsh Indians'. Thus the story flickered.

In the spring of 1805 Lewis and Clark headed west with two boats and six canoes. 'We were now about to penetrate a country at least two thousand miles in width, on which the foot of civilized man had never trodden,' wrote Lewis. Describing the journey as his 'darling project', he added that 'I esteem this moment of departure as among the most happy of my life.' The party was enlarged to thirty-three by the recruitment of a French-Canadian fur trader, Toussaint Charbonneau, and his squaw Sacagawea or Bird Woman, whom he had bought from Indians who had kidnapped her. She was a Rocky Mountains Shoshone Indian aged about sixteen and had just given birth to a son. In time Sacagawea became an American folk heroine, a romanticized princess and decorative stereotype. The stories had it that she played a leading role in the expedition and guided it through dangerous places. One embellishment suggested that she enjoyed a love affair with William Clark. She was certainly useful as an interpreter and go-between in dealings with the Indians the party encountered and she helped to find directions through some of the mountain passes. Her very presence, with a baby slung on her back, smoothed the way among the tribes. As Clark noted: 'A woman with a party of men is a token of peace.'

Sacagawea's rough-and-ready husband, Charbonneau, made admirable sausages, six feet long, from the intestines of a buffalo. He stuffed the skin with meat, kidneys and pepper, boiled it in a large pot and fried it in bear fat until, as Lewis described, 'it is ready to assuage the pangs of a keen appetite such as travellers in the wilderness are seldom at a loss for'. The explorers had taken plenty of biscuit and salt pork but hunted constantly and fed on buffalo, deer, beaver and bighorn sheep. They saw grizzly bears almost every day, 'a terrible looking animal we found very hard to kill', said Lewis.

Everywhere they travelled, along the winding rivers and through hills and prairie, they found animals and birds in abundance. Lewis and Clark discovered plants hitherto unknown, among them Indian tobacco, hoary sagebrush and the Oregon grape, and in time had classes of plants named after them, Lewisia and Clarkia. The expedition had no artist but both the leaders kept journals and Lewis in particular compiled observations more detailed and thoughtful than a mere logging of progress and of things seen. The landscape awed him, and he knew he was privileged to be a pioneer. Crossing the prairie in 1805 he noted that 'the whole face of the country was covered with herds of buffalo, elk and antelopes ... so gentle that we pass near them without appearing to excite any alarm among them'. The earth itself was dense with rabbits, mice, gophers and weasels. On this rolling and almost treeless prairie robins, shrikes, meadowlarks and hawks made their nests on the ground. Lewis saw the grassy ocean of the prairie uninterrupted, intact and innocent under the immense sky, as it had been through the ages. His published descriptions of such beauty and profusion made a thrilling impression on those in the settled lands of the east who longed for images of the big country of the distant West. He was one of the first to make it appear magical and romantic.

Following the Missouri, Lewis and Clark portaged for eighteen miles around the five thundering cascades of Great Falls in Montana and in August crossed the Continental Divide, the watershed of the river systems, through the Lemhi Pass in Idaho. Lewis was conscious of the historic achievement in reaching the headwaters of the 'heretofore-deemed-endless Missouri ... on which my mind has been fixed for many years. Judge, then, of the pleasure I felt in allaying my thirst with this pure and ice-cold water.' Often hungry and freezing the party struggled painfully through the toughest country of their journey, a jumbled mass of high peaks and thick forests. They crossed the Bitterroot Range and descended the Clearwater, Snake and Columbia rivers in five canoes. They reached the Pacific coast on 18 November 1805. 'Men appear much satisfied,' wrote Clark, 'beholding with astonishment this immense Ocean.' The explorers enjoyed the hospitality of the local Indian people, and Clark noted that some of the women arrived 'for the purpose of gratifying the passions of our men' and that 'many of the women are handsome'.

The following year, 1806, given up for lost after two years and four months away, the expedition returned to St Louis. During 8,000 miles of travel one man had died, of appendicitis, and two attacking Indians had been killed. Seaman, the dog, survived the longest walk a dog could wish for. Lewis and Clark were fêted as if they had been to the moon. Although their expedition did not reap the commercial and agricultural rewards Jefferson had hoped for, it nevertheless began the official exploration and mapping of the American West, a task not completed until the end of the nineteenth century. It opened the West to settlement and, above all, to the American imagination.

Three years after his return Meriwether Lewis died violently, possibly murdered but probably, after drinking, by his own hand. William Clark became a benevolent superintendent of Indian affairs and died aged sixty-eight. The Mandan were mostly wiped out in 1837 by smallpox carried to them by the travellers they always welcomed so warmly. Sacagawea returned to her people and died in 1812, although an unlikely story has it that she survived until 1884. In Idaho and Montana there are springs that bear her name.

Sacagawea's son Jean Baptiste enjoyed an extraordinary life as a hunter and mountain guide. William Clark was his mentor and paid for his education in St Louis. In 1823 Prince Paul of Württemberg met the young man on an American hunting trip and persuaded him to travel back to Germany with him to study. Jean Baptiste learned German, French and Spanish and hunted with the prince in Germany and Africa. He returned to the West as a mountain man of culture, traded in furs, prospected for gold in California and Montana and was the most famous guide in the Rockies.

Travelling on his mother's back during the expedition, Jean Baptiste was a natural mascot and William Clark gave him the nickname of Pompey. It was in his honour that Clark picked out a tall column of rock which stands beside the Yellowstone River, east of Billings in Montana, and called it Pompey's Pillar.

10

The punishment of Ixion

> 'You will hear more good things on the outside of a stagecoach from London to Oxford than if you were to pass a twelvemonth with the undergraduates of that famous university.'
>
> William Hazlitt, 1821

During the gale that raged for five days after the Battle of Trafalgar, Vice Admiral Cuthbert Collingwood moved his command to the frigate HMS *Euryalus*. He needed to communicate with the smashed and scattered ships of the British fleet and to send news to London. His own ship, HMS *Royal Sovereign*, was dismasted and her signalmen had no means of hoisting flags. Aboard *Euryalus* on 22 October 1805, the day after the battle, the exhausted Collingwood dipped his pen and began to write one of the most dramatic documents of history. It was addressed to the Admiralty. 'The ever to be lamented death of Vice Admiral Lord Viscount Nelson,' he started, 'who, in the late conflict with the enemy, fell in the hour of victory . . .'

On 26 October Collingwood gave his dispatch to Lieutenant John Richards Lapenotière, a thirty-five-year-old Devon man commanding HMS *Pickle*. This was the second smallest vessel at Trafalgar, a fast copper-bottomed messenger schooner of 125 tons whose fore-and-aft rig gave her the speed and something of the appearance of a yacht. HMS *Pickle* had already achieved a romantic minor footnote in the battle: her sailors rescued a young woman called Jeanette who had stowed away aboard the French warship *Achille* to be with her husband. The *Achille* caught fire in the fighting and her crew jumped overboard. Jeanette was plucked naked from a floating spar, an ever to be celebrated opportunity for an artist to show a nymph amidst the fury of war at sea; and the painting hangs in Fortnum and Mason in Piccadilly, London.

Collingwood told Lapenotière to sail as fast as he could and, if ambushed by enemy ships, to throw the dispatches overboard. To cover himself he sent a copy to Lisbon to be taken overland to England. Collingwood himself would never see his home again: the Admiralty kept him on active service in the Mediterranean and he died on duty five years later.

Lapenotière now had his moment in history. From Cape Trafalgar, at the western approach to the Strait of Gibraltar, he set course for England, 1,200 miles distant, flying *Pickle*'s ensign at half-mast. It was a headlong voyage, the south-west gale thrashing the schooner, her men pumping her around the clock. Off the Cornish coast, Lapenotière gave the news to fishermen who hauled up their nets and sailed at once to Penzance. They informed the mayor who spoke from a balcony to a quickly gathered crowd. *Pickle* reached Falmouth, eight days' sail from Trafalgar, on 4 November. London was 270 miles distant and to get himself there Lapenotière hired a light four-wheeled post-chaise drawn by two horses. At almost £47 the fare was expensive. He set off up the turnpike road, the chaise rattling along at just over seven miles an hour, passing through Truro, Bodmin, Launceston, Okehampton, Exeter, Axminster, Bridport, Dorchester, Blandford, Salisbury, Andover, Basingstoke, Bagshot and Hounslow. The carriage stopped nineteen times for a change of horses and during each brief pause, amidst all the coachyard commotion, Lapenotière gave local people a summary of the battle and left pools of mingled exultation and grief all the way across southern England.

After midnight on 6 November, thirty-eight hours from Falmouth, sixteen days after the battle, *Pickle*'s commander hurried through the smoky fog of London and around Hyde Park Corner to the Admiralty. He found William Marsden, the Secretary, working by candlelight at his desk.

'Sir,' he said, as terse as a headline, 'we have gained a great victory; but we have lost Lord Nelson!'

A messenger rode the twenty miles to Windsor Castle to tell King George III, another ran to Downing Street to inform William Pitt, the Prime Minister. Clerks sharpened their pens, lit their candles and copied Collingwood's dispatch. Printers set it in type and that same day the *London Gazette* published a special edition. Couriers carried copies to Great Yarmouth and put them aboard a ship sailing to the Netherlands. The news reached Ireland and Scotland on 9 November. On 12 January a ship sailed into Cape Town with an account of the battle; and in mid-April, six months after Trafalgar, the story reached the small town of Sydney in Australia.

Twenty-five years earlier, it had taken eleven months for the news of Captain Cook's death to reach London. After burying Cook's remains at sea off Hawaii in February 1779, his men sailed more than 3,000 miles north-west to the Kamchatka peninsula in the far east of Siberia. In June they entrusted their letters and copies of Cook's journals to a Russian traveller who rode across Siberia and the Russian heartland to St Petersburg. There the package was taken to Berlin and at last by ship to London and the Admiralty. The account of the 'ever to be lamented fate' of Cook was published in the *London Gazette* in January 1780.

Seafarers and their families endured the aches of long absence. A wind blowing hard in the night always awakened Elizabeth, Cook's wife, and made her think anxiously of men at sea. Once they had sailed, seamen could expect no news from home. After his first three-year circumnavigation Cook arrived at his house in Mile End, London, to be told by Elizabeth that two of their four children had died while he was away. When he returned from his second three-year voyage another was dead.

News was theatre. The clamour of hooves on cobbles dramatized the spectacle of its arrival. 'Nothing is done in England without a noise,' reflected Robert Southey, watching the coaches arrive at an inn, 'and noise is the only thing they forget in the bill.'

Reports of great events, of war and empire, of Nelson's victories and Wellington's triumphs, came day and night in a stir of rasping horns, shouting ostlers, jostling porters, whinnying horses and clashing pails. Enthroned above the crowd, the red-cheeked mail coach driver preened in his thick Benjamin coat and gloried in the role of the nation's publisher. When the Houses of Parliament burnt down in 1834 the news was scribbled at the foot of the passenger lists of coaches leaving London. Wherever the coaches stopped to change horses the coachman read out the report that 'At this moment both Houses of Parliament are in flames.' By this means the night coach took the news to Birmingham by breakfast time the next day. As a coach bound for London passed through Oxford, students at Christ Church College shouted to the driver and the guard, 'You won't be in time to see the Lord Chancellor's wig on fire!'

Coaches now seem part of a less hurried age; yet speed and urgency, time-clipping competition and the 'expeditious' journeys promised in advertisements were their essence. Timetables boasted of precise arrivals, three minutes to the hour or seventeen after. Tyrant coachmen waited for no one. In brief stops at inns passengers scurried to relieve themselves, bolted food, slurped coffee and chocolate and crammed bread into their pockets. 'Everybody is in a hurry,' said Southey. The velocity of the coach was a marvel but some passengers grumbled that the views of the countryside passed all too quickly when glimpsed from coach windows. In 1798 a traveller thought that speeds of around ten miles an hour were 'highly dangerous to the head', and reported that he had heard of passengers who 'having reached London with such celerity died suddenly of an affectation of the brain'.

Then, as now, the idea that time itself could somehow be shrunk stirred every imagination. From ancient times and into the twentieth century long-distance communication was primarily hand to hand. The courier with his satchel, a man walking or running, riding or sailing, was the essential figure carrying letters between kings, generals, merchants and bishops. The Roman postal system, *cursus publicus*, had shown the way, its riders covering between fifty and a hundred miles a day, pausing every twelve miles or so at post stations where fresh horses waited and carpenters were on hand to repair wagons. The high cost of horses, carts and post staff, however, eventually caused the system's collapse.

Marco Polo's description of the express news network he saw in China fascinated his European readers. It seemed to be a way of taming time. Along the roads spoking from Peking the Great Khan built horse posts twenty-five to forty miles apart; and between these were smaller posts at three-mile intervals for relay runners whose belts of jingling bells warned couriers waiting at the station ahead to spring to their feet. As Polo noted, the Khan received news in a day and a half that would otherwise have taken ten days; and news in ten days that would ordinarily have taken a hundred. From the thirteenth century Italian merchants employed teams of messengers to carry news of harvests, ship cargoes, shipwrecks and the edicts of rulers. In the middle of the fifteenth century horseback couriers delivered mail from Florence to Paris in three weeks. In Cape Town's museum you can see letters bearing the distinctive crease made by the cleft stick carried by runners in the seventeenth century. In northern India, before the railways came, runners crossed in relays from Calcutta to Bombay, taking eleven days in the dry season and three days more during the monsoon rains.

In 1516 King Henry VIII appointed a Master of the Posts to improve the haphazard royal mail deliveries to Berwick and Calais. In Elizabethan times five mail routes radiated from London. The growing importance of Ireland led to the establishment of a postal service from London to Holyhead in 1599. Mail was paid for by the recipient on the basis of distance travelled, two pence for eighty miles, six pence for deliveries over 140 miles; and also on the number of sheets in a letter, counted by a postmaster who held it to the light of a candle. Members of Parliament had the right to send ten free letters a day, a privilege they soon learnt to abuse. They signed blank letters for their friends and also for their servants who sold them in lieu of pay.

The couriers, or post boys, changed horses at post stations; and since inns were convenient stopping places they were the first post offices and innkeepers the first postmasters. A horse, like a man, is fast only in short bursts: the three-mile Derby is run at about thirty miles an hour and the four-and-a-half-mile Grand National steeplechase at around twenty-five. Modern trials show that fit horses can cover a hundred miles on a good surface at twelve miles an hour, with some rests and some cantering, the word derived from the loping of pilgrims' horses to the shrine of St Thomas à Becket at Canterbury.

Well into the eighteenth century road travel of any kind was a penance, usually as slow and mired as in Elizabethan times. Winter weather made many roads too muddy to use. Even in good conditions travellers mounted at dawn to get on the roads before the trains of packhorses and pack mules jammed them. If you walk a surviving packhorse road today, such as the dramatic route through the Rhinog mountains in Wales, you can see clearly that since the track is only three or four feet wide, and frequently steep, strings of twenty or thirty packhorses would be difficult to pass. In fact it is so narrow that in the 1930s a coffin was carried along it slung from a pole on the shoulders of four men in single file.

For much of the eighteenth century sixty miles a day on horseback was good going. In 1746 horsemen took ten days to ride the 600 miles from Inverness to London with news of the English slaughter of Bonnie Prince Charlie's forces at Culloden. Some roads were so bad that stagecoaches sometimes struggled for five days to cover a hundred miles, and proprietors were careful to qualify their arrival times with the words 'God willing'. In the 1730s coaches took a day to reach Oxford from London, and even twenty years later the London–Manchester 'Flying Coach' took three days. Stages ran from most cities by the end of the eighteenth century, but the costs of horses, coaches, taxes and tolls on the turnpike roads produced fares that were beyond the pocket of the majority of people. The lumbering wagon of the local carrier was cheaper. It carried baggage while 'passengers' walked. Travellers leaving England for Europe reckoned on taking a month to reach Rome or southern France by carriage, five weeks to travel from Paris to Naples. On their return, they spent the best part of two days getting to London from Dover. In 1789 news of the fall of the Bastille in Paris reached Madrid two weeks later.

In the 1780s John Palmer, a Bristol theatre manager, considered the system of mail delivery and decided it was time for reform on the roads of Britain. Post boys were unarmed, vulnerable and often robbed. Stagecoaches were irregular and frequently waylaid by highwaymen evidently undeterred by the corpses of their criminal cousins swinging on gibbets.

Palmer called for an efficient public service, a 'mail machine' operated by special coaches that would leave and arrive punctually. They would run at eight miles an hour and carry a guard at the back to deter highwaymen. This would be the Royal Mail and it would be a crime to interfere with its lordly progress. Commercial coaches would have to give way. It would not pay turnpike tolls: indeed, at the imperious blast of a horn, the pikies, the turnpike gatekeepers, would have to fling the gates open. Some postal officials opposed the scheme, and Palmer growled that their attitude was 'don't try it lest it succeed'. Fortunately, his friend William Pitt, the Prime Minister, dismissed objections and, at four o'clock on 2 August 1784, the inaugural Royal Mail coach left the Rummer tavern in Bristol. It arrived at the Post Office in London before eight next morning, the sixteen-hour trip halving the time taken by the commercial coaches and post boys. John Palmer had introduced speed and the certainty of arrival.

The new coaches were decidedly classier than the ordinary stagecoaches; and they were faster and a penny a mile more expensive to ride in. Their bodywork paint was shining black and maroon and the wheels were red. The royal cipher and the words Royal Mail were stencilled in gold on the doors. The coachman and the guard were imposing in scarlet coats and tall beaver hats. Commercial stagecoaches looked different. Their owners painted them in gaudy fairground colours and hired artists to depict scenes on the doors. The commercial coaches bore names like The Times, The Telegraph, Monarch, Rocket, Umpire, Champion, Peveril of the Peak and the comforting Kershaw's Safety Coach. Only one Royal Mail service had a name, the coach from London to Plymouth and Falmouth called Quicksilver. The Post Office ordered frequent changes of horses to keep up a high average speed and stipulated that ostlers should disconnect the horses and back the fresh team into place inside five minutes. Many ostlers prided themselves on making the switch in two minutes. To save time, coaches sped through many smaller towns without halting, the guard picking up mailbags with his stick.

The mail coach rapidly established itself as dependable and admirable, an institution that stitched the nation together. In his essay 'The Glory of Motion', Thomas De Quincey saluted John Palmer as the man who 'had accomplished two things very hard to do . . . he had invented mail coaches and had married the daughter of a duke'. The mail coach, 'as the national organ for publishing mighty events became a spiritualized and glorified object', and 'the connection of the mail with the state and the government . . . gave to the whole mail establishment an official grandeur'. The first Royal Mail from London to Holyhead set off in October 1785. De Quincey recalled a day when he travelled on this service and a Birmingham commercial coach, 'all flaunting green and gold', had the effrontery to overtake; but the Royal Mail coachman 'slipped our royal horses like cheetahs' and the coach swept past the presumptuous Brummie with a raspberry blast of its horn.

The demand for more and faster coaches in news-hungry and rapidly developing Britain led to better roads. The Romans built 54,000 miles of imperial highways, and nothing in Europe rivalled that commitment to communication until the 1780s when the French military started constructing strategic routes from Paris. In Britain the leading road builders were the Scottish engineers Thomas Telford and John McAdam, and the remarkable Jack Metcalf who was blind and tapped his way along his roads with a stick. Telford, a shepherd's son and self-taught engineer, built hundreds of roads, canals and bridges which endure as monuments to his genius. He constructed a thousand miles of roads and a thousand bridges in the Highlands of Scotland, an extraordinary achievement. After the political union of Ireland and Britain in 1801, and the growth in administration in Dublin, there was a clear need to build a good road from Shrewsbury through the mountains of north Wales to improve communication between London and Dublin via Holyhead. Thomas Telford's Menai bridge was the vital link. The first mail coach crossed it on 30 January 1826 and delivered its mailbags and thoroughly stiff passengers to the steamer at Holyhead, twenty-seven hours and 260 miles out of London. Telford built roads with deep and solid foundations.

McAdam found that a shallower foundation was enough, and also cheaper; but both knew the importance of good drainage and of using stones of small size to surface their roads. McAdam told his navvies that a stone was right if it fitted into their mouths.

Telford and McAdam were heroes, not least to the stagecoach companies and drivers. Coaches sped faster and in greater numbers along the new smooth all-weather roads. A service from London left at five in the morning and covered the 186 miles to Manchester before midnight. A man astonished his friends by relating that he was hunting in Sussex on Monday and dining in Dublin on Wednesday. The mail coach reached Devonport in twenty-four hours, with twenty changes of horses on the way. In the eighteenth century it had taken as long as a week.

Mail coaches travelled overnight from London and spectators gathered to enjoy the noisy drama of their evening departure. In the 1830s twenty-eight Mails left each evening. From 1829 most of the Mails, for Dover, Scotland, the North and Holyhead, departed at eight o'clock from the new Post Office at St Martin's Le Grand near St Paul's. The Western Mails started half an hour later from Piccadilly. 'The finest sight in the metropolis is that of the mail coaches setting off from Piccadilly,' wrote William Hazlitt, the essayist.

> The horses paw the ground and are impatient to be gone, as if conscious of the precious burden they convey. There is a peculiar secrecy and despatch, significant and full of meaning, in all the proceedings concerning them. Even the outside passengers have an erect and supercilious air, as if proof against the accidents of the journey . . . as if borne through the air on a winged chariot . . . give me, for my private satisfaction, the mail coaches that pour down Piccadilly of an evening, tear up the pavements and devour the way to the Land's-End.

Four or more 'outsiders' rode on top of each coach with the driver and the guard. Six 'insiders' crammed thigh by thigh inside, believing that the higher fares they paid accorded them an elevated social status, a compensation of sorts for enduring the company of the talkative, the flatulent, the snorers and the vomiting babies.

The mail was brought from the Post Office and distributed among the waiting coaches. At eight-thirty the coachmen of the Western Mails cracked their whips and the coaches set off for Hyde Park Corner and bowled through the villages of Knightsbridge, Kensington, Hammersmith and Kew. They changed horses at Hounslow and ran over the heath now covered by the concrete of Heathrow Airport. Then they took their separate routes for Bristol, Exeter, Poole and Devonport. The Quicksilver for Devonport, the fastest long-distance Mail of all, went by way of Andover, Salisbury, Wilton, Shaftesbury and Honiton and into Plymouth. Passengers and mail were rowed across the River Tamar to Cornwall to join the connecting coach to Falmouth. Later a steam ferry crossed the river, large enough to carry four coaches.

As the Mails hurried north, south, east and west through the night, other Mails were on the roads to London, timed to arrive early in the morning. Meanwhile more than 200 commercial stagecoaches left London daily from such inns as the Spread Eagle in Gracechurch Street, the Bolt-in-Tun in Fleet Street, the Bull and Mouth near St Paul's and the Swan-with-two-Necks in Lad Lane whose manager operated sixty-eight coaches and kept 1,800 horses. Hundreds of stagecoaches plied local routes.

Charles Dickens joked that coach travel resembled the punishment of Ixion, bound by Zeus to a fiery wheel and condemned to roll through the sky for ever. In 1835 he described the departure from London at six o'clock on a winter morning.

> The coach is out; the horses are in, and the guard and two or three porters are stowing the luggage and running up and down the steps of the booking-office with breathless rapidity. The place, which a few minutes ago was so still, is now all bustle. The early vendors of the morning papers have arrived with shouts of 'Times, gen'lm'n', 'Herald, ma'am' ... 'All right,' sings out the guard at last, jumping up as the coach starts, and blowing his horn ...

The coaches reached their peak in the 1830s. In June 1837, wherever they stopped to change horses, the captains of the dashing Mails spread the news that King William IV had died and that Victoria was Queen. It was one of the final hurrahs. The new railways were making their mark and by the end of the 1830s it was plain that they would win. In 1838 a stagecoach raced a steam train from Birmingham and won: a gallant and pointless gesture. By the early 1840s crowds no longer swarmed to see the Mails depart and the yards of coaching inns were left with their ghosts. A last Mail left London for the north in 1847, flying a flag at half-mast. 'Alas!' wrote William Thackeray, the novelist and *Punch* journalist, 'we shall never hear the horn sounding at midnight or see the pike-gates fly open.'

William Hazlitt had come to terms with the inevitable decline of the coach in 1831. 'The picturesque and dramatic do not keep pace with the useful and mechanical,' he wrote. 'The telegraphs that lately communicated the intelligence of the new revolution to all France within a few hours are a wonderful contrivance; but they are less striking than the beacon-fires which, lighted from hilltop to hilltop, announced the taking of Troy and the return of Agamemnon.'

Yet the coaches had done more than their bit for progress. In their heyday they doubled speeds on the road and spread the feeling among people that theirs were times of excitement and progress. The Royal Mail began a tradition of punctuality, public service and guaranteed delivery times. The records are full of admonishments from postmasters, urging drivers to maintain their schedules and rebuking them for any delay. Speed on the roads was obviously limited by the endurance of horses, but the postal service's rapid and frequent changes eliminated fatigue and made deliveries fast. The coach timetable foreshadowed its railway equivalent and established a framework of life governed by clocks and the pressure of time. You could set your watch by the mail coach. It pointed the way to a standard time.

The stagecoach hurrying through the snowy landscape, the driver enveloped in a high-collared coat, the outsiders cheery and cherry-faced, remains the favourite transport of Christmas card artists. In the 1850s the coach was already part of Britain's past. But on the world's frontiers it still had a future.

11

The Atlantic ferry

'The brisk blue-peter beckons; and at last
Our souls shall ride full-sailed before the blast
Into the perilous security
Of strife with the uncompromising sea.'
Wilfrid Wilson Gibson, 'The Blue-peter'

It was coming up to ten o'clock in the morning. Eight men swallowed their coffee and hurried over the cobbles of South Street in Manhattan, tugging up their collars against the snowy squalls.

Boarding the 424-ton *James Monroe* they could hardly fail to be impressed. Brand new, piquant with varnish and bright with brass, she was a thoroughbred built for speed, her gleaming black hull copper-sheathed and copper-fastened in the modern fashion. On deck there was all the ordered activity of departure. Stevedores had stowed cotton, wool, apples and potash into the hold. Hens, cows, sheep and pigs had been coaxed into coops and pens on the deck to provide fresh eggs, milk and meat. Stewards conducted the eight men to comfortable passenger cabins. These were panelled in satinwood with white marble pillars at each door; and rather than the customary straw palliasse, the notorious 'donkey's breakfast' of sea travel, each berth had a hair mattress. Discreet in a locker was a blessed chamber-pot decorated with flowers.

At the last moment, with the shore lines ready to slip, leather mailbags that had hung in the Tontine Coffee House were brought aboard. The company flag, a black ball on a red square, was hoisted. At ten o'clock, true to the advertisements, Captain James Watkinson ordered the sails set. That was why the eight passengers quickly drained their coffee cups. The Black Ball Line, let alone the tide, would wait for no man.

It was 5 January 1818, a big day in the history of the Atlantic. Steering down the East River towards the sea, Captain Watkinson was inaugurating a transport revolution, new standards of speed and reliability, the beginning of scheduled voyages across the ocean. Atlantic services were ripe for improvement. Until then merchant ships had not run to scheduled sailing times. Shipowners on both sides of the Atlantic operated according to the practices that had suited them for centuries, ordering their captains to sail only when the holds were filled with freight and the cabins crammed with passengers. A captain might wait for days or weeks to accumulate a full cargo, and days more for a fair wind. Businessmen and manufacturers were hamstrung by delivery dates that at best could only be vague. Britain's laws of the sea protected its large merchant fleet from foreign competition and there was no pressure on owners to sail regularly or make their ships faster and more comfortable.

Isaac Wright, a New York merchant who exported cotton from plantations in the American South to the mills of Lancashire, reflected on the damage done to his business by unreliable ship deliveries. Three Englishmen, importers of English wool in New York, shared his complaint and joined him in doing something about it. In the autumn of 1817, Wright, his son and the Englishmen put $25,000 each into a shipping company, the Black Ball Line, to run a timetabled service from New York to Liverpool. The critical element was the company's promise that a ship would sail on the first of every month, fair wind or storm, whether or not its holds were full. Another innovation was passenger comfort rather than the squalor of seagoing tradition.

It was a risky enterprise. A ship lacking a full cargo might sail at a loss; and sailing by timetable defied the realities of North Atlantic weather. When the *James Monroe* left New York on its maiden voyage it had room for twenty more passengers and space in the hold for more cargo; but the schedule ruled. Captain Watkinson established the pattern of competitive transatlantic travel, pushing his ship and crew hard, night and day. He made the most of the eastward set of the Gulf Stream and the prevailing westerly wind on the run from Sandy Hook to the Fastnet Rock off the southern coast of Ireland and reached Liverpool in twenty-eight days. Soon the Black Ball ships averaged twenty-three days on the eastward crossings to Liverpool and forty on the westward or 'uphill' voyages to New York. One of them set a record of thirteen days and seven hours eastward, and seventeen days and six hours westward. In general, Black Ball ships crossed the ocean in two-thirds of the time taken by British ships and needed only two-thirds of the crew to manage them, between twenty and forty men. Their lives of hard labour prompted the rueful shanty:

> On a trim Black Ball liner I first served my time
> And in that Black Baller I wasted my prime.

Within a few years the Black Ball Line's success spawned competitors: the Red Star Line in 1822, then the Swallowtail Line in the same year, and later the Black X Line and the Dramatic Line whose ships were given names like *Shakespeare* and *Garrick* and *Sheridan*. Ships berthed in ranks at the waterfront below Wall Street in Manhattan, their bowsprits thrust across the road. Passengers and merchants could now count on a weekly sailing from New York with services to Liverpool, Le Havre and London.

In time the advances in ship design and gear reduced the average run from Liverpool to New York to thirty-five days and the return voyage to nineteen, the ships sometimes sailing at twelve knots. There was heavy betting on three packets which sailed from New York in July 1836 and raced to Liverpool. Two of them took seventeen days and arrived within three hours of each other, and the third arrived the following morning. The profitable American packets dominated the Atlantic traffic for more than thirty years and carried most of the transatlantic mail. They were the main means of communication between the Old World and the New.

The mail was the leather satchel in which postal riders carried their letters. It was also called a *bougette*, a word shaped into English as budget. Bundles of official letters were called the packet and the word devolved to the ships that carried them. Early packet services ran from Dover to Calais and Dunkirk. The Harwich packet started its service to the Netherlands in 1661, sailing on Wednesdays and Saturdays, and in 1675 the Harwich postmaster cordially sent his Dutch counterpart a Christmas present of two cheeses, a Cheshire weighing a hundred pounds and a Cheddar of ninety pounds. Mail and passenger traffic on the Holyhead-to-Dublin packet increased heavily after the Union with Ireland.

In the frequent periods of war from the middle of the seventeenth century to the early years of the nineteenth the packets were marked out as legitimate prey. Privateers, privately owned vessels licensed by government to attack enemy commerce, seized the mails, held them for ransom and robbed them of any gold or silver they contained. The ships were sunk or taken as prizes. In 1689 the threat of French ambush in the English Channel persuaded the government to open a packet port at Falmouth in Cornwall, well clear of the narrow confines of the Channel and with direct access to the Atlantic. It was the chief port for ocean mails until the 1840s and the title of its local newspaper, the *Falmouth Packet*, salutes its history. The Post Office sent the mailbags from London by way of Exeter, and the packets put to sea as soon as the bags were aboard. This was a businesslike improvement on the informal postal system operated by merchant captains who hung postbags in taverns and took them aboard only when they had loaded full cargoes.

The first Falmouth packets served Corunna in Spain, 450 miles to the south, and later supplied Lisbon and Gibraltar. A packet service started to the West Indies in 1702, completing the round trip in a hundred days, but it was ended after nine years by high charges and attacks by privateers. Warships and merchantmen carried the mails after that and the packet was not re-established until 1763. Similarly, packets ran to New York from 1709 but were withdrawn after two years because of high costs and piracy; and forty-five years passed before a permanent service was set up.

Falmouth packets were built for speed. Benjamin Franklin, the American scientist, crossed the Atlantic in three weeks in 1757, and although his ship was several times chased by French privateers he reported that 'we outsailed everything'. The packets were fast but the American privateers hunting in the American War of Independence were famously faster and, with the French privateers, were dangerous predators. They sank or captured forty-three British packets and badly damaged nineteen. To the horror of the Post Office, privateers seized two Holyhead packets in the Irish Sea. As was the practice, the packet captains weighted the mailbags with iron and threw them overboard.

In nine years of war with France after 1793 forty-six Falmouth packets were captured. Sometimes the crews fought back. The twenty-three men of the *Antelope* killed or wounded most of the sixty-five sailors of a French privateer who tried to board them off Cuba. They took the marauder as a prize. Off the French privateering base of Guadaloupe in 1796 ten passengers in the packet *Portland*, five of them army officers, grabbed muskets and beat off a privateer, shooting many of her crew. In 1798 fifteen passengers in the *Princess Royal* joined the crew in taking up guns and beating off a French attacker. The captain's sister and her maid calmly filled paper cartridges during the action.

Packet crews were small, usually fewer than thirty men, poorly paid and meanly fed. They made money by small-scale smuggling of cheese, tea, gin and tobacco and were consequently harassed by customs officers. They were also preyed on by press-gangs seeking sailors for the Royal Navy. The packet men's exemption papers made little difference. They simply showed the impressment officers that they were dealing with that valuable commodity, trained seamen.

Falmouth packets maintained vital lines of communication when war with France resumed in 1803. Again the packet men fought back when they could and the Post Office made many awards to crews who defied the privateers, spreading the idea that the mails were to be defended. The *Lady Hobart* won a fight with a French privateer off Newfoundland in 1803. Four days later she hit an iceberg and the twenty-nine people aboard took to the boats and reached Newfoundland after six days. The packet *Windsor Castle* refused to submit when a privateer attacked near Barbados in 1807: her men scrambled aboard the enemy and killed her captain and thirty-one crew. During the Peninsular campaign of 1808–14, in which the British armies under the Duke of Wellington drove the French out of Portugal and Spain, packets ran mail to and from Lisbon, the British base. The voyage took an average of two weeks but on one occasion a ship almost flew the distance in four and a half days. In 1814 the *Francis Freeling*, with £30,000 of gold aboard, was chased by a privateer for six days but outsailed her and reached the safety of Falmouth. The pressures on the Atlantic packets intensified when Britain, still fighting France, went to war with the United States in 1812.

Everyone who trusted letters to the overseas mail knew the risk of attack by privateers and pirates, and the danger of shipwreck. Mail was also delayed by the quarantine procedures meant to prevent the spread of disease. Postal officials at Falmouth fumigated letters from plague areas in the Mediterranean or the Caribbean by piercing the envelopes with holes and exposing them to the smoke from smouldering sulphur and charcoal. Letters were also handled gingerly with tongs and dipped in vinegar.

By the time of Napoleon's defeat American sea traders and shipyards had long experience of building vessels with fine, fast lines and great spreads of canvas. These traded along the American east coast and among the islands of the West Indies and fished on the Grand Banks. Both fishermen and slavers dealt in perishable cargoes and needed the fastest ships. During the Napoleonic wars American owners ran a lucrative business getting through the Royal Navy blockade of the Continent; and during the war with Britain of 1812–14 their ships frequently outstripped even the fastest British frigates. They drew on their cheap and plentiful supplies of home-grown timber to build numerous ships and were well placed to profit from the growth in the post-Napoleonic transatlantic trade and European emigration. Isaac Wright chose exactly the right time to launch his innovative Black Ball Line and make the ships run on time.

The press loved the American packets for their glamour. The *Edinburgh Review* in 1837 called them 'the best sailing vessels between this country and New York'. Charles Dickens similarly endorsed 'the noble American vessels . . . [the] finest in the world', while the writer and social reformer William Cobbett, who sailed in a Black Ball ship, wrote an irresistible recommendation. 'The Americans sail faster than the others, owing to the greater skill and greater vigilance of their captains and to their great sobriety. They carry more sail than other ships because the captain is everlastingly looking out . . . if I were going to cross again, nothing should prevail on me to go on board of any ship but an American one.'

The packet captains were celebrities of the sea, an elite of respected professionals, often better educated than the usual run of masters, and as well known to the public as any political figure or army hero. Their rewards were handsome, up to a quarter of the passenger fares and 5 per cent of cargo charges. They usually had shares in the ships, too, and invested their money in land and business. Ashore they lived and dined well, dressed in style and were noted for their coloured ties, gold watch chains and 'a certain smartness in their boots and general shore rigging'.

If, as one respectful writer had it, a packet ship was the noblest work of men, her captain was 'the noblest work of God'. It was not enough that a commander was the ablest of seamen: he had to have the conversation and manners to charm the first-class passengers. His graciousness with the ladies and gentlemen, however, sometimes concealed a brutal manner with the crew. A number of captains had their own followings, the forerunners of the celebrity groupies, people who made a point of sailing only with a particular trusted master. At sea they had a close-up view of their hero in action, the wise, rugged, tireless and awesome figure on the quarterdeck urging his ship through a gale; and, when conditions permitted, the urbane host presiding over dinner in the mahogany-panelled cabin. At the end of a voyage appreciative passengers sometimes presented the captain with a piece of silver or a handsome gratuity and placed an announcement in a newspaper saluting his courtesy and skill.

Newspapers cast the captains as men of renown, daring masters of speed who sailed their ships to the limit. An especially fast Atlantic crossing was a journalistic sensation. 'A reputation for being very fast brought cargo and passengers to a ship; fame for her captain and owner, but hell for the crew,' observed Howard Chapelle, the American historian of sail.

The ordinary sailors, however, were not to be pitied, according to Arthur Clark, an Atlantic captain. He saw the forecastle hands as more or less subhuman creatures 'amenable to discipline only in the form of force in heavy and frequent doses. To talk about the exercise of kindness with such men would be the limit of foolishness.' Yet he had a grudging good word for these scarred and hardened 'packet rats', the seamen who worked the ships in the relentless struggle for ocean records. 'For all their moral rottenness, these rascals were splendid fellows to shorten sail in heavy weather on the Western Ocean . . . a snow squall whistling about their ears, the rigging a mass of ice. They worked liked horses at sea and spent their money like asses ashore.'

A few packets sailed for many years in the Atlantic business. One made 116 round trips in twenty-nine years and carried 30,000 passengers to the United States. Two hundred couples were married on board and 1,500 babies born. But in general wooden ships so relentlessly driven usually lasted only a few years until, worn down by the Atlantic, they were sold off.

'Although they are the very best of sea-going craft and built in the best manner,' wrote Herman Melville,

> a few years of scudding before the wind seriously impairs their constitutions – like robust young men who live too fast – and they are soon sold to Nantucket, New Bedford and Sag Harbour for the whaling business. Thus the ship that once carried over-gay parties of ladies and gentlemen to Liverpool or London now carries harpooners. And the mahogany cabin which once held rosewood card tables and brilliant coffee urns, in which many a bottle of champagne sparkled, now accommodates a bluff Quaker from Martha's Vineyard who entertains a party of naked chiefs at dinner, in place of the packet captain doing the honours to the literati, theatrical stars and foreign princes who talked gossip, politics and nonsense across the table in transatlantic trips.

The packet trade wore out captains as well as ships. Sailing a vessel as fast as she would go was dangerous work. Masters pitted their experience and nerve against the ocean, especially at night, looking for every advantage in the wind, risking serious damage if they pressed too hard. 'I never knew an American captain to take off his clothes to go to bed during the whole voyage,' William Cobbett wrote. 'The consequence of this great watchfulness is that advantage is taken of every puff of wind, while the risk from sudden gusts is in great measure obviated.' Such seamanship saved lives. Scores of ships were wrecked every year, but only three packets foundered in the first twenty years of the service.

It was a job for men with the right blend of youth and experience. The stresses drove many packet captains into quieter commands in their mid-thirties, after five years or so at the top. The exceptional Black Ball captain, Charles Marshall, was forty-two when he quit the quarterdeck for a desk and became manager of the company. He had sailed the Atlantic ninety-four times.

For the ladies and gentlemen in the first-class section the heart of the voyage was the dining saloon. Aboard the packet *Pacific* this was a panelled room, forty feet long, one end decorated with an arch and pillars of Egyptian porphyry. A mahogany table ran down the centre of it. Seven state rooms on each side were entered through doorways supported by marble columns. Once they had recovered from seasickness, passengers met four times a day for substantial meals of mutton chops, pork, beef, ham and fowl with plenty of liquor. The ships' rats, it was reported, enjoyed champagne corks as a delicacy. The chickens, geese and ducks kept aboard were served for dinner very shortly after their necks were wrung. Sheep and pigs kept in the forecastle or in a boat on deck awaited slaughter. A passenger once complained to her diary that 'a sheep has jumped overboard and so cheated us of some of our mutton'.

Passengers had to find ways of coping with the inevitable tedium, their own ennui and the frequently oppressive company of others. They wrote their journals, played chess and whist, bet hotly on the ship's daily run and gambled at cards: the packets brought the game of poker from America to England. They sang songs and hymns, danced, played shuffleboard and complained pettily of each other's cheating. They quarrelled when drunk. They joined in the hilarity of rat hunting on the decks. On tranquil days they fished and tried to catch sharks. From time to time they sailed close to drifting mesas of icebergs. If, as they did on rare occasions, they spied a polar bear they naturally ran for their rifles and fired at it.

'You cannot imagine, dear Mother, what a droll life one lives in this great stagecoach called a ship,' the philosopher Alexis de Tocqueville wrote on his way to America. But there were inconveniences. 'Our cabins are so narrow', he said, 'that one goes outside them to dress.'

For all the fun in the satinwood class there were not enough rich people to make transatlantic travel pay. Shipowners responded to the swelling crowds of Europeans yearning to make a new life in America and willing to travel in no comfort at all if tickets were cheap. Liverpool, whose merchants had long experience of packing slaves economically, now stuffed migrants into dark and awful dungeons, three or four hundred or more in each ship. 'These emigrants', wrote a packet captain,

> were the rakings and scrapings of all Europe. Men, women and children were tumbled into the 'tween decks together, dirty, saucy, ignorant and breeding the most loathsome of creeping things. The stench below decks aggravated by the sea sickness and the ship's poor equipment for the work, placed us far below the civilisation of the dark ages. It was not uncommon in mid-winter to be fifty or sixty days making the passage. In gales, which were frequent, hatches had to be battened down and men, women and children screamed all night in terror. Fever, smallpox and other contagious diseases were common and it is a wonder that so many survived the voyage.

As well as the storms, seasickness and bad food there were the irritations caused by other passengers. Two Welsh preachers who sailed to America in 1836 complained of the Irish crowded aboard. 'We were afraid many times that their ungodliness would move the Lord to sink us all not simply because they were papists but also because they were barbarians,' they wrote.

Between decks, in spaces six feet high, carpenters fitted three tiers of rough unplaned shelves for bunks. In the early years the emigrant's ticket of hopes and dreams bought only the cramped and stinking sleeping crevice, a ration of water and a share of a galley fire. These passengers had to bring some straw bedding for their allotted plank as well as their own food, pots, cups and cutlery. The Black X Line asked people to provide themselves with two hams, potatoes, biscuit, flour, tea, sugar and treacle; and a cooking pot, a frying-pan, a mug and cutlery. Since passengers could not know how long the voyage would be it was difficult for them to calculate how much food they would need. Many Irish migrants took only oatmeal and had porridge every day. The brig *Credo*, which sailed for Quebec from Aberystwyth in April 1848, advertised an emigrant fare of £3 5s. for adults and simply told passengers to 'find their own provisions'. A migrant sailing from Liverpool in 1856 wrote of the conditions in the cheapest berths. 'As for the £4 place, that is the lower deck, the home of the Irish. Oh! hole of pity, darkness, barbarism, dirt, flies and stench!'

As their ship yawed and surged passengers took turns to cook in a tiny and filthy galley, usually open to the weather, between the cow byre and the sheep, pig and poultry pens. The ship's carpenter doled out daily allowances of water and fuel for the fires and was on hand to extinguish the flames when they got out of control. In many ships, as a sideline, he performed dentistry. The shipping companies' insistence that passengers supply their own food led to such hunger and distress that the British government passed laws making the companies feed them with a weekly ration of biscuit, flour, oatmeal, rice and molasses – 'just enough', said a captain, 'to keep starvation away'.

Herman Melville saw hundreds of German migrants embarking at Liverpool for New York – and 'they gathered on the forecastle to sing and pray . . . fine ringing anthems'. Every evening at sea, he said, they 'sing the songs of Zion to the roll of the great ocean organ'. Irish emigrants, on the other hand, were, according to one account, quite different:

> harum-scarum, and happy-go-lucky . . . most of them had an idea that America was only two or three days' sail from Liverpool, and often when the coast of Ireland was sighted after perhaps several days of head wind, they crowded to the

rail under the impression that they were looking at the coast of the promised land and would hardly believe that it was old Ireland, which they had left only a few days before.

First-class passengers rarely saw the foul and degrading conditions in which most migrants travelled. But sometimes they could not avoid being assailed by the terrible smell of unwashed bodies, vomit, faeces, urine, rotting food and wet straw. In such conditions smallpox, cholera and especially typhus, known as 'ship fever', carried off passengers in their dozens. But as shipping companies themselves sometimes pointed out, in an attempt to apply a cheerful gloss, births exceeded deaths during the majority of voyages.

No year on the Atlantic was worse than 1847. The great famine in Ireland caused by potato disease in the 1840s killed a million people, an eighth of the population. They died chiefly of starvation, typhus and dysentery. The potato crop, the staple and healthy food of the country people, was seriously depleted by blight in 1845. It failed totally the following year and totally again in 1848. People fled in their tens of thousands and many of them sought refuge in escape across the Atlantic. Thousands died in the attempt because they were weakened by malnutrition and disease and in no fit state to make an ocean voyage aboard overcrowded, dirty and meanly provisioned ships, and those with typhus rapidly passed it on to their fellow passengers.

The burial records and mass graves at the old immigrant quarantine station on Grosse Ile in the St Lawrence River, twenty miles from Quebec City, tell the tragic story. In the six-month shipping season of 1847, the river being closed by ice in the winter, 8,745 migrants heading for Quebec City died during the voyage or in the quarantine station, a twelfth of the 105,000 who set out for Quebec that year. Most were Irish. The average age of those who died at sea was twenty, and of those who died in quarantine twenty-four. The ships with the highest death rate were those which sailed from Liverpool and Cork. Liverpool vessels carried an average of 375 passengers of whom fifty-eight died, 16 per cent of the total. Ships departing Cork had an average of 312 of whom fifty-eight died, a rate of 19 per cent. In comparison, the death rate that year aboard ships from Bremen and Hamburg varied from one-hundredth to one-fiftieth of the passengers.

Some voyages were particularly horrific, with death rates up to 56 per cent. Of the 476 passengers who left Liverpool in *Virginius* on 28 May 1847, 159 died during the three-month crossing to Quebec and 109 in quarantine, a total of 268. Of the 334 who sailed in the *Naomi* from Liverpool 196 perished. Of 552 in the *Avon*, out of Cork, 247 died, and in the *Bee*, also from Cork, 165 of 373 men, women and children were buried at sea or died in quarantine or in hospital soon after their arrival.

The ships anchored in a line off Grosse Ile to unload their dead, dying and sick. Robert Whyte, an Irish emigrant who arrived off Grosse Ile in July 1847, spoke to priests who came aboard his ship to perform the last rites. 'The account they gave of the horrid conditions of many ships was frightful. In the holds they were up to their ankles in filth ... corpses remaining long unburied, the sailors ill and passengers unwilling to touch them.' He saw 'a continuous line of boats carrying its freight of dead, an endless funeral procession, some tied up in canvas, others in rude coffins, constructed from the boards of their berths'.

In the twenty-four weeks of the shipping season that year, 478 emigrant vessels sailed for Quebec, mostly from England, Ireland, Scotland and Wales, and twenty-two of them lost a hundred or more passengers, each case in modern terms the equivalent of a major air disaster: a disaster almost every week. Thirty-one ships lost more than fifty passengers. More than 5,000 of those who died in Grosse Ile lie buried in mass graves, the death toll of 1847 so high that the coffins were stacked in threes.

Jonathan Edwards, who sailed from Liverpool to New York in the *Virginia* in 1866, described in his diary the horror of cholera aboard. 'Death is using his sword very easily. Two died last night and six today.' Then six died, then five, then seven. The ship's doctor declared a steward dead but as some men were wrapping him in a shroud to throw him overboard 'the man opened his eyes and asked what they were doing. They started rubbing him with brandy until he revived. Our hearts bled to see our friends dying and groaning. There was a woman crying bitterly having seen her baby thrown into the sea.' Cholera killed fifty of the passengers, and when the ship reached New York the survivors had no food for two days. And then 'food arrived and the men were like a pack of wolves leaping over each other and snarling. Many were nearly crushed to death; others were cursing and swearing and hitting each other.'

Once cleared by the doctors, the fortunate survivors could remake their lives. For many years to come, however, the Atlantic voyage was as often as not an ordeal of terrifying gales, disease and death.

12

By Vulcan's Art my ample belly's made

'What I ha' seen since ocean-steam began
Leaves me na doot for the machine: but what about the
  man?'

Rudyard Kipling, 'McAndrew's Hymn'

In 1801 a fifty-foot boat nudged its way into the history of shipping. The *Charlotte Dundas* towed two barges for twenty miles along the Forth and Clyde canal in Scotland. It was the first working steamboat. The engine drove a single paddle wheel mounted on the stern and was built by William Symington and designed by the Scottish craftsman James Watt, one of the great talents in the development of steam power. According to the romantic story about his boyhood, he observed the escape of steam from a kettle and saw in his mind's eye not only the invention of the steam engine but the marvellous steam-powered industrial era as well.

He did not, however, invent the steam engine. In the progress of machines there was no single originator or flash of inspiration or overnight success. Inventions had forefathers. The steam engine's distant ancestors included Chinese scientists of the second century BC. In 1631 David Ramsey, a Scot, obtained a patent on his dream of building machines 'to make boats, ships and barges to go against strong wind and tide', but he did not explain how he would do so and no evidence exists that he ever built an engine. In the 170 years between his patent and the reality of William Symington's steam-powered boat a number of mechanics and assorted tinkerers in Britain, France and America bolted together crude cylinders, spheres and pistons of brass and copper in their attempts to create steam power. Thomas Savery demonstrated a pump to the Royal Society in 1699, 'raising water by fire' to make waterwheels turn, and began building steam devices near Fleet Street in London in 1702.

The significant advance was a massive steam engine constructed in 1712 by Thomas Newcomen, a Devonshire ironfounder and blacksmith, to pump water from Tipton colliery in Staffordshire. In this engine, the condensation of steam created a partial vacuum under a piston, and the weight of the atmosphere pushed the piston down and raised the pump rod. More than 300 Newcomen engines working on this principle were manufactured to evacuate water from mines in Britain, mainly in Cornwall and north-east England, and also in France, Germany and Hungary. In 1755 a copper mine owner in New Jersey ordered a Newcomen pump from England and inaugurated the steam age in America.

James Watt, however, cannot be denied his breakthrough. While repairing a Newcomen engine in 1764, when he was twenty-eight, he saw that its action could be improved by getting steam to operate on both sides of the piston. This was the double-acting engine, the great step forward; although it was twenty years before he developed a thoroughly reliable model. Other engineers meanwhile were working on similar lines. Watt moved to Birmingham, then prospering as a centre of engineering and invention, and in 1774 joined Matthew Boulton, a visionary industrialist. Boulton had started out as a button-maker and by the 1770s employed 800 people manufacturing metal goods. He and Watt were at the very heart of the new and expanding world of engineering. Boulton noted in 1781 that 'people in London, Manchester and Birmingham are steam mill mad'. The double-acting rotative engine that Watt designed in 1784 was so dependable that it set the standard for steam engines for the ensuing half century; and in time the internal combustion engine evolved from it. The firm of Boulton and Watt grew into the chief British manufacturer of engines and made 496 of them for ironworks, cotton factories, flour mills and breweries. Watt graded the output of his engines by horse-power, based on his observation of the work done by dray horses, and he defined and popularized the term. He also invented the vital regulator, or governor, to control the speed of an engine.

Britain's primacy in steam was undisputed. British mechanics were the best in the world and spread some of their knowledge as they travelled in Europe and Russia installing Boulton and Watt engines. But the British government banned any export of textile machinery to the United States so that Britain could retain its command of steam technology. Nor were there any construction manuals. It was not until after the Napoleonic wars that the know-how of steam seeped into America with the arrival of migrant British engineers.

Watt's epitaph in Westminster Abbey salutes the engineer who 'increased the Power of Man'. Thomas Carlyle, the essayist and biographer, described him as 'this man with blackened fingers, with grim brow . . . searching out the Fire-secret'. Some thought that Watt's genius did for steam what Shakespeare did for poetry. Like everyone else, poets were caught up in the fervour of the 'gospel of steam' and the way it was transforming the world and propelling the leap into the future. They revelled in what the poet Ebenezer Elliott called the 'Tempestuous music of the giant, Steam'. Erasmus Darwin, in 1792, thought that 'unconquer'd Steam' would eventually propel flying machines, that it would

> Drag the slow barge, or drive the rapid car:
> or on wide-waving wings expanded bear
> The flying-chariot through the fields of air.

In the early part of the nineteenth century steam led the triumphalism of technology; and Watt was the hero of it all. 'Engine of Watt!' cried Ebenezer Elliott, 'unrivall'd is thy sway.'

It would be some years before poets would stand apart from the pounding pistons and cast a more sceptical eye on the 'ruthless King' called steam and on the smoke that masked the sun and sooted the streets and downs. Then they would begin to conceive of the machines as tyrants and conclude that steam-driven factories were dungeons of the poor.

While James Watt had little interest in applying his engines to land vehicles or boats, several engineers in France, America and

Britain developed steam vessels. Robert Fulton, an American mechanic, went to Scotland and saw Symington's engine driving the pioneering *Charlotte Dundas*. He himself had been experimenting with paddle-wheel craft for twenty years. While living in France in 1797 he designed a submarine, *Nautilus*, driven by a hand-cranked propeller when submerged and by a sail on the surface. He demonstrated it at Brest to some of Napoleon's admirals but they were not impressed. He then tried to sell the contraption to the Royal Navy whose admirals agreed with French opinion.

Fulton returned to America and designed and built a far more practical craft, his great contribution to transport history. The *Clermont* was 150 feet long and driven by paddle wheels on each side. Her Watt engine gave her up to five knots. In 1807 she steamed 132 miles of the Hudson River between New York and Albany in thirty-two hours, 'a monster ... breathing flames and smoke', and initiated the age of the American river boat. Fulton launched twenty-one such steamers and in 1814, during the Americans' war against the British, he constructed the first steam-driven warship and modestly named her *Fulton the First*.

In Scotland, meanwhile, Henry Bell built the paddle steamer *Comet* to carry passengers on the Clyde between Greenock and Glasgow in 1812. 'Stinkboat', the river boatmen called her; but her smoke and noise were exciting and she also worked reliably and was immediately popular. Within two years five steamers were running regular services on British rivers and canals.

The open sea, though, was the critical test. In 1814 the steamboat *Thames* set out from her builders on the Clyde and reached London by way of Dublin, Wexford, Milford Haven and Plymouth, a total of 758 miles at more than six knots. She was soon put into regular service, carrying thousands of passengers between London and the resort of Margate on the Kent coast. *The Times* published news of such shipping developments on its new steam presses, installed in 1814. In 1816 the Clyde-built *Margery* was sold to French owners and crossed from Newhaven to Le Havre in seventeen hours, the first steamboat to butt across the Channel. Two years later the *Rob Roy* plied between Belfast and Greenock, and an Englishman started a service between Venice and Trieste. As steam companies proliferated so did public confidence in steamboats.

Steamers were put on the crossing from Holyhead to Dublin in 1819, the ten-hour voyage halving the time taken by a sailing packet. King George IV travelled to Ireland in 1821, the first monarch to use a steamboat. A steam packet service across the English Channel started that year and a service to the Isle of Man from Liverpool in 1822. The Royal Navy's first steamer, *Lightning*, launched in 1823, deposited smoke smuts on officers' noses and made long-haul passages to Algiers and Russia, the beginning of a distinguished fifty-year career as a survey vessel.

For twentieth-century Hollywood the wagon trains pursued by whooping Indians were good box office; but in reality the West was won chiefly by rafts and boats and the men who steered them on the great rivers. Before the settler trails snaked out towards the Great Plains and the Rocky Mountains, and long before the railroads made their début, the Ohio, the Mississippi and the Missouri, more than 6,000 miles of navigable water, were the highways to the frontier.

From the years when America was a British colony the Ohio carried emigrants south and west into their own fables. If, say, you sailed from Liverpool with your bride and arrived in Philadelphia, you bought horses and a wagon and drove 300 miles west along the rough road, staying at taverns if you could afford them, until you reached Pittsburgh in Pennsylvania. Here you sold your wagon and purchased a boat, or perhaps joined forces with like-minded pioneers and bought a share in one. There were plenty of carpenters in Pittsburgh able to knock up a boat quickly, with rotten timbers if they could get away with it. For $35 or so you got a flatboat, often called a broadhorn, an American ark little more than a crude raft about sixty feet long with a rough shed for shelter, a space for some chickens, a stack of hay for a horse or two and a long steering oar to guide the whole clumsy rig down the swirling stream.

At Pittsburgh the Allegheny and Monongahela rivers unite to form the broad and deep Ohio which flows almost a thousand miles to join the Mississippi in the centre of America. Many of the people who bravely pushed out into the river had no clear idea of where they would finish or what they would find. They were bound for the new towns taking root along the fertile Ohio valley, or heading for Kentucky and its promising lands. Some dared to aim for the Mississippi and beyond. After 1801 the river-borne pioneers could buy Zadock Cramer's indispensable guidebook *The Navigator*, with its charts of shoals and landing places. It made sense for people to drift down the river in groups for protection against pirates and Indians who hid in overhanging trees and dropped on to the rafts to kill and rob. If they were spared, the victims were put ashore where they shouted to passing flatboats for rescue. But sometimes the pirates slit their bodies and filled them with stones.

When migrants reached a suitable place to settle they hauled their boats out and cannibalized the timbers to build houses and raise palisades as protection from marauding Indians. Cincinnati was founded on the banks of the Ohio in 1788 by flatboat voyagers who sawed up their boat timbers to build the town's first school and houses and, before long, the chapels of half a dozen denominations.

Flatboats were built for just one downstream journey. Keelboats were more substantial, eighty feet long and twelve feet in the beam, with a cabin for the crew and a mast and square sail. These were the bulk freighters of the rivers and in the early years of the nineteenth century there were 300 of them. They carried cargoes of iron, salt, bricks and timber; and while their crews had an easy time travelling downstream they had to struggle to row and pole the boats back against the current with fifty tons of furs, hides, sugar and coffee on board. For months they hauled on towlines and grasped the branches of riverside trees to inch their way back to their base upstream.

Such heroic work bred a tribe of several thousand rough and brawny rivermen notorious for their swaggering, fighting and drinking. Mike Fink was the best known of them, a roistering keelboat captain on the Ohio and Mississippi. By the time of his death in 1823 his brawling, boozing and shoot-outs had made him a frontier legend, like Daniel Boone and Davy Crockett, although far less noble. In his last scene he boasted he could shoot a tin mug from a companion's head, in the manner of William Tell, but his aim was poor and the man fell dead. Much offended, a friend of the drilled victim drew his pistol and killed Fink in turn.

In September 1811 a steamboat's clangour signalled both the start of the mechanical age on the rivers and the long twilight of the keelboats. Robert Fulton had already proved himself as a steamboat builder on the Hudson and now he and Robert Livingston launched the *New Orleans* paddle steamer, 138 feet long, on her maiden voyage from Pittsburgh to New Orleans. She was a dramatic spectacle. The roar of escaping steam drowned the racket of her engine, the brown water churned under the shuddering paddles, smoke erupted in volcanic plumes and the ding of her bell punctuated the shouts and hubbub of spectators.

The *New Orleans* wheezed down the Ohio, joined the Mississippi at Cairo and turned south for her namesake town. Along the river crowds of gaping settlers and knots of thoughtful Indians watched her pass. Captain Nicholas Roosevelt berthed her alongside the bank every evening and people streamed in from the countryside to cheer and goggle. At Natchez Roosevelt amazed everyone by steaming against the current. When he tied up he invited a crowd on board and showed off the 'stewing, sizzling, whizzing' engine and made everyone jump when he suddenly blew a shrieking blast of steam. The steamer arrived in noisy triumph in New Orleans in January 1812. She was not a true river boat; she drew too much water for one thing, a serious handicap in a shallow river; but she was the forerunner. Within a few years steamboats multiplied. Many of the early vessels were inefficient and slow, and the jaunty keelboat men laughed at them and wagered that they could sail faster; and sometimes they won barrels of whiskey by beating them in races.

The rivers, meanwhile, were a riot of free-for-all life. Flatboats, keelboats, canoes, barges and rafts took cargoes of timber, immigrant families, merchants, peddlers, prostitutes, parsons and adventurers down the Ohio and up and down the Mississippi. Farmers clubbed together to build flatboats, loaded them with prairie produce and drifted down to New Orleans, a city bursting its buttons with trade and people. Here they sold their goods and the timbers of their boats, bought a steamer ticket and cruised home with their pockets full. In this fashion the young Abraham Lincoln

lacable highway: a sailing ship runs before a gale in the Southern Ocean

'To sail as far as I think it possible for man to go': James Cook (1728–79), the greatest oceaneer. Portrait by John Webber, artist on Cook's third voyage

'My Grand Tour shall be one round the whole globe': his botanic majesty Joseph Banks (1743–1820), lover of Tahiti, enthu and impresario. Portrait by Thomas Phill

The adoration of Omai: the Pacific islander as the idealized and exotic noble of the South Seas, after Sir Joshua Reynolds's. *(Inset)* Omai revisited: portrait by William Hodges, who always sought more truthful images, painted around 1775–7

Sex and breadfruit, seamanship and coura William Bligh (1754–1817) of the *Bounty*. Portrait by John Smart

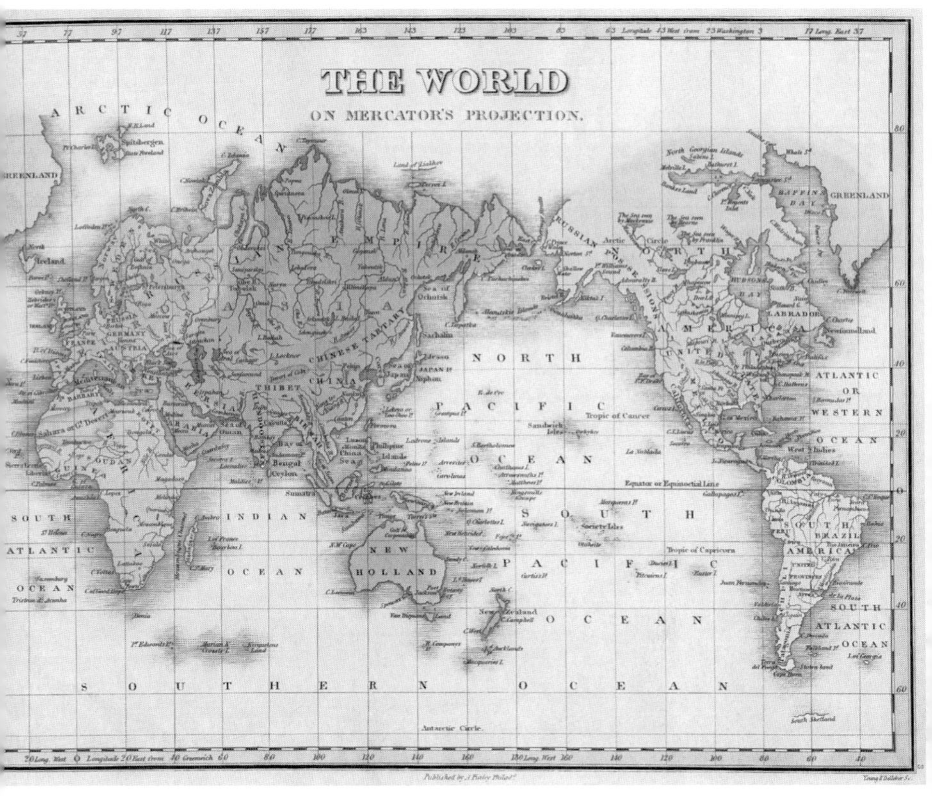

e truth of navigation': for all its inevitable distortions the projection of Gerardus Mercator
–94) was important for navigators and one of the most influential maps of history

what was wanted: the classic pictures of India by Thomas and William Daniell, this one
he Taj Mahal across the Yamuna River, met some of the huge demand in Britain for
ian scenes and shaped perceptions of the Indian empire

Negroes for Sale.

A Cargo of very fine ſtout Men and Women, in good order and fit for immediate ſervice, juſt imported from the Windward Coaſt of Africa, in the Ship Two Brothers.— *Conditions are one half Caſh or Produce, the other half payable the firſt of January next, giving Bond and Security if required.*

The Sale to be opened at 10 o'Clock each Day, in Mr. Bourdeaux's *Yard, at No. 48, on the Bay.*

May 19, 1784. JOHN MITCHELL.

Thirty Seaſoned Negroes

To be Sold for Credit, at Private Sale.

AMONGST which is a Carpenter, none of whom are known to be diſhoneſt.

Alſo, to be ſold for Caſh, a regular bred young Negroe Man-Cook, born in this Country, who ſerved ſeveral Years under an exceeding good French Cook abroad, and his Wife a middle aged Waſher-Woman, (both very honeſt) and their two Children. *Likewiſe*. a young Man a Carpenter.

For Terms apply to the Printer.

Above Shrinking the ocea fast American ships driver hard by daring and sometimes idolized captai revolutionized transatlanti travel from the 1820s

Left Iniquitous business: slave-sale advertisements i newspaper, 1784

Hero of the north, intrepid,
ane, observant and an eclectic
: Samuel Hearne (1745–92)

America the beautiful: Sacagawea
ng the Lewis and Clark expedition to
'acific (1805–6)

Flag-planter: Sir John Barrow (1764–1848), second secretary at the Admiralty, biographer, writer on China and Africa, and backer of Arctic exploration

Sir John Franklin (1786–1847), sent b John Barrow to find the North West Passage, perished in the Arctic ice wit his men

In the jaws of the ice: HMS *Hecla* and *Griper* during Lieutenant William Parry's search the North West Passage in 1819–20

gle on India: William Tayler's 1843 sketch of
orge Everest (1790–1866), Surveyor-General
ndia, who completed William Lambton's
asurement of India, the Great Arc

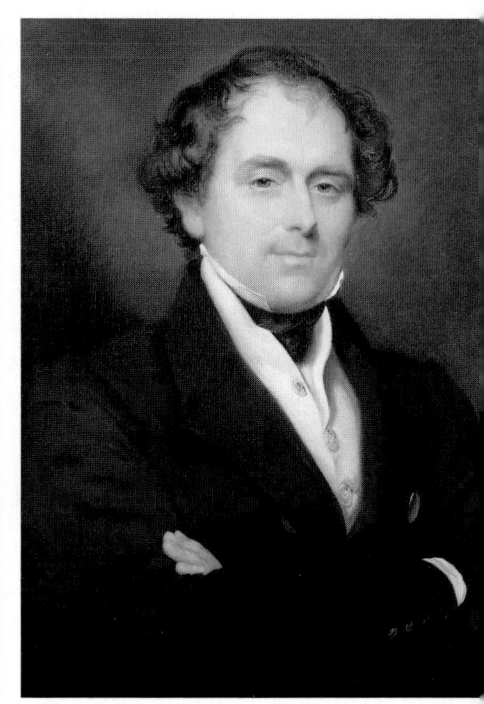

'Care of Mr Waghorn, Alexandria': Thomas Waghorn (1800–50), amazing Mercury of the East and pioneer travel agent. Portrait by Sir George Hayter

ve parasol, theodolite and sword, will travel: British surveyors in the mountains of
itan

The Great Stomp: troops run twelve abreast across the Great Exhibition floor in 1851 to prove it safe for crowds

'We are capable of doing anything,' wrote Queen Victoria after seeing the Great Exhibiti

brought his harvest to New Orleans and gazed at the amazing waterfront of the capital of the Mississippi.

Booming river trade demanded larger and more powerful steamers. In 1819 the *Independence* left St Louis and was the first steamboat to enter the Missouri, America's longest river. In 1820, only nine years after the *New Orleans*'s maiden trip, sixty steamboats plied the Ohio and the Mississippi. In the next eight years the shipyards at Cincinnati, Marietta, Big Bone and similar Ohio river ports built 118 steamers and baptized them with names like *Red Rover* and *Velocipede*. The inhabitants of the infant riverside towns took a proprietorial pride in the clanking boats, seeing them as the juggernauts of civilization, pushing the Indians aside as they steamed upriver towards the new footholds of hamlets, forts and mining camps.

Unruly steamboat captains sometimes raced their rivals and collided with them or ran aground on the sandbanks. Passengers shot buffalo as they passed by grazing herds, and the captains stopped while the beasts were skinned and butchered. Shipbuilders made the vessels more comfortable for travellers who could afford a cabin. For the poor, who could not raise the money for such luxury, the boats at least had the advantage of being very cheap. A dollar or two took them 500 miles, into the heart of the new lands. These deck passengers were accommodated near the horses and cattle and slept where they could and took their chances by drinking the river water. They were close to the roaring and shaking high-pressure boilers which often exploded with horrific consequences: newspaper artists who illustrated reports of these disasters could not resist showing heads and legs flying through the air.

Contemplating the traffic of river boats and wagons George Catlin realized that the tide of European settlement was unstoppable. There was a moment at his home in Philadelphia when he watched a group of Plains Indian chiefs passing through on their way to negotiate treaties in Washington. They looked magnificent in their robes, eagle-quill headdresses and tasselled tunics, 'arrayed in all their classical beauty', as he put it, and the spectacle moved him to the central decision of his life. It was obvious that the tribes would be broken and scattered. As the United States expanded they would be reduced; and as it became prosperous they would grow poor.

Catlin resolved to go to the frontier and paint the Indians before they vanished, their faces, their costumes, their weapons, food, homes and ceremonies. Before their twilight he would celebrate their freedom and humanity. Catlin was a lawyer and had no training in painting the human figure and only hazy ideas of perspective. But he was a fast sketcher. In 1832, when he was thirty-five, he heard that the steamboat *Yellowstone* was leaving St Louis to make the first steam voyage to the upper Missouri and into the heart of Indian country. This was his chance. He got himself a berth and joined a raucous crowd of trappers and fur traders well stocked with whiskey to trade for pelts. It took three and a half months for the steamboat to cover more than a thousand miles along the border of modern Nebraska, through South Dakota and North Dakota to Fort Union on the Montana border where the Yellowstone River meets the Missouri.

The steamboat stopped frequently and Catlin painted with a furious energy. He was an oddity, roaming the frontier without any desire to settle or trade or gather souls. All he wanted, 'with the aid of my brush and my pen', was to rescue the Indians from oblivion. Among the Sioux encamped in hundreds of tepees around the stockade at Fort Pierre he was shocked to see how readily warriors slaughtered and wasted the buffalo on which their lives depended. Offered a few gallons of whiskey for the tongues, they killed 1,400 of the beasts in a few hours, cut out the tongues and left the rest of the animals to rot. No doubt they believed that the herds, hundreds of thousands of animals stretching darkly across the plains, would last for ever. Catlin painted everything he saw, the buffalo hunts, the 'tall, straight and graceful Sioux', and the 'fine-looking' Indian women who lived with white men. Indians had no tradition of portraiture in their own art and were sometimes frightened when they peeked around the easel and saw what Catlin had painted. But many were entranced, too, and Catlin described how 'an Indian often lays down from morning to night in front of his portrait admiring the beautiful face and faithfully guarding it from harm, owing to the belief that there may be life in the picture'.

As well as painting and sketching the Sioux, Blackfoot, Chippewa, Assiniboine, Comanche and Crow, Catlin filled his notebooks with details of their diet and clothing, and the foot-stamping dances through which they expressed themselves. 'No one can appreciate their nobility without seeing them in their own country. I will do all I can to make their looks known to the world. I will paint and scribble . . .' He shrank from nothing as he painted, however strange and seemingly barbaric the spectacle, and made no judgements on the mutilations and tortures he saw.

From Fort Union Catlin and two fur traders canoed through a landscape where buffalo, antelope, bears and wolves were plentiful. Among the Mandan he painted some of his most astonishing pictures. He noted their blue and grey eyes and fair colouring and was struck by their 'peculiar ease and elegance', so much so that 'I am fully convinced that they have sprung from some other origin than that of other North American Indians'. He was charmed by Mink, an attractive young woman whose portrait he painted, but decided it would be wrong to take her as a mistress. He noted the etiquette of the chiefs who served their guests at dinner, but would not themselves eat until later.

The Mandan allowed him to paint the extraordinary ceremony of *okeepa*, the supreme test of a warrior's courage and never before witnessed by a white man. The participants fasted for three days in a dome-shaped sacred lodge and then stood without flinching as sharp sticks were thrust through their breasts and shoulders and tied to rawhide cords hanging from the lodge roof. Bows, arrows and buffalo skulls were hung from the sticks and, streaming blood, the youths were raised several feet in the air and spun rapidly before being dropped to the ground unconscious. When they revived, their little fingers were chopped off. Finally they ran in circles until the heavy buffalo skulls and other objects tore out the wooden skewers from their breasts and ended the ordeal. No one was allowed to help the staggering sufferers in their agony. Catlin's paintings of this ritual created a sensation. His collection provided a unique record of the 'supple and graceful' Mandan in their final flourish before smallpox killed most of them a few years later.

Catlin continued his painting all the way down the Missouri to St Louis. He and his companions watched from a hiding place as Arikara Indians danced in frenzied celebration around the fresh scalps of two white men. He made more journeys into Indian country in Wisconsin and Minnesota and was the first white man to see the pipestone quarry in south-west Minnesota where Indians mined the soft stone they fashioned into tobacco pipes. Warriors warned Catlin not to approach this forbidden place, but he gently persisted, sketched the quarry and picked up pieces of the red stone. A geologist later identified it as a new mineral and named it Catlinite. Today the quarrying is reserved by treaty for Indians. Catlin included peace pipes in his travelling exhibition of Indian artefacts, tomahawks, bows and arrows and clothing as well as his Indian scenes and portraits. Many who saw the show were disturbed by his portrayal of Indians as 'honourable and religious human beings'. Catlin hoped the government would buy his paintings but it did not and he died disappointed in 1872. It was many years before he was vindicated. Today his paintings hang in Washington, New York and elsewhere. Catlin intended them to be his 'perfect monument of an extinguished race'.

13

The great god tripod

'When you can measure what you are speaking about, and express it in numbers, you know something about it.'

Lord Kelvin, 1883

The Greek geographer Strabo said in the early years of the first century AD that 'we must hear accounts of India with indulgence, for not only is it very far away, but even those who have seen it saw only some parts of it'. Eighteen hundred years later there was still much force in his warning. Most of India was a blank to Europeans. In the 1750s a French cartographer could only guess at the route of the River Ganges across the northern plains and drew a speculative serpentine line through what he called 'country of which there is no particular knowledge'. The merchant venturers of the East India Company occupied coastal enclaves in Calcutta, Bombay and Madras and knew little of the hinterland. Great rivers, mountain ranges and tiger-ruled forests were missing from their maps and minds. The Himalayas were barely known: no European had yet sighted the convocations of the world's gigantic peaks.

Britain's gradual conquest and annexation of the provinces of India created a demand for maps to enable the rulers to understand, manage and tax the country. Surveys were needed to bring coherence to the chaos of landscape.

After winning the Battle of Plassey in 1757 Robert Clive became Governor of the province of Bengal and founder of Britain's Indian empire. He commissioned twenty-one-year-old James Rennell to make the first scientific measurements of lands owned by the East India Company. Rennell began with a survey of the plain of the

Ganges, fixing the latitude and longitude of strategic places. He drew much of his information from the notes and sketches made by army officers. Although military expeditions usually found their way with scouts and local guides rather than with the aid of maps, cartography and drawing were becoming an increasingly relevant military skill. Officers who could turn in an accurate pencil drawing or watercolour sketch of a landscape were men of value. By the beginning of the nineteenth century military academies increased their emphasis on teaching cadets topographic drawing and surveying, and a small cadre of army surveyors was established in India.

Although he was Surveyor-General of Bengal, Rennell looked beyond the province and drew maps of the entire subcontinent. India at that time was known as Hindustan, the name applied to it by the Moguls, the Muslim conquerors. Rennell called it India and established the idea that this huge culturally and linguistically diverse mosaic was a geographic entity and an area of British interest. He set out its now familiar shape and borders and in time this carefully mapped and defined British India would be adopted by Indian nationalists as the true and historic India.

Rennell's pioneering cartography opened the way to the detailed mapping undertaken by nineteenth-century British surveyors. In 1800 William Lambton, a military surveyor, persuaded the East India Company to fund a survey of southern India. He had no plan at that time to extend the work into the central and northern parts of the country. In 1818, however, his project grew into the complete charting of the subcontinent and was called the Great Trigonometrical Survey of India. The principal feature was its 1,600-mile backbone, the Great Arc, hundreds of painstakingly measured triangles stretching up the centre of India along the seventy-eighth meridian from the southernmost point at Cape Comorin to the farthest bulwarks of the Himalayas.

The survey occupied the lifetimes of many surveyors and cost many lives, too. It took half a century to complete. The word trigonometry hardly resonates with danger, yet the Great Trigonometrical Survey was an always arduous and frequently perilous adventure in which men found death and crippling disease. A few found a tincture of glory. Theirs was no ordinary survey. As they stumbled through jungles and trudged over deserts and mountains and shook with fever they believed they were bringing rationality to what seemed inchoate. In their minds the Great Arc was a noble endeavour, the mastering of nature. In the interstices of the Great Trigonometrical Survey local topographical and revenue surveys created a Domesday Book of the Raj, an anatomy of India. Through the grit and diligence of the surveyors India ceased to be space and its bewildering immensity was brought to order. The large map on the wall of the district officer, the tax collector, the police chief and the garrison commander was the essential aid to management and the emblem of dominion.

The immediate ancestor of the Great Trigonometrical Survey was the eight-year survey of Scotland started after the uprising of 1745, Prince Charles Edward Stuart's attempt to seize the British throne. After the defeat of the rebellious clans at Culloden the British army called for reliable maps to help them understand the land and bring the disaffected Highlands under control. General David Watson and his quartermaster William Roy were commissioned to make the surveys. Roy was Britain's first major surveyor and his work was the foundation of the Ordnance Survey, the government mapping agency. He designed an advanced theodolite, the chief instrument for measuring angles, which weighed 200 pounds and had a brass circle a yard in diameter. Roy's technique inspired William Lambton in his work in India. Lambton, the son of a Yorkshire farmer, spent thirteen years as an army surveyor in New Brunswick and taught himself astronomy and geodesy, the study of the shape of the earth. He was a shy boffin, a learned forty-year-old lieutenant, when he arrived in India in 1797. He shone as a brigade major under Arthur Wellesley in the war against Tipu Sultan, and Wellesley supported him when he sought backing for his southern India survey.

Lambton began in 1802. He ordered a large theodolite from England, but on its way to India it was seized as a prize by a French ship and landed in Mauritius. In the interests of science the French governor of the island chivalrously forwarded the instrument to Lambton with a letter of good wishes. Thus the legendary Great Theodolite, which survives in the Survey of India offices in Dehra

Dun, began its career. It was one of the most modern scientific instruments in the world. No doubt it was often cursed. It was necessarily robust and heavy, a half-ton monster, but it could not ride on a bullock cart because the jolting on rough tracks would have damaged its sensitive adjustment screws, micrometers and microscopes. Instead it was carried everywhere by a dozen men trained to treat it as if it were a fragile jewel borne on a velvet pad.

The survey was achingly slow and repetitive as Lambton and his men compiled their series of triangles with a measuring chain, rods and calibrated wheels. First, they measured a baseline between two landmarks, say between a fort and a tower. From each end of this baseline they took bearings on a third point and employed trigonometry to fix the lengths of the two remaining sides of the triangle. One side of the triangle formed the baseline of the next triangle. Lambton marked out his first baseline at Madras. It was seven and a half miles long and took twelve days to measure. It had to be accurately done. If it were wrong the error would be multiplied throughout the survey. Lambton rechecked his calculations hundreds of times and spent sixteen nights observing a single star until he was sure that his position fixes were right and his baseline accurately calculated. It was a vigil that set the survey's pattern of assiduous and constant verification. Lambton insisted on checking all three angles within a triangle as the only way to uncover mistakes. Even the most simple device, the steel measuring chain a hundred feet long, produced errors because it expanded in the heat. For this reason much measuring was done in the cool of the early morning. The chain's temperature was constantly monitored, for even a small variation could make a difference, and its length was verified time and again against a standard bar. The chain was set out in five wooden boxes mounted on supports fitted with adjusting screws to make them level. After the first satisfactory measurement the chain, the boxes and the supports were lifted, carried forward, repositioned and re-levelled for the second measurement. It took Lambton almost two months to complete these initial steps.

The survey parties toiled like tortoises across the land, methodically marking their lines and angles. Chronometers were expensive and too fragile for such rugged work and, in any case, could not be relied upon. The surveyors found their latitude with sextants and calculated their longitude by observing the eclipses of Jupiter's moons. Throughout their observations they wrestled with the difficulties of refraction, the bending of sight lines by the atmosphere, as well as the effects of heat, the problems arising from the curvature of the earth and the fact that the earth is not a true sphere but a slightly squashed spheroid.

These men were science on the march. Officers, assistants, medical staff, porters, cooks, water carriers, messengers and the mathematical clerks known as computers were accompanied by a special corps of soldiers for protection. Sometimes, when they halted for a long time to measure a baseline, the surveyors were joined by their families and the parties swelled to 200 and more, a tented trigonometrical circus complete with the elephants, camels, bullocks and horses that carried the chains, barometers, chronometers, thermometers, compasses, spirit levels, plumb lines and heliographs.

As often as not, the arrival of such forces of men and animals caused dismay in villages along the way. The surveyors asked for large quantities of food and water and a force of local labourers. Since they often waited for many days for clear weather and good visibility, they were not a brief inconvenience in village life. One of their requirements was an uninterrupted view from one hill to the next and in many places lines of sight had to be cut through scrub, trees and even houses. Many local rulers were unhelpful or obstructive because they feared that the survey would reveal the extent of their land ownership and crop revenue and open the way to heavier taxes. Some of them suspected that the surveyors used their telescopes to spy on their wives and daughters.

Lambton constructed his lines of triangles westward across the southern peninsula to Mangalore, northward towards the state of Hyderabad and southward to Cape Comorin. He was in Tanjore in 1808, moving south, when his progress was abruptly halted. The chief temple in the city had a tower more than 200 feet high, an ideal platform for observations, and Lambton ordered men to haul the box containing the Great Theodolite to the summit. During this operation a rope parted and the box with its precious half-ton instrument fell and splintered. Everyone stared in shock at the bent and twisted brasswork, the buckled circle. Lambton rebuked no one and drew on his well of patience. He put the wreck into a tent and shut himself away with it. He dismantled all the broken parts and laboured day and night to repair them. He had a watchmaker's skill and a single-minded devotion. The task took him six weeks.

By 1818 Lambton had completed more than 700 miles of the Great Arc, and after sixteen years on the survey trail he was a revered figure. He had the rank of Superintendent of the Survey and at last received the accolades he deserved. Seemingly tireless and contented in his work, he pushed his triangles northward until he died in central India in 1823 at the age of seventy.

The ambitious Lieutenant George Everest, who had joined Lambton as his chief assistant in 1818, succeeded him. Everest was born in 1790, either in Greenwich where he was baptized, or at Gwernvale, his father's mansion near Crickhowell in Wales, now a hotel with an Everest restaurant adorned with a picture of the surveyor's thickly forested face. At sixteen, Everest sailed for Calcutta to join the Bengal Artillery, served in Java and returned to India to build a 400-mile line of semaphore stations from Calcutta. Like Lambton he was a self-taught surveyor; and like his revered chief he felt that the Great Trigonometrical Survey was what he had been born for.

Everest was proud and pleased when Lambton sent him off in 1819 to lead his first survey expedition in the Krishna River region south of Hyderabad. He suffered all the horrors of the survey trail, the discomfort of working in jungles drenched and humid in the monsoon. One day some of the Indian troops detailed to guard him threatened to desert. He coolly ordered them to surrender their drawn swords or he would shoot them. They did; and he had three of them thrashed.

He was not by nature compassionate and usually took a severe view of his men's work. He nevertheless wrote warmly of the way they hacked through forests 'infested with tigers and boa constrictors . . . without water or provisions and with jungle fever staring them in the face'. He saw that surveying during the monsoon led to 'a reckless waste of life and health caused by exposure to the pitiless pelting of the rains'. It was the worst possible time for malaria. Everest fell ill, then the expedition's doctor, then most of his 150 men. It was a lesson. 'Buoyed up by the vigour of youth and a strong constitution,' said Everest, 'I had spurned the thoughts of being attacked by sickness, against which I had foolishly deemed myself impregnable.' He fancied that the spirit of the jungle was punishing those who dared to violate its sanctity. His devastated party, 'baffled and crippled', as he put it, abandoned the survey and limped the 200 miles back to Hyderabad. A tenth of his men died. Everest himself, carried on a stretcher, described his fellow survivors as 'a crowd of corpses torn from the grave'.

It was a harrowing and salutary first expedition; and the fever never left him. Desperately ill in 1820, he sailed for the Cape of Good Hope to recuperate and went back to India in 1821. Fever struck again, leaving him partly crippled, and with abscesses of the hip and neck 'from which fragments of decayed bone have repeatedly been abstracted'. Worn down by heat, struggling through turbid swampy forests, hauling themselves up mountains, the surveyors knew that their chances of living to enjoy retirement were especially small. Quite apart from the fevers, they sometimes felt the terrifying presence of tigers. Everest's doctor once watched helplessly as a tiger killed his servant. Sick, soaked and sweating, dreaming of sea breezes in Sussex, the surveyors wrote in anguish to their masters and begged for leave. Their letters usually fell into the bureaucratic muddle at base; and like many men who worked in the field they soon discovered that head office was indifferent.

Everest was thirty-two when he took charge of the Great Trigonometrical Survey. Another bout of fever nearly killed him. Out on survey he became so weak that two men had to hoist him into his seat so that he could make his observations. To reach the adjusting screw of the theodolite he called on one of his assistants to support his flaccid left arm. 'Without being held up,' he wrote, 'I could not have stood to the instrument.' He drove his assistants as he drove himself, believing that if he faltered they would simply give up. But in 1825 even he surrendered, so ill and corroded by anxiety that he retreated to England for five years. He recovered and returned as chief of the Survey and as Surveyor-General of India, the supremo. He brought with him the latest surveying instruments, including a large theodolite. He also had new compensation bars, ten feet long, for measuring the ground. Made of brass and iron clamped together, these reduced the errors caused by expansion. Before sailing for India he tested them at Lord's cricket ground.

Everest was determined to finish what Lambton had started. The sturdy trunk of the Great Arc grew steadily northward and also sprouted branches, latitudinal lines of triangles forming immense crosses over the broad middle of the subcontinent. Leading his survey forces over the flat northern plains, Everest had to solve the problems caused by the landscape's lack of hills. Surveyors needed high vantage points on which they could set up their instruments to get a view of twenty miles or more across the land. They tried climbing trees at first, but trees swayed too much to be useful, and the tiger-fearing labourers far below fled at every rustle in the jungle. Solid masonry towers, sixty feet high, were the answer and Everest designed and built a series of them for triangulation across the plains. They were expensive, but surveying could be nothing but costly and time-consuming. Staff spent days and weeks waiting for haze and dust to clear so that they could make observations. Everest hastened the process by introducing observation at night, when the atmosphere was clearer, and inventing a system using blue lights made from animal bladders stuffed with chemicals and dye. These were hoisted on masts ninety feet high, but the fact that they moved in the wind always caused difficulty. The calculations, verifications and elimination of errors ate up the hours and days. They could never be hurried. A few thousandths of an inch mattered.

Everest on the move, with his strings of pack animals and hundreds of porters and labourers, was a formidable spectacle; and his temper was a phenomenon, too. He readily found fault in his inferiors. 'You seem to me to be right stark staring mad,' he wrote to an assistant, threatening to dismiss him. He was just as sharp in dealing with his governmental masters and never shrank from acrimonious correspondence. He had a low opinion of the British soldiers attached to his survey parties and growled that they drank too much and could not stand the climate. He also had a keen idea of his own importance. Many Indians knew surveyors as 'compass walas', but Everest took offence when a man wrote to him and addressed him, picturesquely and politely, as Kumpass Wala Sahib. He was the Surveyor-General, he thundered, not some compass wala. When his party reached the Chambal River on the border of Gwalior state he raged at the British Resident's apparent lack of respect and the absence of an escort of the Maharaja's horsemen. He held up the work of the survey, keeping his men at the river for two weeks, until a cavalry escort trotted up to satisfy his ego. The British government angrily rebuked him for his arrogance.

Everest's outbursts were aggravated by his frequent illnesses, fevers and discontented bowel. The old malaria kept him in bed for half of 1835, paralysing his left leg and causing him to endure grim and life-threatening medical treatment. A doctor attached hundreds of leeches to draw out his blood in the belief that this would remove the infection.

In India, as elsewhere in the empire, the practical application of science was considered a strictly European province, an extension of the manner in which the educated elite in Britain distinguished themselves from the lower orders in their own society early in the nineteenth century. Many astronomers and cartographers believed that the rigorous intellectual endeavour of the survey demonstrated European superiority: the theodolite mightier than the sword, science against superstition. The very processes of surveying, with expeditions pushing relentlessly through villages, building towers and making observations from high places with mysterious instruments, were all assertions of the scientific ruling power.

Like most of his contemporaries, Everest disdained Indians whose 'minds are uninformed and bowed down under the incubus of superstition'. Some villagers, he noted, saw the Great Theodolite as the manifestation of a deity. Women prayed to it for the blessing of fertility and others made pilgrimages to it seeking cures for sickness. Advancing over the northern plain, Everest often encountered opposition when he wished to clear trees and houses from his sight lines; and it was not always a matter of financial compensation. People considered the trees the homes of gods; and to underline the point they turned out 'in battle array armed with clubs'. They were suspicious of the surveyors' ritual of hoisting blue lights at night and, thinking that the survey towers caused drought, they tore them down.

For some years the British excluded Indians from surveying and kept the geometry and science mostly to themselves, reinforcing their own loftiness and their belief that Indians were irrational. But there was so much surveying and calculation to be done at the detailed level that Eurasians, the sons of British fathers and Indian mothers, steadily took over much of the work during the nineteenth century, just as they became predominant in running the railways. Everest said that they were his most reliable team members. He later employed Brahmins, the highest caste, as mathematicians to work on the observations. From the 1860s the British trained Indian surveyors to travel in Tibet and China and penetrate areas forbidden to the British. Disguised as merchants or Buddhist pilgrims, they walked thousands of miles in the high mountains, taking bearings with compasses secreted in prayer wheels and measuring angles with pocket sextants. They concealed plane tables for mapping in hollowed books and measured the land by pacing it out, counting each step on prayer beads. The patience of these men brought the British valuable knowledge of the geography of the Himalayas.

In 1843 Everest oversaw the completion of the Great Arc and retired. He had worked on it for more than twenty exhausting years, and stress and fatigue had driven him several times to the edge of mental collapse. Under his command more than 1,400 miles of the country were surveyed, from Hyderabad to Dehra Dun in the foothills of the Himalayas, and this was the heart of his life's work and the foundation for mapping all of Asia. It was described as 'perhaps the greatest geographical achievement on any continent in any age'.

A knighthood was offered but Everest said he was worth a better honour. He was turned down. In 1846, aged fifty-five, he married and fathered six children. In 1861 he accepted a knighthood. Sir George Everest, Kumpass Wala Sahib, died five years later at seventy-six and was buried at Hove.

In the year before Everest's death the Royal Geographical Society approved the suggestion that the world's highest mountain should be given his name. Sir George Everest himself never saw it.

The peak was an enigma for many years, distant and elusive. In 1803, three years before Everest arrived as a cadet in India, Captain Charles Crawford, the pioneer surveyor of Nepal, located the source of the Ganges in the Himalayas and drew maps which showed the distant heights he called 'the snow mountains'. These maps intrigued Robert Colebrook, the Surveyor-General of Bengal, who had already seen the mountains from a distance of 150 miles. In 1807 he saw them again and calculated that at least two were five miles high and therefore higher than the pinnacles of the Andes which were then thought to be the supreme peaks of the world. In the 1830s attempts by British surveyors, including George Everest, to get closer to the high Himalayas were barred by Nepalese suspicions of British intentions. Nepal was an autonomous state which had escaped British rule and would not allow the surveyors to approach nearer than eighty miles.

The thwarted British continued their investigations from the Terai, the Indian hills south of Nepal, and built observation towers to do so. Weather limited the work to the final three months of each year. In 1847 J. W. Armstrong observed Everest, then known as Peak XV, from a place 200 miles distant, and calculated its height at 28,799 feet. Two years later James Nicolson measured it from five different stations, the closest 108 miles off, and calculated its altitude to be 30,200 feet. Andrew Waugh, who succeeded Everest in 1843, supervised the mapping of the Himalayas between 1846 and 1855 and in that time seventy-nine of the greatest peaks were measured. Waugh worked on Nicolson's observation data at Dehra Dun, worrying over the problems of light refraction which could produce errors of 500 feet, and in 1856 concluded that Peak XV rose to 29,002 feet and that there was probably no higher mountain. Waugh's estimates endured. Modern surveys have produced a figure of 29,028 feet, remarkably close to his calculation a century earlier and a tribute to his skill.

What to call it? Waugh said there were so many local names that it would be invidious to choose one of them and he persuaded the Royal Geographical Society that it would be a fitting honour to call it after his old chief. Everest it was; and even for Everest the mountain was monument enough, though he actually pronounced his name Eve-rest. Any other pronunciation would have drawn one of his incendiary rebukes.

14

Get me long distance

'As cold waters to a thirsty soul,
so is good news from a far country.'
Proverbs 25, 25

Sometimes shouts and waving arms were enough. Fires, drums, smoke, flags, guns, mirrors, lights, rockets and wigwagging mechanical semaphore limbs also went some way to meeting the need for rapid long-distance communication. In ancient Greece and along the North African coast chains of fire signals warned of the approach of enemies. In 1588 beacons blazed from hill to hill and told Queen Elizabeth's England that the Spanish Armada was at sea and heading up the Channel. Church bells warned coastal towns of prowling pirates. The Scots kept bundles of firewood ready to tell of the approach of the English over the border. In northern India drummers sat on top of stone columns along the highway from Agra to Srinagar in Kashmir and passed the message that the Mogul emperor was on his way.

Like the man marooned on a tiny desert island and the missionary in the cooking pot, the smoke signals of North American Indians are a favourite joke of cartoonists. But smoke was part of an efficient sign and gesture language developed by the Plains Indians, the Sioux and others. It involved the use of the arms and the pose of the body on foot and on horseback. Given the keen eyesight of its users, it was a practical means of communication far beyond the range of the voice across the flat expanses of the Great Plains. Nor was it merely rudimentary. Indians used it to convey useful information about the movements of deer and bison and also the details of treaty negotiations. Explorers, migrants and frontier troops of the United States army were impressed by this long-distance talk and borrowed some of the gestures themselves.

Communication between ships at sea was advanced significantly by the telescope, introduced in 1609 by the Dutch spectacle-maker Hans Lippershey and improved by Galileo, the Italian astronomer. Galileo's fourth model, designed in 1610, magnified thirty times and enabled him to see the craters of the moon and the four satellites of Jupiter. Simple signals using five flags supplemented shouting between ships during the middle of the seventeenth century. Later, more flags were added and in the 1770s the Royal Navy's *Fighting Instructions* listed fifty-seven signals. *The Signal Book for Ships of War* of 1799 had 300 and was revised and expanded in the *Telegraphic Signals or Marine Vocabulary* devised by Sir Home Riggs Popham in 1803. His mother's twenty-first child, he entered the navy at sixteen and took part in the seizure of the Cape of Good Hope from the Dutch. He was later reprimanded by a court martial for leaving his post in South Africa to support a raid on Buenos Aires. London businessmen thought the raid beneficial to British commerce and honoured him for the same exploit. He died an admiral in 1820 and his grave at Sunninghill in Berkshire is ornamented appropriately with a bas-relief of his flag vocabulary. With improvements his system remained in force throughout the nineteenth century and beyond, essentially from the Battle of Trafalgar to the Battle of Jutland in 1916.

Popham's code allotted the numbers 0–9 to ten distinctive flags and gave these numbers to the letters of the alphabet. Number 1 stood for A, number 2 for B. Number 9 stood for both I and J so that the alphabet had twenty-five letters. Double figures were signalled by hoisting two flags, for example Number 1 and Number 9 represented Number 19, the letter T. Thus any word could be spelled out. But to save time and laborious multiple flag hoists in spelling out every word Popham compiled a lexicon of 999 common words and gave each a number. A group of three flags, Numbers 2, 5 and 3, represented Number 253 in the code book and that was the word 'England'.

Always the warrior, looking for the winning edge, Nelson was a signals enthusiast and took copies of Popham's code with him when he rejoined HMS *Victory* before Trafalgar. As he walked the quarterdeck in the tense prelude to the battle he told Lieutenant John Pasco, the ship's signal officer, that he wished to 'amuse' the fleet with a message: 'England confides that every man will do his duty.' Pasco told him that 'confides' was not in the flag vocabulary and would have to be spelled out letter by letter, whereas 'expects' was in the book and could be sent more quickly as a single hoist of three flags representing Number 269. Perhaps Pasco was a discerning editor, too. 'England expects . . .' has a more striking ring to it than 'England confides . . .' Nelson agreed. The first eight words had their own numbers, 253, 269, 863, 261, 471, 958, 220, 370; and the word duty, not in the vocabulary, had to be spelled out: 4, 21, 19, 24. In the early nineteenth century the letters U and V were reversed in the alphabetical order, hence the U in 'duty' was Number 21 in the twenty-five. A few hours after he ordered the signal Nelson died in *Victory*'s cockpit thanking God that he had done his duty. The severely wounded Pasco was close by. In 1846, forty-one years after Trafalgar, Pasco was appointed captain of *Victory*. The ship was by then a naval shrine based permanently at Portsmouth.

Flags at sea were one thing. Facing the Napoleonic threat, the British Admiralty wanted an overland means of communicating rapidly with the fleet at the principal naval base in Portsmouth more than seventy miles away. A post-chaise could take seven hours and ran the risk of being bogged down in Haslemere or held up by a highwayman at Hindhead.

The answer came in the first place from France and the ideas of the five Chappe brothers, Claude, Ignace, Pierre-François, René and Abraham, pioneers of mechanical signalling. In 1793 Claude Chappe demonstrated a system of movable arms pivoted on towers and, reaching into his Greek, called it semaphore (*sema*=sign, *pherein*=to bear). He transmitted a message using three relay towers over a distance of twenty miles, the signal being read through a telescope and relayed to the next tower. A French civil servant showed off his Greek, too, and coined the term telegraph for the apparatus (*tele*=far, *graphein*=to write).

Chappe's robot sent messages at a speed of one and three-quarter words a minute and in August 1794 transmitted an important piece of war news, the recapture of Le Quesnoy, over 130 miles between Paris and Lille. Napoleon used it to direct his forces, and between 1794 and 1807 lines were built from Paris to Strasbourg and to Brest, Brussels and Lyon. They were, of course, for government business only, although the authorities allowed the national lottery, which helped to fund the telegraph, to transmit the winning numbers.

The British military lagged behind the French and relied on the traditional runners and gallopers; but in 1794, the British captured a French soldier carrying a drawing of the Chappe system. The Reverend John Gamble, the Duke of York's chaplain, studied and improved it, inventing a system first tried in 1795. But the Admiralty preferred an even better device developed by another clergyman, the Reverend Lord George Murray, later to be Bishop of St David's in Pembrokeshire. Piqued at being bettered, Gamble erected his signal mast on Westminster Abbey where their Lordships at the Admiralty could not fail to notice it. They pretended, however, that they could not see it.

Lord Murray's telegraph was a relay of signal stations built on high places about nine miles apart. On the roof of each stood a twenty-foot frame housing six shutters, each of them three feet square and operated by the tug of ropes, providing sixty-three combinations of letters and numbers. On the next hilltop operators with telescopes read the signals and rapidly passed them down the line. Obviously the telegraph was restricted by darkness and foul weather; and during fog a horseman took messages to the first station with a clear view of the next one up or down the line.

In 1796 signal relays ran from London to Deal on the Kent coast and to Portsmouth. The Admiralty post was built on the roof of the home of the First Lord of the Admiralty, and the first repeater was at the Royal Hospital in Chelsea, one and three-quarter miles away. A message from the Admiralty passed through fifteen signal stations and reached Deal in seven minutes. The pardons granted to the Royal Navy seamen who took part in the Spithead mutiny of 1797, over poor pay and conditions, were telegraphed and received long before the pardon document itself, which riders carried the seventy miles from London to Portsmouth in four and a half hours. The telegraph also had its footnote to the mutiny at the Nore, in the Thames estuary, also in 1797. The news that the twenty-five ringleaders had been hanged at the yardarm was received in the Admiralty seven minutes after the execution.

Telegraph stations gave their names to numerous Telegraph Hills, Telegraph Inns and Telegraph Roads throughout Britain. For all their importance to the Royal Navy and national security, however, they were resented by some landowners who did not want them intruding on their property and possibly frightening the horses. Each station was equipped with a stove, beds, a clock, and two brass telescopes made in London by John Dollond and his son Peter. Each station also had a coal and candle allowance, and these expenses were included originally in the Admiralty odds-and-ends budget covering chimney sweeping and rat-catching. The stations were staffed by Royal Navy veterans, usually a lieutenant, a midshipman and two seamen, some of whom had been wounded at sea. Henry Garrett, for example, who served aboard HMS *Belleisle* at Trafalgar, commanded the station at Petersfield in Hampshire in the 1840s.

In 1814, when it appeared that the Napoleonic threat had subsided, the improvident Admiralty closed the telegraphs to Deal, Portsmouth, Plymouth and Yarmouth and demanded that the staff return their expensive Dollond telescopes. No telegraph was working, therefore, when Napoleon, in his last throw, returned to France from Elba and resumed his hostilities in 1815. The French telegraph reported the outcome of the Battle of Waterloo; the British had no means of doing so. Eleven days after Waterloo, Parliament ordered the re-establishment of the telegraph and an improved system was built, employing a semaphore invented by Sir Home Riggs Popham.

Commerce demanded its own telegraphs. It was typical of the prosperous shipowners of Liverpool, anxious to know when their ships were coming in, that in 1827 they should commission the first business telegraph. Ten semaphore stations with tall masts were built along the coast of north Wales from Holyhead to

Liverpool between three and twelve miles apart. Some are still standing. The arrival of ships off Anglesey, safely home from the earth's corners and bound for the Mersey, was reported in minutes to the waiting owners and merchants in Liverpool and gave them time to arrange sales and markets for their cargoes. Holyhead also reported wind strength and direction, so that owners could calculate accurately when their ships would dock. The semaphore operators used numerical codes to send phrases quickly. The number 49, for example, meant the wind was west-south-west. Ships often identified themselves to the signal stations by flying flags in the code devised for merchant shipping by Captain Frederick Marryat, naval officer and novelist. In fine weather the Anglesey-to-Liverpool line was highly efficient, transmitting a signal in under five minutes, rapid enough on its first day for a Liverpool newspaper to describe it proudly as 'faster than the wind'.

15

From Liverpool to the moon

'All of Europe is crossing the ocean.'
Philip Hone, 1836

In the 1830s the fast American sailing packets seemed to be the most wonderful ships on the Atlantic. If the weather were fair and the passage swift, travellers in the saloon class, with all the comforts of the captain's table, four meals a day, and their eyes and noses closed to the suffering of those in the cheapest accommodation, could fancy that they were blessed.

Composing his sea journal in a lyrical hour, one of them wrote that 'time flew on silken wings'. For others, however, that was not fast enough. Calms and storms alike were a punishment. Many travellers never lost their terror of the sea. Embarkation was often a miserable and fearful event and usually a final farewell. In their reports of shipwreck and mass drowning the newspapers spared no horror or gruesome detail. Everyone knew of the wretchedness of sea travel, and those who had made even a short passage needed no instruction in the facts of seasickness, tedium and bad food.

The advent of ocean-going steamships promised at least to shorten the misery and to guarantee arrival times. But it took years for steam to challenge the haughty swans of sail. It was thirty-seven years after William Symington launched his pioneer steamboat in 1801 that the first ship crossed the Atlantic under the sole power of steam. Paddle steamers ran satisfactorily enough on lakes and rivers and in sheltered coastal waters where journeys were short and coal was easily obtained. They were not powerful enough to confront the weather and heavy seas of the ocean.

When the paddle wheeler *Savannah* set out to cross the Atlantic from Georgia to Liverpool in 1819 the confidence in her venture was measured by the ticket sales, none, and the amount of cargo loaded, nil. She was a hybrid, a steamer assisted by sail, and she crossed the ocean in twenty-six days, mostly under sail. *The Times* reported that another ship pursued her for a day in the belief that because she was belching smoke she was on fire. Most of the return voyage, too, was not steamed, but, for the sake of appearance, her paddles were turning as she entered her home port of Savannah. In 1833 the *Royal William*, carrying pine spars, furniture, stuffed birds and ten passengers, used steam and sail to cross from Nova Scotia to Cowes in the Isle of Wight. On some days she used sail only, and her master stopped the engine every four days so that his men could clear the boiler of salt encrustation. In these faltering early voyages, engineers gained ideas of what was possible.

The first steam vessels were built by men accustomed to constructing sailing ships according to the traditions of their grandfathers and their own intuitive feel for line. They had to grope their way, trying and erring, through the new skills of marine engineering. The shape of a paddle ship's hull was distorted by the bulk of the paddle wheels and made her an ungainly and inefficient sailing vessel. In rough water she naturally rolled so that one paddle was deeply immersed and the other churned the air. When the ship was heeled over by the wind blowing on her beam the leeward paddle dug more deeply and tended to force the ship to turn into the wind. It required firm hands on the steering to prevent her making the progress of a one-legged duck.

Designers saw that the paddle steamer's hull required a fairly narrow shape so that the bow wave would be thrown outside the paddles and not on them; and since she could not be a fat ship she could not carry bulk cargo. She was also difficult to stop; and because there was no central slipstream to act on her rudder she was not easy to steer at slow speeds.

With a boiler pressure of around five pounds to the square inch, her engine was inefficient. This would not improve significantly until the advent of steel boilers permitted higher pressure. To meet the problem a ship had to load a huge store of coal for a long journey; and the coal encroached on cargo space. When the ship was laden with coal her paddles were deeply and inefficiently immersed. As the fuel was consumed the ship grew lighter; but the higher she rose the less efficiently the paddles dipped. Only at the mid-point of a passage, when the ship was neither too heavy nor too light, did the paddles dip effectively.

The sheer quantity of coal needed to feed a ship's furnace led many to doubt that a steamer could ever carry enough to fuel an Atlantic crossing. The leading and influential doubter was the Reverend Dr Dionysius Lardner, professor of science at University College, London, and editor of an encyclopaedia. He had warned that trains going through tunnels would consume so much oxygen that passengers would suffocate. Dr Lardner never denied the feasibility of 'steam intercourse' between Britain and America, but he calculated that a ship would consume all its coal in fifteen days, or just over 2,000 miles. A transatlantic steamer would therefore have to depart from Valentia in the far west of Ireland and take on coal either in the Azores or in Newfoundland; and the ship would be so laden that there would be no room for passengers or cargo.

Speaking in Liverpool in 1835, Dr Lardner declared: 'As to the project announced in the newspapers of making the voyage directly from New York to Liverpool . . . they might as well talk of making a voyage to the moon.' For good measure he warned the public against accepting the ideas of steamship builders. They were 'more highly supplied with zeal', he said, 'than with knowledge'.

His opinions carried some weight among cautious marine engineers and sceptical investors. They were also a direct attack on the plans and visions of Isambard Kingdom Brunel, at that time the ambitious chief engineer of the Great Western Railway and reaching his prime as an engineering giant of the nineteenth century. He was designing a ship, the *Great Western*, to do what Dr Lardner said was impossible, to steam across the Atlantic. In Brunel's grand concept passengers would take the Great Western Railway train from Paddington station in London, itself an architectural wonder he had designed, and thence to Bristol where they would be shown to their cabins aboard the steamship. Thus they would travel directly to New York in one fluid steam-propelled Brunelian sequence.

Dr Lardner repeated his doubts about Atlantic steam at the 1836 meeting in Bristol of the British Association, the synod of scientific brains. Brunel rose to rebut him but failed to persuade the audience that Lardner was wrong. By then the *Great Western* was growing in the shipyard at Bristol. Brunel was certain she would steam to New York with fuel to spare and win the argument.

In July 1837, the first year of Queen Victoria's reign, Brunel launched the *Great Western* at Bristol to the applause of 50,000 people. She was towed to London to be fitted with her engine and on 31 March 1838 was ready for the Atlantic. Advertisements for the maiden voyage from Bristol to New York trumpeted her virtues: she was 212 feet long, timber-built and immensely strong, coppered and copper-fastened, with the best possible engines. At 1,340 tons she was also by far the largest steamer of her day: Brunel never built small. A former Royal Navy officer commanded her. The prospectus said the ship had 'coal stowage for twenty-five days of constant steaming and therefore will not be required to touch at Cork for coal'. As an additional enticement to passengers, servants' tickets were sold at half price.

As the *Great Western* steamed down the Thames, Brunel anxiously stomped her deck, a short and intense figure given a little more height by his battered stovepipe hat. He chewed relentlessly on one of the cigars he kept in a sort of cartridge pouch slung around his neck. Without any warning the lagging around the boilers caught fire, choking smoke enveloped the ship and she ran aground on a sandbank. Some stokers dived overboard and swam for the shore. Brunel lost consciousness in the fumes and fell eighteen feet from a ladder into the engine room. Fortunately, he landed on top of his friend, Christopher Claxton, who, having broken his fall, quickly pulled him out of the bilge water and saved him from drowning. Brunel was taken ashore badly injured. Alarmed by the fire, fifty prospective passengers cancelled their 35 guinea tickets for the maiden voyage. A tug pulled the ship from the sandbank and she steamed around the south coast to Bristol at top speed to start her Atlantic crossing.

The *Great Western* was not alone in her venture. Rival shipping interests in Liverpool and London looked sourly at Bristol's bid for glory and chartered a ship to snatch from under Brunel's nose the honour of being first across. It was a stunt. At 703 tons, *Sirius* was half the size of the *Great Western* and was built for the Irish service not the Atlantic. Extra space was found for coal and she topped up with fuel at Cork and set off on 4 April 1838 with forty passengers. The race was on, the shipping world agog.

The *Great Western* completed her trials and left Bristol on 8 April with only seven doughty passengers. A reporter dipped into the purple ink and likened her to to a war horse pawing the ground. 'And see, how the fierce breath of a giant defiance pours out of her eager nostrils!' She began to close the distance, but not without difficulty. No one had foreseen how hard it would be to keep up the pressure of steam. In rough weather the stokers became fatigued. The trimmers, whose job it was to keep the coal stocks balanced and ensure that not too much was shovelled from one side at the expense of the other, could not ferry the coal quickly enough from the bunkers to the furnaces. Tempers disintegrated and one of the crew sought to throw the captain overboard. The sailor was swiftly manacled but the stokers downed shovels and demanded his release. All was smoothed with extra pay and common sense.

The commander of the *Sirius*, meanwhile, ran into bad weather and rejected the pleas of his frightened passengers that he should turn back. He ploughed on and, according to journalistic embroiderers, threw furniture and even a little girl's doll into the furnace to maintain speed. There was a desperate finish. The *Sirius* reached New York on 23 April, down to her last fifteen tons of coal. The city declared a festival. Boatmen crammed their craft with sightseers. And then, razzle upon dazzle, cheers augmented by gunfire, the *Great Western* steamed in a few hours later beneath a glorious cumulus of smoke to trump her upstart rival. Not first over the line, but the winner by more than four days, she had a comfortable 200 tons of coal in hand, the perfect riposte to Dr Doubting Lardner, and Brunel's vindication.

Certainly she had used sails to help her along when the wind was right, but the *Great Western*'s engine had kept turning all the way and she was the first true ocean-going steamship. James Gordon Bennett wrote in the *New York Herald* of the magnificence of her arrival. 'The Battery was filled with the human multitude, one half of whom were females, their faces covered with smiles and their delicate persons with the gayest attire. Below, on the broad blue water, appeared this huge thing of life, emitting volumes of smoke. She looked black and blackguard, rakish, cool, reckless, fierce and forbidding.' Another writer declared that 'Rome in all her glory never witnessed a scene like this'. And yet another reflected that England was now nearer to New York than were many parts of America. In this moment of triumph the *Great Western*'s chief engineer was scalded to death in the engine room.

The future had arrived. Steamship fever raged. To drums and blaring brass the bigwigs of New York led the municipal delirium and stuffed themselves with beef and champagne. Crowds rushed to lower Manhattan to stare at the barely believable spectacle of two vessels which just over a fortnight earlier had been in Europe and which were now in the New World, driven by a power made by man. A newspaperman charged with summarizing the brilliance and significance of the occasion swiftly splashed five triumphant words to fit the newsboys' posters: 'Annihilation of Space and Time'.

The *Great Western* opened the way to the age of the transatlantic liners. She steamed back to Bristol in twelve days with sixty-eight passengers and 20,000 letters, each envelope earning the captain a fee of twopence from the Post Office. The American sailing packets had advanced Atlantic travel by guaranteeing their passengers a starting date for their voyages; but they could not, of course, tell them when they would arrive. The steamship owners could. The *Great Western* was a financial success and made sixty-seven voyages across the Atlantic between 1838 and 1847, averaging fifteen and a half days to New York and just over thirteen days back, carrying about eighty passengers on each journey. Her triumph, however, did not silence Dr Lardner. Nor was he abashed by the scandal of his elopement with a married woman to whose husband he had to pay £8,000 damages. He remained a public figure and doomsayer and crowed that, for all the vaunted supremacy of steam, sailing ships proliferated on the Atlantic.

Certainly the *Great Western*'s builders failed to construct sister vessels which would have enabled them to run a shuttle service and win the Atlantic mail contract and its buttressing government subsidy. Perhaps Dr Lardner's gloom deterred Bristol investors. In the event the shrewd Nova Scotian Samuel Cunard seized his opportunity and commissioned four paddle steamers. To the Great Western Company's dismay he won the mail contract that was the foundation of his Atlantic success and the making of his name.

Steam was progress but there were good reasons for many shipowners to stay with sail. Winter weather in the Atlantic was a menace to the early steamers, while sailing packets operated the year round. Ton for ton, it cost three times as much to build a steam vessel as a sailing ship. A steamer's engine occupied a lot of space and so did the coal. A ship with 800 tons of cargo capacity had to allot 600 to coal; and in the 1840s a steamship burned sixty tons a day, almost three times as much as the efficient steamers of the twentieth century.

Steamers also seemed to many observers to be slow, lumbering, smoky and dirty. 'Let jack set plenty of canvas,' wrote a critic, 'as his good ship pleases, and three to one the liners will do at all times as much as the smokers.' Many passengers liked the way their splendid sailing ships outstripped the wheezing 'tea kettles'. They had style. A captain recalled that 'when a packet ship running before a westerly gale in mid-ocean overhauled a wallowing side-wheeler, the joyous shouts and derisive yells of the steerage passengers on board the packet, as she swept past the tea kettle, were good for the ears of sailormen to hear'. In those days, he said, a packet captain overtaking a steamer passed as close as possible so that 'the dramatic effect might not be lost upon the passengers of either vessel'.

Typically, Brunel did not rest on his achievements for a moment. His restless mind vaulted to the future. He had demonstrated that a large steamer could carry an ample supply of coal to cross the Atlantic as well as enough passengers to make a profit. Now he drew the lines of what would be the largest ship in the world. In 1839, the keel was laid. She would be called *Great Britain*. Twentieth-century historians would call her the first modern ship.

16

Taller than my giant

'If any man can be at two places at once, Waghorn is he.'
William Makepeace Thackeray, 1844

In October 1837, refreshed after four years of leave in England, Captain Seymour Burt of the Bengal Engineers set out to rejoin his regiment in the Himalayan foothills. Being young and modern-minded, he chose the fastest means available. Rather than undertake the usual passage to India, the long voyage under sail around the Cape of Good Hope that might last six months, he left London as a pioneering passenger in a steamship. He would travel through the Mediterranean, cross Egypt overland by horse and camel, and board another steamer to take him down the Red Sea and across the Arabian Sea to Bombay. What a story he would have to tell when he arrived at last in the officers' mess in Simla.

Burt bought his ticket from a new steamer company founded that same year by Brodie McGhie Willcox and Arthur Anderson. Willcox had started as a shipbroker in Lime Street, London, in 1815. Anderson was a Shetlander who had served in the Royal Navy during the Napoleonic wars. When peace was restored he was paid off in Portsmouth and walked to London to find work. He was a hungry young man when Willcox took him on as a clerk. In 1822 the two became partners and tried to make a living in the uncertain and bankruptcy-riddled business of chartering ships. By 1837, in common with many operators, they were running at a loss.

Indeed, they were on the edge of going broke when their fortunes turned dramatically. Into the 1830s the Post Office sailing packets carried all the mail to Britain's expanding empire and other places abroad. This arrangement was persistently criticized by James

MacQueen, a former sugar estate manager in the West Indies, who badgered the government to give contracts to private companies to carry the mails and to promote steam-driven vessels as 'the mail coaches of the ocean'. He succeeded. In 1834 the government awarded its first mail contracts to shipping lines, a form of official subsidy that provided the British merchant fleet with the financial foundation of its nineteenth-century prosperity.

One of the first beneficiaries was the struggling partnership of Willcox and Anderson. Just in time to enable them to beat bankruptcy, the government gave them a contract for £29,600 to carry mail by steamship every Friday from London by way of Falmouth to Spain, Portugal and Gibraltar. Hitherto the mails had been carried by sailing packets, which often took three weeks over the voyage. Willcox and Anderson promised the government regular sailings and deliveries: Falmouth to Vigo in fifty-four hours, Lisbon in eighty-four hours. They relaunched themselves as the Peninsular Steam Navigation Company.

Only three years later, in 1840, by which time their service to Spain and Portugal was a success, they won a £34,200 contract to transport mails to India by way of the Mediterranean and the Red Sea. By a royal charter of 1840 their firm became, more grandly and ambitiously, the Peninsular and Oriental Steam Navigation Company, P&O for short. From that time it grew into a great shipping line of the world and an imperial giant.

Meanwhile, on 13 October 1837, Captain Burt paid his fare of £18 and embarked aboard the *Tagus*, one of two 'splendidly fitted up' paddle steamships of Willcox and Anderson's new company and claimed in the advertisements to be the 'largest and most powerful that have yet been put afloat'.

Burt was thrilled by the ship's modernity. Although he and nine other passengers were confined together in a cabin only twelve feet by eight, he noted happily that he 'never saw such splendid dinners placed on any table on board ship'. The trip from London to Falmouth took only thirty-six hours, a satisfying improvement on the sailing ships that typically waited for days off the Kent coast for a wind to blow them down the English Channel. The *Tagus* clanked and corkscrewed its way at ten knots to Vigo, Oporto, Lisbon and Cadiz, and arrived in Gibraltar eight days out of Falmouth.

Here Burt took another ship to Malta. He gave the island a bad report because of a local custom calling on sinners to toll bells at night, the greater the sin the longer the tolling. He boarded yet another steamer to cross the Mediterranean to Alexandria and found to his dismay that the standard of catering did not compare with what he had enjoyed aboard the *Tagus*. Instead of generous quantities of ginger beer and champagne, he complained to his diary, there was but a meagre wine ration and coffee served too soon after dinner by an inconsiderate steward.

When he disembarked in Egypt Burt hired a camel which delivered him and his baggage to a Nile boat for the trip to Cairo. It was a stinking vessel, overrun by fleas and rats, and Burt slept in his clothes and kept them tightly buttoned, his kid gloves pulled up over his wrists and his face covered. After five days of squirming and grumbling he reached the capital and the mercies of Hill and Raven's recently completed hotel.

In Cairo Burt had a meal with Thomas and Harriett Waghorn. Mrs Waghorn was the only Englishwoman in the city and she was married to a remarkable and driven man. Her husband was well on the way to becoming a legend as a transport pioneer as well as a menace to bureaucrats. He was thirty-seven, had joined the Royal Navy as a midshipman of twelve, had been wounded in action and had worked for five years as a pilot on the River Hooghly at Calcutta. A portrait shows an open, clean-shaven face with a determined set to it. He was a tall man, so much so that a fairground showman once refused him admission to an exhibition of a giant, saying to Waghorn's companion: 'Sir, take that gentleman away, he is two inches taller than my giant.'

Captain Burt and Lieutenant Waghorn had much in common. As servicemen and old India hands, as well as enthusiasts of speed and progress, they were interested in the fastest possible travel to Bombay. The passion of Waghorn's life was the overland route. He wanted to race the clock and calendar and bring Bombay closer to London. He studied maps and timetables to see how hours and days could be shaved off journeys across Europe. He juggled the possibilities of dashes by stagecoach, Rhine boat and railway, schemed to cross the Mediterranean by the fastest ships available and dreamed of shrinking the miles over the Egyptian desert to the Red Sea to catch the steamships to Bombay. He was one of those buzzy bees dreaded by officials. 'A wild scheme', wrote a Post Office executive of Waghorn's enthusiastic campaign for the overland route and steamships. 'This gentleman', conceded another, 'has at least the merit of perseverance.'

Waghorn was no mere theorist, but an adventurer prepared to see for himself how fast he could speed mail and people to India. He was Mercury. In Europe and Egypt he seemed to be everywhere at once, ever bustling, going boldly. The East India Company at last agreed to try him out and gave him letters to deliver to Bombay. At 7.30 one October morning in 1829 he boarded the stagecoach from the Spread Eagle in Gracechurch Street, London, crossed from Dover to Boulogne by steamer, and reached Paris by post-chaise. Learning that the Simplon Pass in the Swiss Alps was closed by avalanche, he took a series of carriages along the Mount Cenis road to Padua in Italy. Here he heard that the Venice–Trieste steamer had broken down and he had no choice but to continue by carriage. Nevertheless, he reached Trieste nine and a half days out of London, and this, as he boasted, was four and a half days faster than the British Post Office service.

At Trieste, however, he just missed a sailing ship bound for Alexandria. Dismay changed to elation when he learned that it was becalmed down the coast, and he called a carriage and hurried off in pursuit. It seemed within his grasp when a favourable wind sprang up and filled its sails. The frustrated Waghorn could only stand on the shore and watch. Back in Trieste he found a berth in another ship, but it took seventeen days to reach Alexandria. He travelled the 120 miles to Cairo by donkey and Nile boat and there chartered a camel for the two-day desert journey to Suez.

When he arrived on 8 December, the steamship he had expected to find was not there. His resource of patience was always slender and at Suez it lasted only two days. Ever the action man, he hired an open boat and a crew of six to sail down the centre of the Red Sea in the hope of meeting the steamer as it came up from Bombay.

Without benefit of compass or chart, his Arab crew mutinous, he sailed 620 miles to Jiddah in just under a week. Here he was told that his steamship had not yet left Bombay.

This news did for poor Waghorn. He collapsed with fever and was ill for six weeks. He finally reached Bombay in March, four months and twenty-one days out of London. His mail-running journey was hardly a success but he refused to be beaten and continued to campaign for the overland route and for steamships. In his dealings with bureaucrats he would not be rebuffed. 'Mr Waghorn', wrote a Post Office official, 'is a very old friend of ours and rather a troublesome one.'

In 1835, well aware that there was 'a sort of stigma on my sanity', he set up his own unofficial mail service to India. Through a co-ordinated system of forwarding agents, letters received at key points were sent on as quickly as possible aboard the next vessel. Soon the service was handling hundreds of dispatches stamped 'Care of Mr Waghorn, Alexandria'. The Waghorn overland agency could carry letters out to India in seventy-four days, but once, with perfect ship connections, a delivery from Bombay reached London in a record forty-seven days. In 1837 Waghorn himself took letters from Suez to London by way of Marseille in a remarkable ten days. By the time he dined in Cairo with Captain Burt in 1837 Waghorn was the East India Company's mail agent in Egypt, in charge of the overland route from Alexandria to Suez. His 'wild scheme' was a reality and his vision vindicated. He was the great popularizer of the overland service, and more than 500 passengers a year were using his facilities to travel to and from India.

The crossing of the isthmus from Alexandria to Suez took about sixty hours. The first part of the journey was in a barge towed by horses along a forty-four-mile canal, built in 1819. Its construction cost the lives of 20,000 labourers who died of disease and brutal overwork. After nine hours in the barge the passengers reached the Nile and transferred to a boat, at first a sailing vessel, later a small and verminous steamer, for the one-day journey to Cairo. Here they waited in a hotel for news transmitted by semaphore across the desert that a steamer had reached Suez, eighty-four miles and twenty-four hours away.

They travelled to Suez by various means: in coaches drawn by four horses, in two-wheeled vans, on the backs of horses and camels and in sedan chairs slung between two donkeys. They stopped at rest-houses every ten miles to change horses and to eat meals of chicken, pigeon, mutton, beer and wine. According to one account they found it difficult to see what was served for breakfast because each dish was cloaked darkly by a layer of flies; but in any group of British travellers there were always those who grumbled about the food, the natives and the general bloodiness of abroad.

After P&O won the mail contract for India in 1840 it gradually took control of the desert transit and pushed Waghorn out of the business. He had no chance against such a large and well-funded company and returned to London. But he remained restless and obsessed and spent the remainder of his life finding ways of making mail services faster. In 1845 we find him personally carrying dispatches, written in India for *The Times*, from Alexandria to London. Couriers picked up the mail at Suez, galloped it across the desert and gave it to Waghorn in Alexandria. He sailed across the Mediterranean for Trieste, travelled through Germany by coach, took the fastest possible ship to England and hurried into *The Times* at 4.30 one morning carrying dispatches that were twelve days from Suez and an amazing twenty-nine from Bombay. It was the freshest possible news.

Waghorn wore himself out with all his activity and was often beset by money trouble. He died aged forty-nine in 1850. But he had made his name and made a difference. He was a doer. William Makepeace Thackeray, one of his admirers, wrote a sketch of his energetic activities:

> The bells are ringing prodigiously and Lieutenant Waghorn is bouncing in and out of the courtyard full of business. He only left Bombay yesterday morning, was seen in the Red Sea on Tuesday, is engaged to dinner this afternoon in Regent's Park and I make no doubt he is by this time in Alexandria, or at Malta, say, or perhaps both. Waghorn conquered the Pyramids themselves, dragged the unwieldy structures a month nearer England than they were . . .

We last saw Captain Seymour Burt dining in Cairo with Waghorn in 1837. Waghorn arranged the camels and food and wine for Burt's onward travel and charged a modest guinea. As soon as the meal was over Burt left by camel for Suez and two days later boarded the 411-ton East India Company paddle steamer *Hugh Lindsay*, bound for Bombay.

The ship was more suited to trips on the Thames estuary. Her bunkers gave her a range of five and a half days, but she needed ten or eleven days to cover the 1,710-mile stretch across the Arabian Sea. On her first voyage from Bombay to Aden her saloon and cabins were crammed with coal, only one passenger was carried and the ship was so low in the water that her decks were awash. She used her sails as much as possible and reached Aden on the last shovelful of coal. It took six days to refill her bunkers.

Captain Burt's voyage from Suez to Bombay took thirty-two days, the ship stopping for two days each in Jiddah, Mocha and Maculla to refuel. While the coal was loaded, all the hatches and windows were stopped up and the cabin doors sealed with canvas to keep coal dust out, but without success. 'We appeared in a short time more like chimney sweeps than white linen gentlemen,' said Burt. He was comforted, however, by the captain's hospitality and washed the coal grit from his throat with plenty of good wine.

Within a few years of starting their services to the East, P&O recruited hundreds of Indian seamen, or lascars, a word derived from the Persian *lashkari* meaning a soldier. The East India Company employed them during the late seventeenth century as crew for their London-bound ships. They filled the gaps left by the high rates of death, sickness and desertion among British sailors and they developed a reputation for skill and reliability. The demand for lascars increased during the Napoleonic wars when the Royal Navy's need for seamen depleted the ranks of merchant sailors.

At the end of their voyages lascars were paid off and left stranded in British ports, especially London. Because they were outside the poor laws and abandoned to beg or perish, the Society for the Protection of Asiatic Sailors was founded in 1814 to care for them. A missionary pointed out that evangelists spent a lot of money taking Bibles to India but ignored the plight of Indian sailors in Britain.

The expansion of Britain's trade made lascars crucial to merchant shipping, and the growth of the steamship fleet increased the demand for them even further. Engine rooms needed large numbers of men to feed the furnaces. In a four-hour shift a fireman shovelled two tons of coal fetched from the bunkers by the trimmers. It was always gruelling labour. Many European seamen hated it. Lascars did not complain, and their willingness to work in hellish conditions in the engine room kept British merchant shipping running in the tropics. A myth grew of the lascars' ability to stand heat better than whites, an idea that they were genetically suited to stoking furnaces.

Basil Lubbock, an historian of nineteenth-century shipping, gave his version of the problem:

> At first it was thought that stoking in the Red Sea was more than any white man could endure. Thus men from West Africa were signed on ... but [they] collapsed in front of their furnace doors. Experiments were tried with other tropical races with the same result. Then yellow men were tried, and finally steamship owners were compelled to fall back on their own colour. Here we come to the triumph of the Liverpool Irishman ... the only man who could submit to this near approach to the fires of hell and survive.

The truth was, however, that for all Lubbock's rugged regard for 'the Liverpool Irishman', many ships' officers saw lascars as desirable crew because they were disciplined, quiet and sober compared with British sailors whose temperament and drinking habits often made them difficult to handle. In a highly competitive industry lascars were also cheap. They earned about a quarter of the wage of their British shipmates, put up with food of inferior quantity and quality and were allotted less space on board. The law in 1876 gave a British sailor sixty cubic feet of room and a lascar thirty-six cubic feet. These allowances were later doubled in each case, against the opposition of the shipowners.

In 1847 P&O pioneered the employment of all-lascar crews on their ships sailing to Britain. Officers communicated with them through Indian petty officers who spoke English. The lascars kept vessels in 'magnificent order', officers said, and were preferable to the European crews who were always 'growling and quarrelling'. By 1914, 51,000 lascars formed nearly 18 per cent of the total number of seamen in British ships; and in spite of discriminatory legislation lascar communities settled in London, Liverpool, Glasgow and Cardiff.

Arriving in Bombay on New Year's Eve 1837 Captain Burt arranged to be carried to Simla, 1,200 miles distant, in a palanquin. Until the middle of the nineteenth century and the advance of the railways this bunk-sized wooden box with projecting poles, carried on the shoulders of bearers, was a common form of transport for long journeys in India, and many European households owned one. Fanny Eden, a sister of Lord Auckland, the Governor-General of India, had her first ride in a palanquin that year 'and felt very much as if I was getting into my coffin'. Captain Burt bought a second-hand model and had it fitted with curtains to keep out the sun and two iron hoops to hold the essentials, a flask of water and a bottle of brandy. His baggage included an oil lamp, saucepans, cutlery, an egg cup, candles, beer, cigars, sherry, two loaded pistols, writing paper, a penknife, a clock, a mirror, blankets, a nightcap and his clothes.

To carry the palanquin he hired sixteen *hamals*, or bearers, paying them half their fee in advance and complaining in the usual British way that it was a 'rascally charge'. He grumbled to the Governor of Bombay about it and was advised that any attempt to get the bearers to reduce their charges would fail. The British in Bombay were always griping about the palanquin bearers who were well known for their trick of putting the conveyance down in the road and refusing to go on until the passenger agreed to pay a much larger fare. The bearers were especially adept at discovering if their passengers were jurors heading for duty in the court: they knew that late jurors were fined and so they put down the palanquin and demanded a ransom twice the size of the fare.

Captain Burt was granted a travel permit by the Bombay police and, with a supply of bread, tea and roast chicken aboard, he set off on 9 January, his large cocked hat resting on the palanquin's roof.

The *hamals* worked in relays, six at a time, and travelled thirty miles a day. When they grew tired they took it in turns to lie prone while their comrades walked up and down on their backs, providing a refreshing massage. They liked to sing as they went along, and Burt related that he had a supply of cotton wool to stuff his ears against the noise, a story he no doubt told in the officers' mess. The traveller Sir Richard Burton once complained about the noise of palanquin bearers. 'After a day or two,' he grumbled, 'you will hesitate which to hate most, your bearers' monotonous melancholy grunting, groaning chant, when fresh, or their jolting, jerking, shambling, staggering gait when tired.'

Burt often left his palanquin to walk, for exercise and to relieve the bearers' calloused shoulders. He sometimes stayed in the simply furnished government postal bungalows, in those days the stepping stones of India, built ten or fifteen miles apart on well-frequented main routes and perhaps fifty miles apart elsewhere. At other times he stayed overnight in his palanquin, and his bearers found him water, milk, eggs and chickens. At night he was puzzled to see dull lights moving and swinging until he learnt that these were lighted sticks carried by village watchmen and postmen to scare off tigers.

Burt's lurching bearers carried him over the rough roads by way of Cawnpore, Aligarh, Delhi and Ambala. Like Fanny Eden he passed through countryside ruined by one of the worst famines of the century. Scores of bodies lay in the fields and hundreds of skeletal figures tottered along the road. Fanny Eden saw starving women selling their babies for a rupee. Such devastation continued to be a periodical spectacle in India until the improvements in grain yields brought about by the 'green revolution' of the 1960s.

For his journey from the northern plain into the hills Captain Burt left his palanquin and hired a *jampan*, a chair lifted on two bamboo poles and carried by four bearers. Extra men were posted on each side to prevent the chair from toppling over the cliffs.

It was not the safest vehicle for travelling the steep and narrow mountain paths, he thought, 'for if either the front or rear men fall the machine and its burden must come down also, unless the side men contrive to keep it up'. In Simla and other hill stations the British, particularly women, were carried about in *jampans* until about 1914. They were superseded by two-wheeled rickshaws which, given the steepness of the streets, had to be pushed and pulled by four men.

On 3 May 1838, six and a half months after leaving London, and almost four months after setting off from Bombay, Captain Burt arrived in Simla and reported for duty.

17

The path to my fixed purpose

'. . . is laid with iron rails. Naught's an obstacle to the iron way!'

Herman Melville, *Moby Dick*

Isambard Kingdom Brunel shared his father's genius. Marc Brunel, who beams with Pickwickian geniality in a portrait painted in his middle age, was born in Normandy and trained as an officer in the French navy. In Rouen he fell in love with Sophia Kingdom, a Plymouth girl of sixteen who was there to study French. During that time of revolutionary Terror in the 1790s his royalist opinions put his neck at risk and he fled to New York. After his escape Sophia, too, was in danger but was spared execution and imprisoned in a convent for nine months. She was freed in the amnesty that followed Robespierre's fall from power in 1794. In 1799, after six years in America, Brunel left for London, was reunited with Sophia and married her.

Having been a citizen of France and America Marc Brunel now became British and set up in Portsmouth dockyard as an engineer working for the British navy against Napoleonic France. He saved the Admiralty a fortune by installing a production line to manufacture wooden rigging blocks, the pulleys used to hoist and trim sails and spars aboard ships. Each ship needed hundreds of them and hitherto they had been hand-crafted and so were expensive. Brunel used steam-powered woodworking machines, invented by Henry Maudslay, to make blocks rapidly and cheaply. Subsequently Brunel mass-produced boots for the British army and developed steam tugs in spite of the Admiralty's opinion that such vessels were fantasies. Starting in 1825 he built a tunnel beneath the

Thames, his grandest engineering achievement, and he did so with his invention of a shield, the forerunner of all tunnelling machines, which he based on his observation of the wood-gobbling action of that old enemy of wooden ships, the teredo.

Isambard was born in Portsmouth in 1806. Recently the Isambard Brunel Junior School in the city discarded the logo which showed him smoking his habitual cigar. In keeping with modern ideas the school said the picture set children a bad example and it replaced Brunel with a set of cogs. It was not the first time that censors had stubbed out the cigar; but, jammed in the corner of the determined mouth, it was a familiar accessory, like the heavy sideburns and knocked-about hat. It was all of a piece with what the photographs show: a man at work, a Victorian big-thinker in the empire of iron.

Unlike many of the engineers of the Industrial Revolution, Isambard Brunel was neither poor nor poorly educated. He grew up in a comfortable home and was educated in two cultures. His progressive father gave him a solid scientific grounding, sent him at fourteen to learn mathematics at a college in France and took him on at seventeen as his engineering assistant. He brought him into his Thames project, and Isambard was nearly killed when water rushed into the tunnel and drowned six men.

The younger Brunel thus combined the more cultivated and academic approach of the French tradition with the energetic, involved and practical British way of doing things. His schooling was obviously important, but he rose because, in common with many inventors and engineers more humbly born, he was a visionary and an almost fanatical worker. He found it difficult to delegate and was so driven that he eventually burned out. He was born at the right time, fortunate to have an exciting arena in which to perform, the frontier of engineering. In that extraordinary time no one had his mastery of so many forms of technology, and he built ships, bridges, railways, docks, tunnels and harbours.

In 1831, when he was twenty-five, Isambard went by stagecoach to Manchester and there made his first journey on a train. The experience planted the seed of a dream of creating great railways. 'The time is not far off', he wrote during the trip, 'when we shall be able to take our coffee whilst going smoothly at forty-five miles per hour. Let me try.'

In 1804, two years before Isambard Brunel's birth, the Cornishman Richard Trevithick built the first locomotive to run on rails, the portable steam engine as it was called. Trevithick was one of the engineering titans, another who started out with little schooling, but from an early age he was trained by his father and uncle, both mine managers, to build and repair the pumps and engines that kept underground tunnels free of water. He grew into an uncommonly large and powerful man and, for fun, would throw a sledgehammer over the top of an engine house or upend a friend and plant his bootprints on a ceiling. He designed his rail locomotive to work with steam at high pressure, what he called 'strong steam', an improvement on James Watt's excellent engine. Watt was jealous enough to suggest sourly that Trevithick should be hanged for developing it.

Trevithick built his locomotive for Samuel Homfray, owner of an ironworks at Pen-y-darren near Merthyr Tydfil in south Wales. Merthyr had erupted over forty years into the greatest iron-making centre in the world, a fire mountain which lit up the sky with sparks and flames. The four ironworks there were headed by pharaohs of the furnace: ironmasters could hardly be of the diffident sort. On the mountainsides they had all they needed for industrial power: iron ore, coal, timber, limestone and plentiful water. On the slopes today there is still a sense of distant uproar, of a land sucked dry by primitive industry; and in the tangled woods above Merthyr you may wander among the fallen stones of great buildings, evocative remnants of ironworks which bring to mind the ruined temples of Mexico, raised long ago to insatiable gods.

Merthyr's ironmasters shipped their iron and coal by canal down the valley to the port of Cardiff. Trevithick's job was to build a tramway linking Homfray's ironworks to the canal. Homfray's rival, Richard Crawshay, wagered 500 guineas that no locomotive could haul a ten-ton load on iron rails. He lost. At four miles an hour Trevithick's locomotive pulled trams filled with ten tons of iron and seventy passengers for nearly ten miles. The stone blocks on which he secured his rails are still embedded in a quiet glade. His locomotive was not seen at the time as an advance, for no one thought much about railways. Pragmatic industrialists believed that canals were the modern and profitable way of transporting freight and that horses were more efficient and cheaper than locomotives.

The rising cost of fodder, however, began to make horse power more expensive. Coal in the colliery districts was a cheap fuel dug out by meanly paid labour. The cost advantage, as well as the increasing demand for more efficient haulage of coal and iron, sharpened the interest in steam locomotion. George Stephenson was a local genius on Tyneside who grew up breathing steam and repairing colliery engines, and although barely literate he knew how to assemble reliable locomotives. He seemed cut from granite, an exceptionally stubborn, difficult and determined man; and, when they met, he was in tune with Trevithick, the Cornish Samson, who certainly matched him in toughness.

Stephenson inaugurated the railway age in 1825 by building the Stockton and Darlington line, the first public railway to use locomotive power, in this case his engine called *Locomotion*. Mostly the wagons hauled freight, which is what the pioneers thought railways were for. Horses were employed to draw the passenger carriages. The first modern passenger railway, with stations and signals and services that ran to timetables, was Stephenson's thirty-four-mile line connecting Liverpool to Manchester. Its opening in September 1830 was a festival and a frenzy visited by misfortune. Tens of thousands of people crowded bridges and embankments to watch. The chief guest was the Duke of Wellington, the unpopular Tory Prime Minister, whose prejudices included the belief that railways would 'encourage the lower classes to move about'. (It was in the same conservative frame of mind that the Duke also rebuffed a promoter of steamship travel with the message that he had no time 'to receive the visits of gentlemen who have schemes in contemplation of the alteration of the public establishments'.)

On the inaugural run from Liverpool two trains set off on parallel lines and stopped halfway to take on water. Some of the passengers got out to mill around and chat, one of them William Huskisson, railway enthusiast, Tory Member of Parliament and President of the

Board of Trade. Down the other track came the locomotive *Rocket*, built by George Stephenson's son Robert and winner the year before of a locomotive competition with a speed of twenty-nine miles an hour. Huskisson tried to scramble aboard his own train but fell under the *Rocket*'s wheels and his leg was mangled. He was carried to a rectory where he died nine hours later, the first railway victim, the first blood on the tracks. The Duke, who saw it all, did not ride a train again for twelve years.

'The incredible speed' of the inaugural train 'burst upon the world', said the progress-minded Samuel Smiles; but railway doubters worried that the human frame would not withstand high-speed travel. Thirty miles an hour seemed frightening to many. Captain Thomas Hardy, who commanded HMS *Victory* at Trafalgar, refused to get on a train. Thomas Carlyle shuddered to recall what he termed a railway 'flight', writing that 'In the nervous state I was in it seemed to me certain that I should faint from the impossibility of getting the horrid thing stopped. Snorting, roaring, we flew: the likest thing to a Faust's flight on the Devil's mantle.'

Many early railway travellers, however, revelled in velocity and wrote of the sensation of skimming the earth like birds. Charles Greville, Clerk to the Privy Council, wrote that he experienced 'a slight degree of nervousness and a feeling of being run away with, but it entirely renders all other travelling tedious by comparison'. In prose and poetry some women expressed the erotic experience of the railway's throbbing rhythm. The young actress Fanny Kemble was enchanted by 'the magical machine with its flying white breath and rhythmical pace . . . I stood up with my bonnet off . . . the sensation of flying was delightful and strange beyond description'.

Riding on the engine with George Stephenson himself, she located the thin vein of charm in the old crag and admitted that 'he certainly turned my head'. Queen Victoria had her first train ride, from Slough to London in 1842, and an editorial in the *Atlas* reflected a public nervousness in its comment that 'a long Regency in this country would be so tremendous an evil that we cannot but desire that these royal excursions should be either wholly abandoned or only occasionally resorted to'.

George Stephenson foresaw a national rail network, built chiefly to carry iron and coal and the products of the new factories. Businessmen who had invested in canals pursed their lips at the prospect of the new technology proving superior to the water transport they found so profitable. No one could imagine then how railways would transform Britain and the world; but Henry Booth, treasurer of the Liverpool and Manchester, wrote perceptively of the 'sudden and marvellous change in ideas of time and space. Speed – despatch – distance – are still relative terms, but their meaning has been totally changed within a few months; what was quick is now slow; what was distant is now near; and this change in our ideas . . . will pervade society at large.' In a speech he said that the mechanical principle, the philosophy of the nineteenth century, would spread and extend itself. 'The world has received a new impulse.'

From the early 1830s the railways spread like jungle creeper, hand-built by brigades of navvies, tens of thousands of men, each one shovelling fifteen tons of earth a day. Hills and rivers were conquered by tunnelling miners and bridge-building masons. Between 1825 and 1844 the mobile columns of navvies laid 2,000 miles of track and threw up hellish townships of turf huts where they drank and brawled uproariously. Yet they meekly accepted the church workers who believed the navvies could be redeemed and came as missionaries to care for their wild and beery souls.

In 1833, amid all the excitement of a new technology on the march, Isambard Kingdom Brunel won the opportunity to create a masterpiece. Bristol's merchants, anxious to outdo their rivals in Liverpool, appointed him engineer of the Great Western Railway with the awesome task of building a railway from London to Bristol, with branches into Wales and the West Country. He was twenty-seven.

Brunel set out to design a high-speed passenger line, what he defined as 'the finest work in England', and to make himself the master of every stone, iron and timber detail of its construction, every bridge and cutting. One of his achievements was the Box Tunnel in Gloucestershire, almost two miles long, a five-year struggle of blood and gunpowder which occupied thousands of men and killed more than a hundred of them. The story that Brunel aligned it so that on his birthday, 9 April, its gloom was pierced by the light of the rising sun is, I am sad to say, just that. The ubiquitous Brunel surveyed the land, on foot and on horseback, travelling by night in coaches, negotiating with landowners, hiring men, dealing with lawyers, enthusing everyone by the sheer power of his personality. He revelled in it all, forever against the clock, chain-smoking his cigars, working far into the night, sleeping little, often rising at three. To make himself more productive he built a horse-drawn office, nicknamed 'the flying hearse', and fitted it out with a cat-nap couch, engineering instruments and a case holding fifty cigars. In this mobile home he often fell into an exhausted sleep, and when he did so his men loaded it as gently as they could on to a train at Paddington station and sent it off down the line where horses waited to take it to the next section of the construction.

The first Great Western train from Paddington to Bristol ran in 1841, and three years later the line was extended to Exeter. An MP told an astonished House of Commons at 10.30 one evening that he had left Exeter, almost 200 miles away, at 5.20 that afternoon and arrived in London at ten. By then, new railway schemes were spreading in an orgy of greed and speculation. 'The whole world is railway mad,' sighed Brunel in 1844. 'I am really sick of hearing proposals. I wish it were at an end.'

Within a few years rail passengers grew accustomed to travelling at thirty or forty miles an hour. Many felt they lived in an age of rapid change, freed from nature's constraints. The curtain was falling on the stagecoach, to poetic regrets, and distance was shrinking. Now the railways were imposing their rule on time itself, demanding order and uniformity.

Hitherto every town and village had kept its own time. Sundials were accurate enough indicators, the length of a man's shadow on the ground was a fair guide to the approach of the dinner hour, and a sun clock on a church wall told a priest to prepare for matins. Sandglasses measured time at sea. Church and city bells regulated prayer and work.

Trains, however, had to run to schedules, and in 1839 William Bradshaw published the first of his timetables. Railways set the national standard and made Britain a single time zone. The station clock became the national timepiece and the arbiter of a new discipline as well as a rendezvous for spies and lovers. The stationmaster gravely checked his amulet pocket watch as trains arrived and departed. He was the new figure of authority.

Brunel rode the flood-tide of his confidence, powers and precocity. In the 1830s and 1840s he worked hard enough for half a dozen men. In time he would build twenty-five railways in Britain, India and Europe, three great steamships, five suspension bridges and more than a hundred other bridges and harbours. In 1831 Bristol accepted his design for the Clifton gorge suspension bridge; in 1833 he started building the Great Western Railway; in 1837 he launched the *Great Western* to conquer the Atlantic; and in 1838 he created the *Great Britain.*

The story of Victorian technology could quite properly begin with the construction of this vessel at Bristol. Even on the drawing board she was a fluid creature, adapted and reshaped in accordance with rapidly evolving ideas. She started on paper as big, wooden, paddle-driven and fast. She finished as bigger, made of iron, propeller-driven and faster. She led the way in Britain's steam-powered maritime dominance, the first large iron ship driven by the most powerful steam engine in the world.

When Brunel started his design, shipbuilding in iron was in its infancy and no builder was expert. Iron itself had only become available in the right quality and quantity following the late-eighteenth-century developments of coke furnaces, rolling mills and steam hammers; and especially Henry Cort's invention in 1784 of the puddling process in which air was stirred into molten iron in the furnace to make it purer. Until then iron plates had been essentially hand-wrought. By the 1820s a few river barges were made of iron and some small steamers were constructed in the 1830s. Brunel so admired the iron Channel packet *Rainbow*, built in 1837, that he decided on iron for his new vessel, calculating that it would be a fifth lighter, a fifth more capacious and considerably stronger than a timber ship. He ordered iron plates from Coalbrookdale in Shropshire and saw the keel laid in 1839.

At 322 feet and 3,500 tons, *Great Britain* was twice the size of any forerunner and the first large vessel to be driven by a propeller, one of the seminal developments of marine engineering. Brunel originally designed her to be driven by paddles but he saw all the advantages of the new screw propulsion pioneered separately in 1836 by Johann Ericsson, a Swede, and Francis Pettit Smith, an English farmer. Brunel changed his design to incorporate a propeller driven by an engine, designed chiefly by his father, that would give twelve knots.

Prince Albert attended the launching of the *Great Britain*, at Bristol in 1843, and was on hand to grab a bottle of champagne and hurl it at the bow when the baptismal wine missed the target. Afterwards he returned to London by a Great Western train in two hours and forty minutes, so that he was home for dinner with a tale for his Victoria. Newspapers praised the ship as the 'stupendous progeny of Mr Brunel', 'Queen of the Waters', and 'Monster of the Deep'.

Her size, in fact, was a problem. As she was being finished in Bristol it became clear that she would be too massive to negotiate the lock gates into the River Avon unless they were enlarged. They were; and still she stuck and had to be pushed back while more masonry was hammered away, and more cigars champed, as she was squeezed through at the top of the tide. She went to Liverpool and in July 1845 steamed on her maiden voyage to New York, cheered by thousands, but with only forty-five people aboard with enough confidence in screw propulsion to buy a ticket. She arrived in just under fifteen days, averaging almost nine and a half knots, and the *New York Herald* welcomed 'the monarch of the ocean'. The following year, with 180 passengers, she ran aground in Ireland. Although hammered by huge seas her iron hull was an indestructible fortress, so much so that her survival persuaded doubting shipbuilders of iron's great qualities. The passengers were ferried across the sands in carts.

After eleven months aground she was salvaged, sold and fitted out as a full-rigged sailing ship for the growing trade to Australia. On such long voyages sail still had an economic edge over steam. The *Great Britain* was on the Australia run for twenty-three years, a service interrupted by troop transport voyages to the Crimea and

India in the 1850s. In 1887 she became a storage hulk for coal in the Falkland Islands and lay, more or less forgotten, until 1970. Rusted, battered and barnacled, she was raised on to a pontoon and towed to her birthplace at Bristol. I stood above the Clifton gorge and watched her pass beneath Brunel's Clifton suspension bridge as she was towed up the Avon towards dry dock and years of restoration. Her bulk and her lines still spoke of Brunel's vision.

In the end Brunel's dreams engulfed him. In 1851 he sketched out his third ocean-going creation, the *Great Eastern*. A ship of 19,000 tons, almost 700 feet in length and eighty-three feet in the beam, she was six times the size of the largest battleship in the Royal Navy, more than twice the length of the *Great Britain*, a floating city.

'My great babe', Brunel called her, and he acknowledged that she was the most formidable challenge of his life. He put all of his ship and bridge experience into her construction and declared that 'I have never embarked on any one thing to which I have so entirely devoted myself . . . on the success of which I have staked so much reputation'. She was his giant leap into the future. The stiffness and strength of her cellular design was far ahead of her time, and no ship would be larger for the next fifty years, not until the *Lusitania* and *Mauretania* were built in the twentieth century. She was driven both by an immense screw, twenty-four feet in diameter, and by paddles fifty-eight feet across. Her gigantic engines were designed to drive her at twelve knots.

Brunel conceived her as a round-the-world leviathan taking thousands of emigrants to the booming lands of Australia and servicing the growing demands of the India trade. She had cabins for 4,000 passengers, or for 10,000 soldiers if the need arose, and enough bunker space for coal to enable her to ignore coaling stations and steam non-stop around the Cape of Good Hope to India or Australia.

The *Great Eastern* was built on the Thames, rising as a massive black iron cliff above the mudflats and grim little terraces of the Isle of Dogs. She was too large to be built and launched stern first in the normal fashion and so was constructed alongside the river bank, her bow facing downriver to the sea. An army of riveters working at portable forges hammered three million rivets into 'vast plates of iron that seemed big enough to form shields for the gods'. The 10,000 plates were each ten feet long, two feet nine inches wide and three-quarters of an inch thick. The ship was 'made almost as strong as if she were of solid iron,' said the *Illustrated London News*, 'whilst she is rendered as light, comparatively speaking, as a bamboo cane'. Opinions of her varied from deep doubt about such 'a swollen hunk' to exultation over a ship that would 'bring the nations of the earth closer together'.

No one could be sure that she could be persuaded sideways into the river and for Brunel himself the launch was daunting. He was furious to find that thousands turned out for the event in November 1857, although he could hardly have expected anything else. Such excitement was not to be wondered at, a journalist wrote, 'when we consider what a splendid chance presented itself of a fearful catastrophe'.

Hydraulic rams gave the ship a shove but she would not budge, and some of the newspapers had their fun, jeering at what they called 'a monument to folly'. But she was launched successfully the following January. Her builders were bankrupted and another company took her on to put her into a trade for which she was not designed, the North Atlantic. Brunel went aboard her before she put to sea. It was his last sight of her. He was taken home, seriously ill.

The *Great Eastern* steamed down the English Channel in September 1859. An engine-room explosion blew off one of her five funnels and killed five stokers. *The Times* reporter on board described how the funnel rocketed into the air and 'a shower of glass, gilt work, saloon ornaments and pieces of wood began to fall like rain'. Brunel, worn out by the six-year drama of her construction and his years of urgent, anxious, sleepless working, died a few days later. He was fifty-three.

The *Great Eastern* left on her maiden voyage to New York in June 1860. Only thirty-eight passengers were aboard, pygmies amid the opulence, awed by the golden pillars and sweeping staircases, the piano lounges, the spacious cabins with crimson velvet couches.

Each was fitted with the extraordinary luxury of a bathtub with a hot-water tap. Strolling the promenades, passengers listened for a trumpeter who would summon them to lunch. But that was on good days: the ship was not sea-kindly and rolled badly. As for speed, sailing vessels could still match her. The following year, bound for New York from Liverpool, the *Great Eastern* ran into a gale and heaved and pitched violently. Cabins were flooded. A wave smashed the rudder. In the Grand Saloon 'three or four gentlemen were dashed with violence against the great mirror and burst through it, the glass falling about them in slices inflicting cuts and bruises'. The ship was adrift for a week until the rudder was repaired and she put into Queenstown in Ireland, her reputation wrecked.

She seemed to be a lumbering mastodon, yet, as the naval architect Ewan Corlett argued, 'she was a brilliant technical success'; and her design would probably have enabled her to survive the sort of collision that sank the *Titanic*.

Her owners had no great commercial sense, however. They did not use her on the Australia route and put her on the Atlantic. They sought passengers among the wealthy elite, having no wish to carry cheap-fare emigrants. She was too large for her time, but, as it turned out, there would be a giant job for a giant to do.

18

The precipice of time

'Her subjects pours
From distant shores,
Her Injians and Canajians;
And also we,
Her Kingdoms three,
Attind with our allagiance.'

William Makepeace Thackeray,
'Mr Molony's Account of the Crystal Palace'

In 1851 millions of British people loosened their waistcoat buttons and celebrated both themselves and what was evidently evolving as the British century. In January *The Economist* rosily reviewed a fifty-year slice of history that was 'more full of wonders than any other'. It invited its readers to think back to the bad old days of 1800, to recall

> the heavy taxes on the necessities and luxuries, paying twofold for our linen shirts, threefold for our flannel petticoats, receiving our Edinburgh letters in London a week after they were written, travelling with soreness and fatigue by the 'old heavy' at seven miles an hour, instead of by the Great Western at fifty, and relapsing from the blaze of light which gas now pours upon our streets into a perilous and uncomfortable darkness.

At the beginning of the century, it said, the criminal law was in

> a state worthy of Draco; executions by the dozen; the stealing of five shillings punished as severely as rape or murder; slavery

> and the slave trade flourishing in their palmiest atrocity; religious rights trampled underfoot; Catholics not citizens, Dissenters despised; Parliament unreformed. The people in those days were little thought of, where they are now the main topic of discourse; steamboats were unknown and a voyage to America occupied eight weeks instead of ten days; in 1850 a population of 30 million paid £50 million in taxes; in 1801 a population of 15 million paid £63 million.

That was January. In February *The Economist* looked back on a half-century of progress in the sciences: the finding of the planet Neptune, the mapping of the moon's surface, Michael Faraday's discoveries in electricity and magnetism, the start of the electric telegraph and the development of steamships and railways. In medicine there was the 'crowning discovery of chloroform, by which pain is banished at will'. *The Economist* did not mention the opinion of a sawbones British army surgeon in India that the relief afforded by chloroform was 'highly pernicious' and unwelcome in the field hospital. Practical doctors, he said, regarded pain as 'one of the most powerful stimulants known'.

The Economist pressed on with its paean. 'In the days of Adam the average speed of transport was four miles an hour; in the days of Nimrod, or whoever was the first horse tamer, the rate had risen to ten miles; in 1828 it was still only ten miles; in 1850 it is habitually forty miles an hour, and seventy for those who like it.'

And look at the moral progress, it said, the shrinking of corruption in public life, the decline in swearing and drunkenness. 'Intemperance is as disreputable as any other kind of debauchery, and except in Ireland and the Universities, a drunken gentleman is one of the rarest sights in society.'

Such trumpetings formed a frontispiece to the marvel of the year. The Great Exhibition, opened by Queen Victoria in May, collected the achievements of Britain and the world in the cathedral of glass and iron that *Punch* memorably called the Crystal Palace. It was designed by a former gardener's boy, Joseph Paxton, as an assertion of Britain's modernity and industrial supremacy, a view of an empire in triumphant progress, an empire by no means at its apogee.

No one else could have done it. 'We are capable of doing anything,' the Queen confided to her journal. The *Illustrated London News* recorded 'the pardonable exultation felt by Englishmen in the fact that in England alone' was such an exhibition possible. 'Until further notice,' it added, 'London is not simply the capital of a great nation, but the metropolis of the world. John Bull is no longer an ogre, but a genial and courteous gentleman. The old joke about the gloom, smoke and dirt of London and the austerity, inhospitality and semi-lunacy of the English character has been dissipated.'

Of course, there were grumblers. *The Times* bleated that the Great Exhibition site in Hyde Park would become 'the bivouac of all the vagabonds of London'. Colonel Charles Sibthorp MP warned of the bad characters who would arrive and advised local residents 'to keep a strong lookout over their silver spoons and serving maids'. Some did not disguise their hope that Joseph Paxton's soaring greenhouse covering twenty acres of the park would blow down in a gale and that every one of the 293,655 panes of glass would be smashed. In a response to those Jeremiahs who predicted that floors would collapse under the weight of the crowds, a force of 250 strapping soldiers, twelve abreast, came to run and stamp on a section of flooring; and with this hobnailed demonstration put the doubters to flight. By the time the exhibition closed, after five months and eleven days, more than 6 million people had seen it, 17 per cent of the British population, including a woman at least eighty-four years old who had walked all the way from Cornwall.

Queen Victoria visited the exhibition many times and loved every moment, writing that her beloved Prince Albert, who had played a leading part in it all, was for ever immortalized. She was entranced by the monumental evidence of Britain's leadership in steam and iron. Among 100,000 exhibits, half from Britain and the empire and half from other countries, there were locomotives, steam hammers, pumps, turbines and printing machines. A twenty-four-ton boulder of coal, like a newly arrived black meteorite, was a symbol of strength. The venting of steam and the clank, thud and rumble of engines brought prideful tears to the eyes of progressive men. The Almighty, it was said by one observer, had 'vouchsafed to confer upon us so much of His Own Power'.

The frock-coated priests of invention explained their wonders to the wide-eyed Queen. Here, for example, were the spinning machines that dominated industry in Lancashire, manufactured the cotton goods that made up more than half of Britain's exports and undermined the cotton trades of India. Here were machines for making cigarettes, chocolate and soda water. Here at the sharp end of technology were the new steel pen nibs. The Queen watched a demonstration of the electric telegraph. 'The boy who works it does so with the greatest ease and rapidity, truly marvellous,' she said. She saw how sticky opium was harvested from Indian poppies and made into 'curious large cakes and balls and sent to China'. The little balls were symbolic. Britain's victory in the Opium War a few years earlier had compelled the Chinese to accept deliveries of the drug they regarded as harmful. The British saw themselves as the patient apostles of free trade and the Chinese as obstructions to economic progress. Out of this profitable war Britain had won a new colony in Hong Kong, a base in Shanghai and dominance as a trader and drug merchant in the Far East. Opium and Lancashire cotton remained China's chief imports for more than thirty years.

Naturally the Queen admired the Koh-i-noor diamond, the Light of the Universe, brought from the Punjab as a crowning imperial trophy and mounted in a robber-proof golden cage. She inspected hats made in Australia by convicts, anachronisms really because by then transportation was virtually at an end. There were Staffordshire glazed water-closets; an ingenious postage stamp slot machine to help make the new postal age even more convenient; photographic devices; and an array of Sheffield steel Bowie knives which, the Queen noted, were 'made entirely for Americans, who never move without one'. Possibly she inspected some of the mad-inventor oddities, a physician's walking stick containing an enema syringe and test tubes, a corset with a quick-release device to free the wearer in case of emergency; Mr Shillibeer's expanding hearse; and a bed designed to tip its occupant on to the floor beside a waiting cold bath. The American section, barely hinting at future industrial might, exhibited a sewing machine, some of Samuel Colt's revolvers and a vacuum coffin guaranteed to preserve a loved one from decay.

Nothing seemed more futuristic than the amazing electric clocks on display; but leading engineers tended to disparage electrical devices as playthings. Brunel, the colossus of steam and iron, thought that electric machines hardly warranted a place in the exhibition, for they 'can as yet only be considered as toys'.

While the Crystal Palace provided a window on Britain's economic prowess and imperial expansion, the ticket offices gave a view of its hierarchy of class. Only the well-off could afford the admission charge of £1 during the first ten days. Tickets were five shillings after that, still too much for the working classes who were effectively excluded by price. After three weeks the admission sluice-gate admitted people for a shilling from Monday to Thursday, two shillings and sixpence on Friday and five shillings on Saturday. The middle classes could thus ensure that they would not have to mingle with or hear the vowels of the lower orders or the picturesque corduroy folk up from the country. The years of Chartist agitation and riot of the 1830s and 1840s were receding in the memory and there was relief that the sort of people associated with such trouble seemed well-behaved. The exhibition was closed on Sunday. Smoking and alcohol were forbidden at all times. Purity was buttressed by buns and lemonade. The profits from admission fees paid for a wonderful legacy, the land on which were built the Victoria and Albert Museum, the Royal Albert Hall and the Imperial College of Science and Technology.

Perhaps the most welcome amenities at the exhibition were the public conveniences, the first of their kind. Like any other city, London was a place of much discomfort, especially for women, because it lacked lavatories, apart from grim medieval riverside latrines. In the Crystal Palace George Jennings installed flush lavatories of the latest design, what he called 'monkey closets', pioneering London's network of public conveniences, many of them sanitary temples made handsome by ornate ironwork. In the 1850s much of London stank of excrement and the Thames was a sewer, 'a Stygian pool', as Benjamin Disraeli said, 'reeking with ineffable and unbearable horrors'. Sheets soaked in chlorine had to be hung in the House of Commons to offset the river's odour.

As for relief, men at least had their clubs and pubs or the ability to be relatively discreet in an alley or in the bushes of parks. Women were expected to endure. It would be some years, into the 1870s, before women's lavatories were introduced, against the objections of the comfortably-off, xenophobic and misogynist residents who had their own facilities at home and viewed street lavatories as 'German abominations' used by foreigners and tarts. But lavatories had commercial value: shopping grew as an activity only when hotels, restaurants and the new women's clubs provided civilized stopping places.

The Great Exhibition established Thomas Cook as the inventor of popular travel, the creator of adventures and the realizer of dreams. He made his name with an admired feat of packaging, almost generalship, sending more than 150,000 people to the exhibition and ensuring they were provided with train seats, admission tickets, hotel beds, breakfasts, lunches and dinners.

After that there was no holding him. Cook's Tours entered the language and the national experience. He seemed at first an unlikely travel agent. When he organized his first outing in July 1841 he was the thirty-three-year-old secretary of the Leicester Temperance Society in the heart of an expanding industrial and urban region. He was struck by the obvious, that the new railways spreading over the land needed passengers; and that there would be a public appetite for the excitement of railway trips if someone would arrange them at cheaper prices than the rail companies were charging. You supply a train, he said to the Midland Counties Railway, and I will supply the passengers. He fixed a day out at Loughborough, with games, tea, cricket, bands and rousing temperance speeches, and sold nearly 600 one-shilling tickets to fill nine open trucks drawn by a sooty locomotive.

Inspired by that success he arranged more temperance trips and in 1845 organized his first commercial venture, escorting 350 tourists from the Midlands to Liverpool. The outing included a seaside and mountains tour to Caernarfon and Snowdon. Cook also wrote the guidebook. He repeated the trip two weeks later and walked up Snowdon with his tourists. The following year he took a party of 350 on a rail and steamer tour to Scotland which was for the English the newly discovered land of Walter Scottish romance and royal holidays. Cook's tourists were themselves a spectacle, local people coming out to gaze at them, to cheer them and to call out the local brass bands in welcome.

Cook's logistical experience made him the ideal man to organize travel to the Great Exhibition. He tailored his fares to the working-class pocket. He and his seventeen-year-old son John drummed up business, sold the tickets and travelled to London with the tour parties. Passengers short of cash pawned their watches with Cook to pay their fares. To people who complained that some of his tourists were excited by drink Cook countered that the sight of an inebriated man was a useful warning against excess.

As railways developed in the 1850s Cook found a market in the expanding class of people who had a certain amount of leisure and money to enjoy modest adventures, and he became the impresario of the seaside railway excursion. Indeed he was one of the originators of the Victorian seaside itself. It was on a Cook's tour that many townies saw the sea for the first time in their lives. The railways opened up the south coast, the Isle of Wight, Devon and Cornwall, the coasts of Wales and Lancashire and Yorkshire. The trains took the trippers to the very edge of the sands and straight into freedom, to the pleasures of the promenades and the medically approved benefits of fresh air and salt water. Cook himself inspected hotels and recommended them to his clients. To those who complained that his tours seemed to be purely for enjoyment Cook retorted that there could be nothing wrong in introducing people to the beauty that God had made. He suggested that churches would benefit from sending their clergy on refreshing excursions that would improve their performance in the pulpit.

In 1855 he took the big step of arranging his first tour abroad, escorting passengers to Paris. In the 1860s the regular Cook's tour to Paris left London at six in the morning, stopped for a large lunch in Dieppe and reached Paris at eight in the evening. Cook knew that his foreign trips met a yearning for romantic adventure. He extended his service to Switzerland, his clients crossing the Alps in sledges and stagecoaches because there were no railways, and added Italy to his tourist trails. Many of his passengers were women who no doubt saw the tours as a way of emerging into the air from the stifling confines of strict homes. They regarded Cook as an almost brotherly chaperon, a gallant guide and protector so confident in dealing with ticket collectors, hoteliers and foreigners in general. Abroad, as everyone knew, could be awful. One guidebook advised against soup and wine, warned of French trickery with the bill and suggested that each passenger equip himself with a brace of pistols. It was well known that foreign lavatory facilities were unspeakable and women were urged to buy what looked like a hat box but was in fact a cunning commode.

The restaurateurs, hoteliers and guides of France, Switzerland and Italy were delighted to see Thomas Cook's parties, Cookites as they were sometimes known; but British residents abroad looked down their noses and saw people of the more common sort intruding in their private backyards and undermining the genteel British image.

Within a few years the sense of satisfaction in technological excellence that swelled in the months of the Great Exhibition evaporated in the disillusion, waste and incompetence that characterized Britain's conduct of the Crimean campaign against Russia in 1854–6. Britain went to war to ensure its naval paramountcy in the Mediterranean and to stymie any Russian threat to its communications with India. The British army was unfit to fight a long campaign. Badly managed and undermined by government parsimony, it had become a withered muscle since Waterloo. The conceit of a military superpower was soon badly dented by reports of troops ill-supplied with food, blankets and tents dying in their thousands of cholera, dysentery and fever; a suffering army led by elderly and inept commanders like the well-meaning Lord Raglan and the oafish Lord Cardigan.

The British government and people knew of the failures and horrors from the dispatches of a rising breed of men, the war correspondents emerging as stars in the world of popular newspapers. The most influential of these, the original and greatest war correspondent, was William Howard Russell of *The Times*. Born in Ireland in 1820, a parliamentary reporter at twenty-two, he covered the Irish potato famine, the Great Exhibition and the Duke of Wellington's funeral in 1852. In 1854 Thomas Delane, his editor, sent him to the Crimea where he became a friend of soldiers and a scourge of bungling generals. Like many journalists he was hopeless with expenses and the despair of accountants. He wrote with a quill and sent his reports by steamship to be published two or three weeks after he had penned them.

Russell invented the art of the correspondent. He knew that for a reporter nothing beats going and seeing. He was a colourful storyteller and the unsparing master of horrible detail. He had a sure instinct for writing what was wanted. On the breakfast tables of Britain and the desks of government ministers he served up war undisguised, the swollen bodies, oozing blood and shattered limbs, the 'humane barbarity' of the surgeons. He never ignored the suffering of horses, either, and described 'the multitudes of dead horses . . . piles of remains of wretched animals . . . the country dotted all over with carcasses in every stage of decay'.

'So deep had been the sleep of the military service', says the history of the Royal Army Veterinary Corps,

> that when the Expedition was sent to the Crimea, those in authority preferred the sailing vessel which took a month to reach the Black Sea, to the steamer which completed the journey in a fortnight! The losses at sea were consequently heavy, and advice tendered by the veterinary professions as to the best method of shipping horses, by carrying them on sand and shingle ballast in the hold and allowing them to lie down, was not taken by those responsible for transport.

Losses during the winter of 1854–5 on the heights of Balaclava were scandalously heavy, the animals dying in their hundreds from starvation. Desperate and skinny, tail-less and maneless, the ravenous creatures gnawed at the skins of the animals that had dropped dead, and chewed ropes and blankets for nourishment. Mud froze and stuck to them, pulled out their hair in sheets and ripped the flesh from their bones. It was not until 1855 that the army appointed a Principal Veterinary Surgeon in the Crimea and an effort was made to care for sick horses.

The state of the military veterinary service reflected the condition of the British army itself. It was typical of the army's attitude that each of the few veterinary medical chests sent to the Crimea included a tin lantern that provided the barest glimmer of illumination. Twenty-two years after the war, the official history says, equipment and veterinary arrangements remained hopeless and disgraceful. Regulations specified three bandages per 5,000 animals and the military veterinary kit still contained the lantern that gave no light.

The equestrian statue is the most splendid embellishment of military art. As in life, the mounted soldier assumes more authority by his elevation and, if ugly, may be lent handsomeness by his horse.

Statues and portraits of officers on horseback, and the gloriously vivid paintings of cavalry actions framed on the walls of regimental messes, salute and romanticize the historical relationship of men and horses joined centaurially in the savagery and suffering of war. The partnership endured well into the twentieth century. Thousands of horses went to war in 1914. The last British cavalry charge took place in Palestine in 1917. The mounted frontiersman, John Wayne glued into the saddle, remains the emblematic figure of the conquest and settlement of the American West. If there could be one permeating and evocative smell of history it might be that of horse dung: London's streets were full of horses into the twentieth century, and manure supported squads of cleaners who shovelled it up to sell to farmers and gardeners.

Fox-hunting and the stagecoach assisted the development of British military horses in the nineteenth century. The passion for hunting that grew in the last decades of the eighteenth century bred the fast, fearless and powerful hunters that cleared gates and hedges. The growth of stagecoach traffic on the improved roads also encouraged the breeding of strong horses; and these were perfectly suited to drawing artillery. The big thoroughbred hunter, able to bear the considerable weight of a heavily armed and accoutred soldier, became the British cavalry's much-admired mount of choice.

The load-carrying capacity of horses, their endurance and food and water requirements, were a preoccupation of military authorities until the end of the First World War. A large horse weighing a thousand pounds would live perhaps twenty years and was good for about ten years' service. It needed, according to one army prescription, twelve pounds of hay a day, ten of corn and eight of straw; and, in cool weather, five gallons of water. Some veterinary officers believed it should carry no more than fifteen stone, or 210 pounds, but a soldier and his equipment usually came to more than 250 pounds and, on the march, with a ration of corn and some spare horseshoes, to nearly 300 pounds.

Cavalrymen naturally took the greatest professional pride in their horsemanship and were often fond of their animals. But they also drove them very hard, and were frequently indifferent to their suffering. It took them a century to learn to rest their horses by dismounting and leading them. With few exceptions they counted it a matter of honour to remain in the saddle and saw themselves diminished, emasculated in some way, if they had to go on foot like a common soldier. The question of appearances extended to their uniforms which were tailored to be as tight as possible for the sake of parade-ground swagger. This was a sometimes fatal fashion. Soldiers unhorsed in fighting in Africa in 1852 could not scramble back into their saddles because their trousers and tunics were too tight; and they were killed like hapless peacocks. On service in India, cavalrymen, like infantrymen, wore thick clothing that made no concession to the climate. As if sewn into their uniforms, some could dismount only with difficulty.

In 1796 an army inquiry concluded that heavy losses among cavalry and artillery horses were caused by the ignorance of the men appointed to care for them. These were the regimental farriers, non-commissioned officers who carried axes for killing wounded horses and who applied primitive remedies to those which fell ill. The farriers' chief work, however, was to shoe the horses; and a farrier buried with military honours had horseshoes and pincers placed upon his coffin, just as a surgeon's casket was adorned with an amputating knife and saw. A horseshoe lasted a month, or around 200 miles, and military expeditions included shoesmith wagons carrying a forge, iron and coal so that the farriers could make new shoes and nails. In many cavalry regiments it was the farriers' task to flog miscreant soldiers with a cat o' nine tails.

The army's concern about the treatment of horses led to the founding of a military veterinary service in 1796; and the term veterinary surgeon was coined around that time. Edward Coleman was appointed its first chief and held the post for forty-three years. One of his achievements was simply to clean the stables, and under his command the improvement of hygiene brought an immediate benefit to equine health. Some of his assistants also advanced the understanding and limiting of glanders, a serious infectious disease. But in its survey of Coleman's career the history of the Royal Army Veterinary Corps notes his neglect, indifference and dictatorial manner, his 'inordinate love of money and the large sums gained by him from his position'. When he died he left £47,000, a fortune.

Coleman scorned veterinary officers with medical training and said farriers' sons were the men best fitted to care for horses. Many cavalry officers shared his view that there was little value in veterinary surgeons, believing that their own status and experience as mounted warriors made them expert horse doctors. Among caste-conscious officers veterinary surgeons were often disdained and treated as social outsiders. They were paid less than army doctors and, unlike regimental surgeons, were viewed as men who did not hold a proper commission.

A few veterinary officers served during the Peninsular War in Spain against Napoleon's forces, from 1808 to 1814, but they could do little. The Duke of Wellington called Spain the grave of horses. Similarly, in the Waterloo campaign, the best that could be done for lame and sick horses was to destroy them. The treatment of cavalry horses was, in any case, primitive. The first army veterinary manual recommended the placing of a red-hot iron on the chest of a horse with pneumonia, on its belly if it had enteritis, and on its loins in the case of a kidney infection. Just as doctors favoured the bleeding of their patients, so did veterinary officers. The operation was intended to remove infected blood from the body and had the effect of encouraging the production of fresh blood, a process that could make a patient feel temporarily better. The military manual suggested that a vet should treat a sick horse by bleeding it of a gallon of blood and pouring boiling water twice daily on the affected part. Farriers and grooms had always bled horses regularly in the belief that the procedure made the animals healthier.

Even a small military campaign demanded large herds of horses. In India the British, like the Mogul emperors, travelled with an immense menagerie and entourage of camp followers. A British force of 8,500 men, assembling in Meerut in northern India in 1813 for action against the Gurkhas in Nepal, bought and hired horses, elephants, camels, mules, donkeys and bullocks to support them. Hundreds of camels carried ropes, harnesses, nosebags, horseshoes, ammunition, wine, clothing, bedding, hospital stores and litters, as well as grain for the horses. Elephants and bullock carts transported tents and other bulky supplies. Officers and men were serviced by more than 17,000 servants, water carriers, saddlers, blacksmiths, cobblers, tailors, milk girls and musicians. Each officer had up to twenty-five servants to care for his clothes and equipment, pitch his tent and cook his dinner. A groom looked after his horse and a grasscutter found the animal's food. Up to six camels carried his gear. One man looked after every three baggage camels. Many officers found it too hot to ride in the middle of the day and lounged in palanquins carried by six bearers. Possibly they brushed up their Hindustani as they lolled, perhaps learning a phrase I found in a Victorian grammar: 'We took possession of the city in the very first charge.'

Thirty years after the Nepal campaign the gathering of an army had hardly changed. Ordered to march from India to Afghanistan, a British regiment spent twenty days buying animals, and officers made their own haphazard arrangements for freighting baggage. As soon as the regiment moved off, many camel drivers absconded, taking their fees and their animals. As the numbers of camels shrank, baggage was piled on to those remaining, so much so that many could not rise from the ground. A force of 9,500 men travelled from Bengal with 38,000 camp followers and 30,000 camels. Another force took two months to march eighty miles up the Indus valley. The army then stumbled for sixty miles through the Bolan Pass in Baluchistan; and when I travelled through it I found it easy enough to visualize the horses dying by the dozen of thirst and exhaustion, the stinking air black with flies. How fortunate the army was that when it reached Kabul there was no resistance.

The British belief in their command of technology, the resonant theme of the Great Exhibition, was one of the seeds of a storm that simmered and grew into the Indian Mutiny of 1857. Such developments as the electric telegraph, the railways, steam engines, photographic equipment and surveying instruments were to the British the plain evidence that they were more advanced and civilized; an arrogance that was an aspect of the distance that steadily grew between the British and the people they ruled.

For more than 230 years from its establishment in 1600 the East India Company was a trading organization. Its employees in India were mostly bachelors. Few British women made the long and expensive voyage to India. In Calcutta in the 1780s there were 4,000 British men and only 250 women. The Company encouraged its young men to take Indian women as their *bibis*, or mistresses. From the most senior to the junior they did so; and many raised families with them.

Their letters and diaries reveal their admiration of 'supple and slender well-rounded limbs', hair as dark as monsoon clouds in June and 'enchanting breasts, handsome beyond compare'. They wrote in praise of brown skin, so frequently bathed and always sweet-smelling. Sir Richard Burton reflected on 'eyes large and full of fire with long drooping lashes undeniably beautiful'. A doctor described the 'ladylike manner, modesty and gentleness' of the officers' mistresses he saw in Madras. They possessed 'such beautiful small hands and little taper fingers, and ankles neatly turned, as to meet the admiration of the greatest connoisseur'. It was not to be wondered at, he thought, that British men preferred these girls to their own countrywomen. Indeed, mused Samuel Brown, an East India Company official, a man who enjoys the company of playful native women 'shrinks from the idea of encountering the whims or yielding to the furies of an Englishwoman'. One army officer so desired a Muslim girl for his *bibi* that he acceded to her request and submitted to circumcision.

A story went around that sleeping with an Indian woman enabled a man to keep cool in the summer furnace. Richard, Marquess of Wellesley, Governor-General of Bengal from 1798 to 1805 and a considerable philanderer, wrote to his wife in England saying that the climate had so stoked his desires that he needed an Indian mistress. She sanctioned a liaison, provided that he was careful and tender, and warned him that 'the greatest men are often foolish about wicked women. Do not go around begetting a lot of children, that will be ridiculous and horrible.'

Some men in India of course enjoyed their women in multiples. The flamboyant Sir David Ochterlony, the British Resident in Delhi, reputedly led his harem of thirteen concubines, mounted on their individual elephants, on daily outings. But many men were content to make lasting relationships with one woman. The Calcutta lawyer William Hickey enjoyed such an affair with Jemdanee, his vivacious companion who died in 1796 giving birth to their son. She was, wrote the heartbroken Hickey, 'as gentle and affectionately attached a girl as ever a man was blessed with'. And a portrait by John Zoffany of 1785 shows a devoted General William Palmer gazing at his *bibi* and their children.

Such relationships were a respectable commonplace. A *bibi* ran the household, managed the servants, nursed her man through his fevers and, in short, made a home. No doubt some *bibis* were treated with indifference and cruelty, and some had to endure the hostility of their own families, but as far as we can tell most relationships were affectionate. The intimacy of them brought British men nearer to the manners, customs, people and languages of India. They would never again be so genial and so close.

The children of these unions eventually outnumbered the British in India and some became famous. Sir Robert Warburton, one of the outstanding figures on the North West Frontier, who knew the languages and people well, was the son of a British officer who had married a niece of the ruler of Afghanistan. Lord Roberts of Kandahar, Afghanistan campaigner and British commander-in-chief in the Boer War, was the grandson of a princess of a Rajput warrior clan. Colonel James Skinner, who raised the famous and gaudily uniformed cavalry regiment Skinner's Horse, was the son of a Scottish soldier and a Rajput mother, and had a number of *bibis*. So did his brother Robert who overreacted when he suspected one of his wives of straying: he beheaded her and her servants and shot himself. William Watts, who profited mightily in Bengal, had a *bibi* whose grandson became Lord Liverpool. He was the longest-serving Tory Prime Minister, from 1812 to 1827, a record Margaret Thatcher always hoped to beat.

The long afternoon of the sahibs and *bibis* could not last. At the end of the eighteenth century a new intolerance began to set in. There was a steady social cooling towards those who seemed too close to Indian people, customs and religion. The East India Company, which had been interested chiefly in money, was building the structure of a civil service, and its merchants were becoming judges, bureaucrats and administrators. A feeling grew in London that the British in India should not compromise their status as a small ruling class and should maintain a distance from those they governed. They could not do so, it was reasoned, if they were sleeping with them. Too many British and too many mixed-blood children, the argument went, would erode Indian respect for 'the superiority of the European character'.

Charles Cornwallis, the Governor-General in India from 1786 to 1793, excluded the sons of sahibs and *bibis* from higher-grade jobs in the civil service and from commissions in the army. Discrimination of this kind assisted the evolution of a ruling class that saw itself as an aristocracy.

In the interests of commerce and a quiet life the East India Company had excluded Christian missionaries, but in 1813 the ban was lifted after pressure from the evangelical movement and debate in the House of Commons. The evangelist William Wilberforce, campaigner against slavery and moral corruption in general, said he ranked the conversion of India to Christianity as a cause greater than the abolition of slavery. Hindu gods, he told the Commons, were 'monsters of lust, injustice, wickedness and cruelty' and the religion 'one grand abomination'. He hoped that India would exchange 'dark and bloody superstition for Christian light'.

The evangelicals had the tide with them. Missionaries seeped into India and operated as a sex police. What so many British men found amiable and enjoyable the missionaries found sinful. Judging by the carvings and paintings in the temples, Indian women celebrated lovemaking for its own sake and led white men to perdition under the tropic sun. The envoys of the Church understood that common British soldiers could not be expected, whether in India or at home, to control their beastly desires; but officers were surely different. They were gentlemen who should know that surrender to sensuality was a betrayal of their higher Christian civilization.

The military authorities went to great lengths to accommodate the sexual energy of tens of thousands of young soldiers and to deal with the venereal disease that was the scourge of the army. Military brothels and medical inspections of prostitutes went a long way to meeting the problem. On the one hand the army provided Indian women for the soldiers and, on the other, it instructed officers that liaisons with women undermined British prestige.

The steamship's plangent horn seemed to signal the ending of the long affair between British men and Indian women. With the shortening of the passage to India British women began to arrive in larger numbers. They were the wives of administrators and officers or the single girls, sniggeringly known as the 'fishing fleet', hoping to land a husband. The *bibis* all but vanished as the memsahibs drew a line and created a respectable and Victorian way of life in which there was more formality and, perhaps, less laughter. Single men continued to live with Indian mistresses, but no longer in the old easygoing way.

The fading of the *bibis* was part of the process of estrangement. The encroachment of Western ideas on Indian life and traditions caused great unease. Christian missions attacked Indian religions. Many men of the East India Company had once taken a scholarly interest in languages, religions, customs and temples, and had certainly not interfered; now they were encouraged to see these as the unwholesome evidence of Indian ignorance. Thomas Macaulay, the historian who framed an education policy for India in the 1830s, forecast the fading of Hinduism under the advance of English schooling. Many British liberals came to see India as irrational and chaotic and felt that God had chosen them to raise it from its history.

In 1833 the East India Company ceased to be the great trader and became the ruler of India. It undermined Indian princes, reduced their power and self-respect, annexed their lands, swept away jobs in princely courts and extracted heavy taxes from farmers. The assertion of a racial, intellectual and religious superiority spread anxiety and fears of conversion to Christianity. Their trampling evangelism sprang from a conviction that their duty was to enlighten people trapped in darkness. Arrogance grew with their self-righteousness. India and its native rulers could not withdraw into the past and hope to escape the spread of Western ways: this was a clash of old and new.

The more tightly they drew their control the more distant the British became. Their military cantonments were self-contained enclaves built several miles from the cities they found so jumbled and unhygienic. In their hill stations, too, the British kept their distance, exchanging the dissolving heat of the plains for the breezes of Simla and its sisters for seven months from March to October. It was a prodigious effort to haul up everything that was needed, the snooker balls and rubber stamps and whisky. Seven thousand feet up the rulers were isolated and insulated from the people they governed.

Seventy years after the Mutiny, Rabindranath Tagore, the Bengali poet, reflected on the meeting of Britain and India.

> The West comes to us, not with the imaginations and sympathy that create and unite, but with a shock of passion for power and wealth. The East is waiting to be understood by the Western races, in order not only to be able to give what is true in her, but also to be confident of her own mission: 'Never the twain shall meet' – the reason is because the West has not sent out its humanity to meet the man in the East, only its machine.

Into Indian fears and resentments fell the rumour of the cartridges. The East India Company ran three armies, based in its presidencies of Bengal, Madras and Bombay, and in total these mustered 232,000 Indian soldiers and 45,000 British. The Indian Mutiny concerned the Bengal Army. This was badly managed. Many of its officers had come to India to make money and had little interest in the country and its cultures. They had a poor grasp of the languages and contempt for their Indian troops. They swaggered like conquerors and were bored and distant. The troops they neglected were badly paid, housed, clothed and fed, with no way of advancing by merit, and as men of high caste, they resented the absence of respect. A few senior British officers, old India hands, warned that treating Indians as inferiors was dangerous. They went unheard.

In 1857 there was a rumour that the cartridges of the new Enfield rifle were coated with the mixed fat of cows and pigs, an insult to Hindus who held the cow sacred, and to Muslims who regarded the pig as unclean. The end of the paper cartridge had to be bitten off, the powder poured into the barrel and the ball pushed down on top of the powder with a ramrod. Too late, the British insisted that the story of the cartridge grease was not true. Experienced officers recommended that the sepoys should use their own lubricants for the cartridges. But the grease was an issue. Rumour-makers insisted the British were using it to defile the sepoys and attack their religions. The Mutiny exploded. Look, said the British, we bring the gift of progress and this is how they repay us.

19

Ice and fire

'And pluck till time and times are done
The silver apples of the moon,
The golden apples of the sun.'
W.B. Yeats, 'The Song of Wandering Aengus'

As witnesses of the savage and strange, first-footers through snow, swamp and sand, explorers were actors in the empire theatre, the human interest in the narratives of geography. A few were noble, some were misfits, most were utterly driven and many were more than a quarter mad. Their courage was a wonder. Bleeding, fevered, starving and frozen they sought the missing fragments of the world and starred in the stories of their age, their suffering followed avidly in the newspapers.

In the middle of the nineteenth century two particular enigmas of exploration engaged the British imagination: the source of the Nile and the North West Passage through the Arctic.

Geographers had debated the question of the Nile for more than 2,000 years. It was thought that it flowed from fountains in the core of Africa or from lakes fed by the snows of the Mountains of the Moon, mentioned by Ptolemy. The Scottish traveller James Bruce found the spring of the Blue Nile in Ethiopia in 1770. But in the 1850s the birthplace of the White Nile, the greater of the two rivers, into which the Blue Nile flows at Khartoum, was still a secret; and to British minds an intolerable puzzle that demanded a resolution.

So, too, was the North West Passage. After the Napoleonic wars the Arctic and its mysteries were an area of special interest to Royal Navy officers looking for adventure and recognition. In

1845 Sir John Franklin led 128 men in two ships, lavishly equipped and provisioned, into the Arctic maw. He was perfectly confident of cracking the conundrum. In a short time the expedition disappeared in the ice. Instead of the expected message of triumph there was silence. Soon the Arctic seas and islands were criss-crossed by searchers. The fate of Franklin was a riddle and the hunt for any trace of him an obsession spawned by an obsession.

Both the North West Passage and the Nile were considered in Britain to be quests reserved for the British, as if by the appointment of the Almighty. In 1856 the Royal Geographical Society commissioned two Indian Army officers to find the source of the Nile. Richard Burton, aged thirty-six, was tall and brooding and more than physically formidable: a restless scholar who knew thirty or so languages; a daring traveller who penetrated Mecca in disguise; a fencer who wrote the standard volume on swordplay; a writer who would make a fortune and reputation translating erotic oriental works like *A Thousand and One Nights*. John Hanning Speke, aged thirty-one, was blond, blue-eyed, steady, sober and devoted to shooting. The two of them had faced death together in a fight with Somali tribesmen in 1855. Speke was wounded eleven times and Burton, cutting down the attackers with his whirling sabre, reeled as a spear pierced his cheeks and knocked out four of his teeth. They made a desperate escape to the coast.

When they started their Nile journey in June 1857, striking west from Zanzibar, the immense swathe of east and central Africa that became Kenya, Uganda, Tanzania, Zaire and southern Sudan was mostly unexplored and still part of the white man's African darkness. Burton and Speke rode aloft on camels behind a column of porters, the voiceless extras of African explorations. Horses and mules were too vulnerable to the deadly bite of the tsetse fly to be of much use; and men coped better than animals over rough terrain. The column fell into a routine of breakfasting at five, walking a few miles and making camp before noon, when it became too hot to travel. After seven months Burton and Speke reached Ujiji, an Arab slave town on the eastern edge of Lake Tanganyika, more than 700 miles from the coast. They were the first white men to see it, although only hazily through badly inflamed eyes. Burton shook with fever, and when he realized that the lake lay too low to be the Nile's source he felt 'sick at heart'.

He was too ill to accompany Speke who set off to investigate the reports of Arab slavers of another large lake 200 miles north across the plain. In August 1858, after trekking for twenty-five days, Speke came to the paradisaical blue water he named Lake Victoria. There and then, without further exploration, with no evidence to support his decision, he declared this enthralling vision the mother pool of the Nile. He hurried back to tell Burton.

Here began the irredeemable rift. Burton disdained a claim founded only on intuition. Speke seethed. 'He can never be wrong,' he wrote bitterly. Burton knew that Speke might have guessed correctly and that he, Burton, had missed his chance of discovering the fountain of the Nile. They journeyed wretchedly back to the coast, desperately sick and carried much of the way in litters. Burton noted that as they were jerked along the delirious Speke screamed and barked.

They went separately to London in May 1859. In Burton's version of events Speke had promised they would report together to the Royal Geographical Society; and Burton was, after all, the expedition leader. As soon as he landed, however, Speke was a reporter bursting with a scoop and dashed to the RGS to claim the source of the Nile. It was a sensation, just what the Society wanted, a trophy brought to Britain out of Africa.

Burton arrived twelve days later and was almost ignored. Speke was a celebrity, the courageous young explorer, a 'lion, whose roar was to thrill the ladies', as Burton noted. Money was being raised to send him back to Lake Victoria and follow the Nile into Egypt, to validate and crown his triumph. The defeated Burton was a terrible sight. He disputed Speke's claim. His eyes stared out from dark sockets in his emaciated face. He was, in his wife's words, a mere skeleton.

Around this time, on 25 May 1859, a skeleton was found on the west coast of King William Island in the Canadian Arctic. Shreds of dark blue Royal Navy uniform adhered to the bones. Brown hair lodged in a pocket comb. The remains were found by a search party looking for clues to the fate of Sir John Franklin's expedition. The searchers, led by a navy lieutenant, Francis McClintock, later found a beached boat, two skeletons, boots, books, guns, knives, chocolate, tobacco, bedroom slippers, a copy of *The Vicar of Wakefield* and a sixpence. A rust-stained note in a cairn said that Franklin was alive on 28 May 1847 and that two weeks later he was dead. By the time of McClintock's search it was plain that none of Franklin's men had survived. What had killed them no one could say.

We last met John Franklin in Australia, with his uncle by marriage Matthew Flinders. As a chubby fourteen-year-old midshipman Franklin had fought at the Battle of Copenhagen in April 1801 and two months later sailed from Britain aboard the *Investigator*, commanded by Flinders, to circumnavigate and map Australia. On their way back to Britain, Flinders, Franklin and others were shipwrecked on a coral reef 200 miles off Queensland. A small boat was saved and the intrepid Flinders sailed it to Sydney to get help. A ship rescued Franklin and the other castaways and took them to China. Franklin sailed home in an East Indiaman in 1804 and in the following year was signals officer aboard HMS *Bellerophon* at the Battle of Trafalgar. The cannon fire of that afternoon left him permanently half-deaf. He was also wounded during the Battle of New Orleans in 1814, but he matured into a stout and florid man, charming, brave and religious.

From 1819 to 1822 Franklin led an overland exploration of the Arctic coast. He was a Micawber of the snows, his poor management and lack of experience turning an arduous journey into an horrific adventure. He and his nineteen companions lost their canoes and, without food, nibbled lichen from rocks, gnawed their spare boots, crunched bones left by wolves and sucked the rotting flesh of deer. As men died the party divided. One group included Dr John Richardson, Midshipman Hood, John Hepburn, a sailor, and Michel, an Iroquois Indian. Michel brought what he called wolf meat, but its strange taste led Dr Richardson to suspect that it was human flesh and that Michel had murdered three of the group and was feeding on the corpses. Then Michel shot Hood through the head; and a few days later the doctor, sensing that he and Hepburn were next for the larder, pulled out his pistol and killed Michel. Richardson and Hepburn staggered off in search of Franklin and found him slumped in a stinking log cabin with three companions, two of them dying. Franklin, Richardson and the others chewed old deerskins and ate the grubs of warble flies. It was in this desperate state that Indians found them, nursed them and saved their lives.

Franklin sailed home a hero, to marry, fatten and receive a knighthood. He went to the Arctic on a charting mission in 1825–7, governed Tasmania for seven years until 1843, and on his return volunteered for one last go at the old temptress, the North West Passage. Sir John Barrow, the second secretary at the Admiralty and an Arctic enthusiast, said that with one more heave the Passage, 'an object peculiarly British', would become part of Britain's imperial glory. If you let foreigners do it, he warned, 'England would be laughed at by all the world for having hesitated to cross the threshold.'

Plump John Franklin volunteered to lead the expedition. His determined second wife Jane lobbied hard to get him the job. He was fifty-nine years old.

For his walk from Zanzibar to Lake Victoria and northward down the Nile John Speke chose a companion quite different from the saturnine Burton. Captain James Augustus Grant, a modest Scot, an Indian Army officer decorated for gallantry at the relief of Lucknow, deferred to Speke in every way.

They left the coast in October 1860 with a party of more than a hundred soldiers, servants and porters and took thirteen months to walk 800 miles to the west side of Lake Victoria and the kingdom of Karegwe in modern Uganda, the first Europeans to see it. The welcoming king introduced them to his wives whose beauty lay in their sphericity and who were so fattened by a diet of milk that they could barely stand. Speke, in his scientifically Victorian way, ran his tape measure over one of them, recording a height of five feet eight inches, a bust of four feet four inches and a thigh of two feet seven inches.

When Grant fell ill, Speke walked alone to Buganda and stayed there for three months as the guest of Mutesa, the king. On most days the ruler indulged a Caligulan whim and ordered the head lopped off one of his people. He sent gifts of virgins to Speke who tactfully passed them on to the men in his party. Grant rejoined him and Speke sent him east. Speke himself hurried down to Jinja on the northern shore of Lake Victoria, and there, on 21 July 1862, he marvelled at the spectacle of the Nile, 'old Father Nile', as he called it, spilling from the lake and tumbling to the rocks below.

'It was a sight that attracted one for hours,' Speke said happily, 'the roar of the waters, thousands of fish leaping with all their might, the fishermen coming out in boats, hippopotami and crocodiles lying sleepily in the water.' This was the prize. The Nile was his, and his alone.

He rejoined Grant to walk into the kingdom of Bunyoro. As the entry fee the king took Speke's gold chronometer. Skinny and ragged, the explorers toiled down the river and in February 1863, almost two and a half years after leaving Zanzibar, tottered into Gondokoro, 300 miles from Lake Victoria. To their amazement an old shooting friend of Speke bustled forward to shake hands. Samuel Baker had journeyed up the Nile from Egypt with his wife Florence to look for them. Fed, regaled and restored, Speke and Grant travelled downriver to Cairo and reached London in June 1863.

Baker and Florence continued upriver to see if they could corner some Nile glory. The king of Bunyoro took one look at Florence, an attractive Hungarian with startling yellow hair, and nominated her as the entry fee. At this Baker drew the line and his revolver. He told the king he would shoot him if he dared to repeat such an insult to a lady. The doughty Florence launched into an indignant tirade doubtless more frightening than her husband's revolver, and the king gave way, mumbling that there was no need to make a fuss and that, anyway, he would have given Baker a virgin in exchange. In 1864 the Bakers found Lake Albert and decided that it was a contributory source of the Nile. 'It is impossible to describe the triumph of the moment,' Baker exulted. 'England had won the sources of the Nile!' His book of adventures, with all the human interest of the part played by the remarkable Florence, was a best-seller. But the geographers were not at all satisfied that the Bakers had solved the mystery of the Nile.

Sir John Franklin had left London in May 1845, in command of the steam vessels HMS *Erebus* and *Terror*, intending to find the North West Passage and sail on to the Bering Strait and Hawaii. He had enough canned food for three years, plenty of corned beef and soup. Crossing the Atlantic the officers of *Erebus* enjoyed Sir John's benign rule, his large library and meals eaten with silver cutlery. 'You have no idea', wrote one of them, 'how happy we all feel.'

They steered for glory. There was no commercial value in the voyage. Earlier exploration had proved that any North West Passage would lie too far north to provide a practical route for sailing ships. Even for steamships the route would be severe and limited by ice. The main point was to win knowledge and write the name of Britain across the north. Sir Roderick Murchison, President of the Royal Geographical Society, said Franklin's quest was for 'science and the honour of the British name and Navy'. Geographic discovery, said Sir John Barrow, was Britain's gift to the world.

Most of those who sailed believed that their technology and Royal Navy tradition would see them through the ice. They had the best clothing Britain could make, although it did not keep out the cold. Only three officers in the entire expedition, including Franklin, had any experience of the Arctic and the appalling demands it made on bodies and minds, the way that frostbite took the toes and fingers, how snow-blindness came with the dazzling light, how melancholy crept in with the long northern darkness. The frozen sea made prey of ships and crushed them as a snake squeezes the life from its victim, while fear, cold, darkness and sickness drove men to despair and madness.

At the end of July, two months after leaving London, *Erebus* and *Terror* were sighted by whalers in Lancaster Sound. They were never seen again. The absence of news after two years caused no great alarm because ships habitually overwintered. Sir John Ross, for example, looking for the North West Passage fifteen years before, was ice-bound for four winters until his rescue in 1833. But after three years without word from Franklin the first expeditions set out to find him. Everyone knew his ships had food supplies for three years. The Admiralty offered £20,000 to anyone who rescued him, £10,000 for any searcher who found evidence of what had befallen the ships and their crews. Over the years there would be more than forty official and private searches. Lady Franklin badgered the Admiralty to keep up the hunt. To the public, Franklin was news, and the story of the lost hero was constantly refreshed by lengthily reported searches.

In the spring of 1854 John Rae entered the story. A tough and shrewd Orcadian doctor, who worked as a surveyor for the Hudson's Bay Company, he was the outstanding Arctic traveller of his time. In his twenties he learned from Indians and Eskimos how to be an Arctic winternaut. He wore deerskins and furs and thick moccasins and was a skilled snowshoe walker, canoeist, fisherman and hunter. He travelled Eskimo fashion with sledge and dogs, built igloos as he went and curled up in them with his copy of Shakespeare. He was a rare individual. It was common for white men in the Arctic to condescend to or disparage native people, to see them as children. They certainly did not value them as John Rae did.

During a survey expedition in April 1854 Rae met Eskimos near Pelly Bay in the south of the Gulf of Boothia and saw that one of them wore a Royal Navy cap band. The man told him that it came from white men who lay dead some distance away. In further meetings and conversations Rae heard the first sensational account of the end of Franklin's expedition. Eskimos told him that in 1850, families travelling in the south of King William Island met about forty starving white men hauling a boat over the ice. These desperate men begged for food and the Eskimos gave them seal meat and passed the night with them. But the Eskimos lacked the meat to continue feeding both themselves and the large group of ravenous men, and they went on their way. Months later, Eskimos found thirty bodies at a place that became known as Starvation Cove. Some had been cut and cooked. The dwindling band of survivors had fed on their fellows, as a twentieth-century scientific analysis of the bones would confirm. 'It is evident', wrote

Rae, 'that our wretched countrymen had been driven to the last dread alternative.'

The twenty-five people Rae interviewed had not themselves met Franklin's men or seen the bodies; but Rae's cross-examination convinced him of the truth of what they said. The Eskimos brought him telescopes and watches, forks and spoons bearing officers' initials and a silver plate inscribed 'Sir John Franklin KCB'. Rae bought forty-five of these relics, journeyed 1,200 miles to his base on Hudson Bay and reached London in October 1854. Then he went to the Admiralty to show what he had found and to reveal the horrible story.

It appeared in *The Times*, a terrific shock, the worst of news, and Rae found himself detested for being the bearer of it. It was bad enough that Franklin and his men had died in desperate circumstances, but such a fate could have been likened to honourable death on the battlefield. The idea that British officers and gentlemen had supped the brains and eaten the flesh of their own colleagues inspired revulsion. It showed civilized Christian men reduced to the primitive savagery they despised in other races. Lady Franklin, already furious because the Admiralty had officially declared Sir John dead, was appalled by Rae's revelation. It seemed to stain her husband's noble virtue. *The Times* said the Eskimos were liars, 'like all savages', and they had lied to Rae. Charles Dickens, too, scorned the Eskimos as 'uncivilized blood and blubber' people and suggested they had murdered Franklin's men, since 'we believe every savage to be in his heart covetous, treacherous and cruel'. Rae defended the Eskimos. They did not harm Arctic travellers, he said. They were a 'bright example to civilized people'.

Despite Jane Franklin's opposition, the Admiralty eventually gave Rae the £10,000 reward offered for evidence of her husband's fate. Determined to discover more, she paid for expeditions to continue searching. Lieutenant McClintock's, in 1859, was the last and it found the skeletons on King William Island as well as the note left in a cairn. Written by Captain Francis Crozier, Franklin's second-in-command, the message said that *Erebus* and *Terror* were trapped in the ice of Victoria Strait in September 1846 and abandoned the following April. A number of men walked thirty miles across the ice to King William Island. Another note, dated April 1848, recorded the death of Franklin, on 11 June 1847, and of eight more officers and fifteen men.

Crozier indicated in his note that he would march south. It was presumably his party of forty who had begged food from the Eskimos, as described by John Rae. The bones of some of Franklin's men, scientifically examined 150 years later, showed not only evidence of cannibalism but also high levels of lead, no doubt ingested from the tinned food off which they had lived.

The original Eskimo discoverers of the bodies found guns and plentiful ammunition, yet the party had been unable to hunt enough food and had starved. Rae, a critic of heavyweight and expensive naval expeditions, knew what was wrong. There were too many in the party. The harsh land would not support so large a group; and hunting demanded great skill and experience. That was why Eskimos travelled in small numbers, dressed in furs and using dogs. The British wore unsuitable clothing and leather boots, disdained the dog teams they thought undignified in gentlemen's expeditions and relied on tinned meat.

For all the horrific lessons of the Franklin disaster, traditional British attitudes endured. Clements Markham, an officer in one of the ships that searched for Franklin in 1850–1, later became President of the Royal Geographical Society, a passionate advocate of polar exploration, a promoter of the heroic school of discovery and mentor of Captain Robert Scott, leader of British Antarctic expeditions. Markham disliked dog teams and believed that there was a purity in men dragging their own sledges. He disapproved of travellers like John Rae who advocated Eskimo techniques. An obituary of Rae, who died in 1893, aged seventy-nine, noted that he was never shy with his opinions. 'Putting forward his views with point and insistence, as he did, his remarks were, as a rule, somewhat unwelcome to the naval authorities.'

The question of the Nile was not resolved by John Speke's treks to Lake Victoria. Richard Burton, for one, was sure that the source lay much farther south. The two fought it out in their books. Burton dismissed Speke as 'unfit for any other than a subordinate capacity'. Some Nileographers thought Speke brash and unqualified. The geographer James McQueen sneered that Speke had been too busy dallying with African girls to do a proper exploring job and excoriated him for giving the name of Victoria to a lake in a 'barbarous and degraded' country. A gladiatorial showdown was arranged for September 1864, with Speke and Burton engaged to take the stage at the British Association conference in Bath and argue their case. 'Silly tongues', said Burton, 'called it the Nile Duel.' The day before the debate the two met for the first time since they had left Africa in 1859. They glanced briefly at each other but said not a word.

Next day, when Burton took the stage, his notes in hand, Speke was not there. After his encounter with Burton the day before, he had gone out shooting. Around four o'clock, as he climbed a stone wall, his gun discharged and mortally wounded him. Perhaps the trigger had snagged on a branch. A coroner's jury returned a verdict of accidental death.

Ever theatrical and dark, Burton wrote that 'I saw him at 1.30 pm and at 4 pm he was dead. The charitable say that he shot himself, the uncharitable that I shot him.'

So the Nile dispute lingered. In 1865 the Royal Geographical Society called on the great David Livingstone to be the judge, to settle once and for all 'a question of intense geographical interest'. Livingstone was fifty-two, the best-known explorer in the world, a legendary figure. Since his arrival in Cape Town in 1840 at the age of twenty-seven he had trekked more than 30,000 miles on foot and on the backs of oxen and donkeys, and had opened the interior of Africa to European knowledge. He was the first white man to cross the continent, to follow the Zambezi River to its source and to see Victoria Falls and Lake Nyasa.

Livingstone's heart was full of Africa. As a committed wanderer he preferred to live and walk with Africans rather than whites, although he believed in the white man's superiority. The core of his life was his struggle against the Arab-run slave trade and his determination to plant in Africa the civilizing benefits of commerce and Christianity, the material and the moral.

He agreed to resolve the question of the Nile. As he saw it, the river could be a highway bringing goods and God's word into the continent, a way of driving out the slavers. He sided with Burton and thought that the river began somewhere south of Lake Tanganyika. In April 1866 he marched along the Rovuma River, the border of modern Tanzania and Mozambique, heading for the country south of Lake Tanganyika, now Zambia. Along the way his porters deserted and he lost his medicine chest containing his vital supply of quinine. 'I felt as if I had received the sentence of death,' he said. He continued his journey with four African companions.

Frequently ill as he wandered, he arrived worn out in the slaving town of Ujiji, three years after leaving the coast. He rested for half a year and travelled westward looking for the Lualaba River which he thought was the Nile. He could not know that it was part of the Congo River. When he reached it he was the appalled witness of an Arab massacre of hundreds of villagers. No one would help him to travel down the river to verify its course, and when he stumbled back to Ujiji, sick, defeated and with most of his teeth gone, he needed the charity of the Arabs to survive.

The world had heard no news of Livingstone for three years when, in 1869, James Gordon Bennett, of the *New York Herald*, summoned the reporter Henry Morton Stanley to his hotel room in Paris. He sent him on a great assignment, to cover the opening of the Suez Canal, to report from Jerusalem, Constantinople, the Crimea, Persia and India, and then to find Livingstone. Bennett had a powerful instinct for a newspaper story; and the thirty-year-old Stanley had ambition, confidence and unlimited supplies of his employer's money.

After fulfilling the first part of his boss's orders he went to Zanzibar, organized an expedition of 190 porters and soldiers, left the coast in March 1871 and headed west. As he went he asked for news of the lost white man. Eight months later, on 10 November, under the Stars and Stripes, he marched grandly into Ujiji, into the setting of the famous scene. He felt like embracing the pale, frail, whiskered man who shuffled out to meet him. Instead, he was properly understated and raised his hat; and the man lifted his faded blue cap and agreed, just as Stanley had presumed, that, yes, he was Dr Livingstone. Stanley gave him all the supplies he needed as well as medicine, letters and newspapers. He poured champagne to mark the moment; and savoured his scoop.

The usual story of Stanley's background is that he was born John Rowlands in Denbigh in 1841, the illegitimate son of a farmworker of the same name, and of Elizabeth Parry, a bakery girl. These are the names in the baptismal record. A persistent local version says that the father was a Denbigh solicitor, James Vaughan Horne, who avoided scandal by paying John Rowlands to admit paternity. Horne's memorial is in Llanrhaeadr church, near Denbigh. Stanley wrote a colourful account of his own early life and said he escaped from the St Asaph workhouse after thrashing the cruel master, which was not true. He sailed to New Orleans, where he was adopted by a generous American businessman Henry Stanley, who became his guardian, gave him his name and relaunched him in life. Stanley fought for the South, then the North, in the Civil War, and became a roving reporter for the *New York Herald*.

Stanley and Livingstone became friends. They talked and travelled for three months, circumnavigated Lake Tanganyika and walked 300 miles east to Tabora. Livingstone had no wish to return home. He was a widower, committed to Africa and certain he could find the spring of the Nile. In March 1872, carrying Livingstone's own journals, Stanley left for the coast to write the sensational story that made his name.

With his African companions, Susi, Chuma and Jacob Wainwright, Livingstone wandered 500 miles south, beyond Lake Bangweulu in modern Zambia, searching for the Lualaba. Ravaged by dysentery, he grew sick and weak. On the last day of April 1873 he asked Susi how far it was to the Lualaba.

'I think it is three days, bwana.'

'Oh, dear, dear,' Livingstone murmured.

Next morning his companions came upon him kneeling by his bed, as if in prayer, and found no sign of life.

They removed his heart and intestines and buried them beneath a tree, then salted the corpse and dried it in the sun for two weeks. They wrapped it in bark, tied it to a pole and carried it a thousand miles to the coast, a nine-month journey. Jacob Wainwright faithfully accompanied the body on the voyage from

Zanzibar to Southampton. With Stanley he was a pallbearer at the funeral in Westminster Abbey in April 1874. A surgeon had checked the corpse to be sure of its identity. The shoulder injury caused by the bite of a lion in 1843 was plain to see. Livingstone had written of the attack that the growling lion shook him and crunched the bone to splinters, inducing a state of dreaminess. 'The shake annihilated fear and allowed no sense of horror . . . this peculiar state is probably produced in all animals killed by the carnivora; a merciful provision by our benevolent Creator for lessening the pain of death.'

The Nile was still a mystery. There was the possibility that the Lualaba was the source. Stanley resolved at Livingstone's funeral to return to Africa and answer all the questions, and vowed to finish Livingstone's work and open Africa to 'the shining light of Christianity'. The *New York Herald* and *The Daily Telegraph* put up the money, and Stanley organized an expedition on a grand scale. In November 1874 he started from Zanzibar with three British companions, 350 porters, eight tons of provisions and a forty-foot boat, carried in six sections, called *Lady Alice* after his American fiancée. He himself, with his sun helmet and brown boots, was the picture of the African explorer. No doubt he brooded as he marched. His enemies in journalism and the geographical establishment detested him for his ambition, his lack of a gentleman's upbringing, his Americanness and his success. Some had denounced his finding of Livingstone as a newspaper stunt without a shred of truth.

Stanley sailed his boat around the perimeter of Lake Victoria in two months and showed that it was a single body of water decanting into the Nile. In other words Speke's guess had been right and Burton wrong. Similarly he circumnavigated Lake Tanganyika and found no outflowing stream that could have been the Nile. There now remained the question of the Lualaba: where did it go? Stanley pushed west to the centre of the continent, to modern Zaire. In November 1876 he began the most dramatic of his African adventures, following the Lualaba, determined to find where it went.

He hacked through rain forest so dense he could not read his notes in daylight, then launched the boat and some canoes into the broad river. The journey was always dangerous and frequently terrifying, a chronicle of sickness, hunger, the loss of supplies and many ambushes and fights with riverside tribes. Stanley fought thirty battles along the way and lost forty of his men to spears and arrows; and he was always ready to shoot his way out of trouble. Wherever he travelled he heard the blood-chilling sound of war drums. Three months after setting out he knew that the Lualaba was the Congo, not the Nile; but there was still a long way to go to the Congo's estuary on the Atlantic. A number of men, including Stanley's last British companion, were drowned in the rapids. The party shouldered the boat and canoes and carried them through the snake-infested jungle around huge falls. In August 1877, close to the Congo estuary, Stanley sent messengers ahead with an appeal for help. He was exhausted and starving. Several white traders came up from Boma with food, tobacco, beer and champagne, and carried Stanley to safety in a hammock. Only a third of his original party of more than 350 were still with him.

Nine hundred and ninety-nine days had passed since his departure from Zanzibar. But he had come back with valuable information about the heart of Africa and its lakes and rivers. The Lualaba was not the Nile but the Congo curling through the jungle to the Atlantic. For all practical purposes the source of the Nile was, as John Speke had discovered, in Lake Victoria, although the river's hydrography is complex and its origins lie in the highlands north of Lake Tanganyika. Stanley had completed the last of the heroic Victorian expeditions in Africa. He would record it all in his book *Through the Dark Continent*. Meanwhile, the long ordeal on the river had turned his hair white.

As a road to riches the North West Passage turned out to be a phantom. In the age of the sailing ship no way was found through the maze. A statue of Sir John Franklin in Waterloo Place in London salutes 'the great navigator and his brave companions who sacrificed their lives in completing the discovery of the North West Passage'. Brave, yes, but their lives were squandered and Sir

John never found the Passage. The unexpected bonus of his failed quest was the knowledge of complex Arctic geography gathered by the many men who went to look for him. The first continuous navigation of the North West Passage was made by the Norwegian Roald Amundsen in a voyage from 1903 to 1906. He went by way of the Rae Strait, the only channel free of pack ice, discovered and charted by the remarkable Dr Rae.

Amundsen's interest in polar exploration was inspired by his boyhood reading of John Franklin's account of his desperate Arctic adventure in 1819–22. So it was that as he found his way through the North West Passage, Amundsen made reality of the dream conceived when he read the story of the young Franklin who had sailed with Flinders who had sailed with Bligh who had sailed the great oceans with Captain Cook.

Between 1879 and 1884 the ruthless and energetic Henry Morton Stanley enhanced his reputation as a giant among African explorers. In the personal service of King Leopold of the Belgians he created the immense Congo colony in the heart of the continent and founded the towns of Leopoldville and Stanleyville. He made his last African expedition in 1887–9. He died in 1904 and his autobiography was published five years later. In his examination of it, *Sir Henry Stanley, The Enigma* (1989), Dr Emyr Wyn Jones found that as well as writing a highly coloured account of his boyhood, Stanley described himself as an American naval officer, although he was a clerk who deserted the service. He also wrote that his 'father' died within weeks of his birth, although he died thirteen years later. He claimed that his adoptive mother died in 1860, and his adoptive father in 1861, although they both died in 1878.

20

The ghost of the swamp

'The only time off was when we had a go of fever.'
British colonial officer, West Africa

Disease was bad enough at home, the doctors groping in their ignorance and helplessness. In the stifling air of the tropics the fevers seemed more horrible, terrifying and prolific, wiping out families, scything the soldiers, decimating the sweltering administrators. Every year in West Africa, into the twentieth century, malaria and other fevers killed one in twelve among the British colonial officials; and many more went home in ruined health. Across Africa and Asia the faded tombstones of countless young men, women and infants swallowed up by fever evoke the ache of loss and grief far from home.

To Europeans tropical diseases raised a deadly barricade in the way of civilization, the Christian message and profit. They were the 'gigantic ally of Barbarism'. In particular they kept the white man out of the heart of Africa and thus kept it 'dark'. Sir Patrick Manson, the Victorian physician who was one of the pioneers of tropical medicine, likened diseases to the monstrous three-headed hound that stood sentry at the gates of hell. 'The Cerberus that guards the African Continent,' he wrote, 'its secrets, its mystery and its treasure, is disease.'

In one of its most malignant guises Cerberus was a female mosquito of the *Anopheles* species. The male was innocent. The female was guided to her human target by exhaled carbon dioxide and perspiration. Alighting on the skin she sank her proboscis into it, prospecting for blood. She drew off a quantity three times her own weight and processed it to nourish her eggs which she laid in water. If this blood were infected with malaria she transmitted the disease to the next person she bit. Through her proboscis she injected saliva containing an anticoagulant to make human blood run freely, the easier to drink it. Her saliva also contained the malarial parasites of the former victim; and so the disease was transmitted. Within a week or two of being bitten the victim began to feel a frightening chill, then waves of uncontrollable tremors followed by profuse sweating and a high fever reaching 106 degrees Fahrenheit and, before long, delirium and coma.

Malaria was widespread in ancient Egypt, China, India, Greece and Mesopotamia. Alexander the Great probably died of it in 323 BC. It was native to Italy for twenty centuries and was saluted as a defender of Rome for the virulence with which it felled invaders. It was mightier than the sword. It existed in Britain from Anglo-Saxon times and was perhaps the cause of the death in 1658 of Oliver Cromwell who had grown up in the marshlands of East Anglia. King Charles II suffered from it, too. Samuel Pepys wrote about it in his diaries.

In places where it was endemic and populations were relatively small most inhabitants acquired immunity. It lived in their blood but the symptoms were not usually serious. It struck most significantly at intruders who lacked immunity. Malaria was the mysterious force that devastated invading armies. Hordes of Vandals and Huns were laid low by it when they swept into Italy. Soldiers who survived their encounters with the disease became malarial hosts and took it with them in their blood, just as the Mongol horsemen of Genghis Khan and his successors, riding rapidly through Asia, carried plague fleas in their saddlebags.

When Europeans landed in the ports of West Africa to develop the slave trade they had none of the local population's immunity from malaria and they died in such large numbers that the coast earned its frightening reputation as the 'white man's grave'.

'Beware and take care of the Bight of Benin,' went the sardonic rhyme, 'for one that comes out, there's forty goes in.'

Those who recovered from initial attacks of malaria retained the parasites that caused the typical recurrent bouts of sweats and shivers, hence the description of malaria as 'intermittent fever'. The fevered explorer, prostrate on a litter and racked by violent tremors, was a familiar figure in the accounts of journeys on the tropical frontiers. At the age of seventeen, Nelson was almost lost to the Royal Navy when malaria struck him during his service in India. With scant hope for his survival, his shipmates put him aboard a ship bound for England. He recovered; but feverish shakes dogged him for years just as they did Kipling's British soldier who complained that 'the blasted English drizzle wakes the fever in my bones'.

A tombstone on the North West Frontier of Pakistan honours an officer who died 'gloriously charging the enemy, sword in hand'. But for every man killed in battle scores or hundreds more died of disease, far from the sound of guns. In the seventeen years after 1819 half the British garrison in Sierra Leone were killed by fevers. In India, malaria and other diseases carried off thousands of British soldiers; and, it should be remembered, wiped out millions of Indian children as well. Sergeant John Pearman, who served for eight years in India, recorded that in 1845 two or three men died every day in his barracks of 2,000 soldiers. During a pause on a long march, 'men fell down on parade as fast as they could be carried to the hospital tent. It was a jungle fever that attacked us.' He shrewdly observed that apart from cholera and fever another enemy was the soldiers' prodigious thirst for rum; and that by his observation 'the drink did more for death than the fever'.

No one knew what caused malaria. Some of those who pondered how it spread made the connection between the disease and the stagnant water of marshes and called it swamp fever. In 1827 it was given the name malaria, meaning bad air, because it was thought to be a contagion spread by vapours from the earth or from corpses or sewers, a miasmic fever associated with dirt.

Malaria was distressing enough but the disease caused by another African Cerberus, the *Aedes aegypti* mosquito, was even more terrifying. As they fed on human blood these mosquitoes transmitted the virus of yellow fever, known to sailors and soldiers as yellow jack, a terrible progression of fever, severe pain in the eyes, haemorrhage, liver failure and a consequent yellowing of the skin, and finally the black vomit that signalled the imminence of death. It killed more swiftly than malaria. Africans encountered by the slave-trading Europeans were carriers of yellow fever but were largely or partly immune to its effects.

Slaving vessels took both malaria and yellow fever across the Atlantic to the Americas where they had not been known before. During the voyage from Africa the mosquitoes sucked blood from sailors and the chained slaves, laid eggs on the lids of water barrels and steadily spread disease. The journey was long enough for there to be two or three breeding cycles and the seamen quickly dropped and died. No doubt some of the tales of ghost ships have their origins in true stories of ships swept by yellow fever. Sometimes so many slave-ship seamen fell sick and perished that surviving crew unshackled a number of slaves and ordered them to the deck to help sail the ship.

Sailors who reached the West Indies and America, apparently in good health, were often carrying the diseases in their blood. Relieved to be ashore and anxious to get away from the hell of the ship, they hurried inland and took the fevers with them. The mosquitoes, meanwhile, left the ships to establish themselves and their parasites in the hospitable forests of the New World.

Until the arrival of the Spanish conquerors in the early sixteenth century the American continents were effectively insulated from the diseases of Europe, the East and Africa. The Bering Strait land bridge between Asia and the Americas was inundated 10,000 years ago so that all of the American land mass was protected by the sea. Unlike the population of the Old World, the original peoples of the Americas, north and south, had no experience of diseases like smallpox, cholera, typhoid, typhus, scarlet fever, rubella, measles, mumps and diphtheria. They had no resistance and their healers, of course, were helpless. Within a few years the diseases brought across the Atlantic by the Spanish invaders caused the worst demographic catastrophe in history.

The conquistadors presented an astonishing spectacle to the inhabitants of the New World. Raised up on horses, creatures never seen in the Americas, the armoured, helmeted and black-bearded Spaniards carried strange and potent weapons, their swords and particularly their guns, and they used these with much enthusiasm to slaughter people by the thousand. Yet far more deadly than their gunfire was the smallpox virus in their exhaled breath. It raced apocalyptically across the countryside and overwhelmed peoples like the Aztecs and the Incas. In many regions it killed half of the native population and in some areas destroyed nine out of ten. In Mexico 25 million people were reduced to a million. In the decades after the Spanish invasion more than 50 million native people perished and their civilizations collapsed. 'The Indians die so easily', observed a missionary in 1699, 'that the bare look and smell of a Spaniard causes them to give up the ghost.'

The appalling destruction of so many indigenous peoples and the rise of sugar cultivation in the Caribbean created the early demand for plantation labour and the consequent shipping of slaves. Yellow fever arrived in Barbados in 1647 and killed around 6,000 white settlers. An epidemic more than forty years later took thousands more. Europeans fled the island to America but could not escape the disease. The survivors of yellow fever acquired immunity from further attack, but sailors, soldiers, merchants and others who had had no previous contact were highly vulnerable, perfect victims, so that the British and French armies shipped out to the Caribbean succumbed to it in their thousands. The large French force that sought to regain Haiti after the black rebellion of 1802 slaughtered thousands of people but was itself smashed by yellow fever and defeated.

Yellow fever moved around the islands, a terrifying visitor, as if seeking fresh blood. It seeped into the ports of Brazil and the United States. In 1793 ships from the West Indies arrived in Philadelphia and brought a yellow fever epidemic in which 5,500 people died, a tenth of the population. No doctor could say how or why it spread so rapidly. Some people blamed dirt and squalor. Some believed bonfire smoke would keep it away. Some thought quarantine was the answer. Although a few people began to see that the disease was not spread by contact with a victim, many panicked, abandoned their sick relatives and fled to the countryside. During the nineteenth century yellow fever moved up the Mississippi and tore through cities like a fire in a straw house, bringing death to thousands in New Orleans and Mobile. Five thousand people died in Memphis in 1878, one in seven of the population.

The fear of yellow fever kept many European immigrants away

from the southern part of the United States. They chose to settle in the healthier North or to head into the West. The presence of the disease affected investment in the South and also intensified racial prejudice. Whites blamed blacks for spreading the disease, tried to prevent them moving over the countryside, and argued that since slaves were immune it was plain that the Almighty had fitted them for lives of servitude.

Borne in the blood of westbound pioneers in wagons and river steamers, malaria raged from the Gulf of Mexico to Canada and travelled into the South and the Midwest following the line of settlement up the Mississippi. It was the chief cause of death and illness along the river. Migrants called it ague or bilious fever. Mark Twain, who grew up in Hannibal beside the Mississippi, remarked that the severe shaking associated with malaria enabled people 'to take exercise without exertion'. Charles Dickens, who also saw malaria's effects along the river, described Cairo, Illinois, as 'a hotbed of disease, a breeding place of fever . . . a grave uncheered by any gleam of promise'. In southern California today eucalyptus trees endure in profusion as a reminder of the years in which malaria swept through the state. People believed that the disease could be contained by the planting of what was known as the 'Australian fever tree'.

Malaria, smallpox, cholera, measles, chickenpox and diphtheria were as destructive and terrifying to the Indian tribes of North America as the plagues that attacked medieval Europe. They commonly reduced Indian peoples to a half or a quarter of their previous strength and rendered some tribes extinct. In 1763, Sir Jeffrey Amherst, the British military commander fighting Pontiac's rebellion in Pennsylvania, wondered if 'it could be contrived to send the smallpox among those disaffected tribes', and waged germ warfare against them by ordering his men to supply the Indians with blankets that had been used by smallpox victims. In the 1830s a smallpox epidemic killed two-thirds of the Blackfoot people, half of the Assiniboine and most of the Mandan. Cholera obliterated half of the Cheyenne in 1849.

Settled tribes were always the most vulnerable. The nomadic way of life of the Sioux, the hunters of the Great Plains of America, gave them a greater chance of escaping the worst effects of smallpox;

and one Sioux tribe survived the epidemic of 1837 by agreeing to be vaccinated by missionaries.

From the middle of the seventeenth century, long before they knew how malaria spread, Europeans were aware of a drug that mitigated the fevers. Quinine was made from the bark of the cinchona tree that flourished between 4,000 and 9,000 feet on the eastern slopes of the Andes in Peru, Bolivia, Ecuador, Venezuela and Colombia. The story of its discovery relates that in 1638 the wife of the Spanish Viceroy of Peru was saved from dying of fever by a potion made from stewed cinchona bark. Indians called the bark *quinquina.* Because it was introduced into Europe by Jesuit doctors it was known as Jesuit bark, a good enough reason for stout Protestants, like Oliver Cromwell, to reject its use.

King Charles II, however, was treated for malaria by an apothecary who used the bark as a secret ingredient in his medicine and did not disclose it to the King. Restored to health, the King spoke highly of the cure, and its reputation spread. During the eighteenth century the trade in the bark grew, but doctors began to find that the efficacy of their medicines depended on their securing the right kind of bark. The quality was unreliable and, inevitably, those who supplied the doctors often adulterated it. Cinchona had to be transported hundreds of miles from the Andes slopes to the sea and it was expensive. Nevertheless the market grew. James Lind, the naval surgeon, recommended in 1788 that an infusion of the bark should be taken every morning by 'persons subject to agues'. Nelson ordered that men landing from warships and entering marshy areas should be protected by a drink of Peruvian cinchona bark soaked in wine. Thousands of British soldiers attacking Napoleon's forces at Walcheren in Flanders in 1809 were routed when the enemy breached the dykes and surrounded them with water infested by mosquitoes. There was not enough of the magic bark to treat the sick troops and 4,000 of them died of fever.

In 1820 French researchers identified quinine as the active agent in cinchona bark. It became clear to the East India Company that it needed a cheap source of quinine if it were to fight a highly dangerous enemy and expand and manage its territories in India and elsewhere in the East. Sir Joseph Banks suggested that the seeds of Andean cinchona trees could be brought to Kew Gardens for trials and propagation and then transplanted in the hills of India. It was not until the 1850s, however, thirty years after his death, that British botanists made their first exacting and dangerous journeys into the South American forests to search for the elusive bark and the vital seeds.

Still no one knew what caused malaria. For much of the nineteenth century the belief persisted among medical men that it somehow arose from swamp vapours; and there were always those stern observers who saw it as divine punishment for corrupt ways of living, for laziness, drunkenness and immorality.

21

Alarms in distant places

'The Atlantic is dried up.'
The Times, 1858

In the first decades of the nineteenth century the public knew that electricity was an amusing phenomenon. It generated sparks and transmitted a shock through the linked hands of a line of people. Experiments in the 1740s had shown that current passed along a wire. A Scottish doctor writing in 1753 envisaged an electric telegraph. A Spanish inventor, Francisco Salva, built one in 1795. Francis Ronalds demonstrated an electric wire device in London in 1816 and the Admiralty rejected it because it was no improvement on the mechanical hilltop-to-hilltop telegraph. In 1824 the *Encyclopaedia Britannica* published a discouraging review of the latest research. 'It has been supposed that electricity might be the means of conveying intelligence,' it said, 'but, ingenious as the experiments are, they are not likely ever to become practically useful.'

Thinkers and tinkerers pressed on. The mathematical genius Karl Gauss constructed an electric telegraph in Germany in 1833. Two years later in Bonn the Russian Pavel Schilling transmitted a message over three miles of wire. William Cooke, an Englishman with no deep knowledge of electricity, saw Schilling's contrivance and thought it had commercial potential. He built a telegraph with a needle pointing to letters of the alphabet. Then he joined forces with the physicist and musician Charles Wheatstone, professor of experimental philosophy at King's College, London, and they took out a patent in 1837 on an instrument for 'giving signals and sounding alarms in distant places'. Cooke was the businessman and Wheatstone the boffin who, among other things, invented the concertina and a sound amplifier he called a microphone.

Isambard Kingdom Brunel encouraged them to install a telegraph alongside his Great Western Railway from London to Slough. It was already clear that railways would not work efficiently without a means of rapid communication to control train movements, although the early telegraph, transmitting at four or five words a minute, was hardly fast. Nevertheless, the railways and the telegraph grew together and the railway companies financed the telegraph's early development.

It was Cooke and Wheatstone's fragile wire that hanged a killer. In January 1845 John Tawell poisoned Mrs Sarah Hart with cyanide at her home in Slough and was seen boarding the Great Western train at Slough for London. A telegraph message to London warned that 'the suspected murderer in the garb of a Quaker, with a brown greatcoat which reaches nearly down to his feet, is in the last compartment of the second first-class carriage'. When he arrived at Paddington station Tawell was recognized from the description and arrested; and in March he was hanged at Aylesbury. The story was a newspaper sensation and opened minds to the possibilities of the telegraph. There would be an echo of Tawell's arrest sixty-five years later when the poisoner Dr Harvey Hawley Crippen, crossing the Atlantic with his mistress, was identified in a message transmitted by the infant wireless technology. Arrested aboard ship, he was in due course tried and executed.

Samuel Morse, a portrait painter and professor of art at New York University, heard of the electric telegraph while touring Europe. On his return he built a transmitter, demonstrated it in 1838 and improved it in the early 1840s with the help of Joseph Henry, a pioneer in electromagnetics. Morse also expanded existing telegraphic codes to produce his eponymous system of dots and dashes that became the world's chief telegraph medium, almost a new way of speaking. With a $30,000 grant from the United States Congress he built a fifty-mile line from Baltimore to Washington and in 1844 tapped his first telegraph message: 'What hath God wrought!'

Morse himself wrought a company and made his fortune as the telegraph network multiplied. By 1848 most of the United

States east of the Mississippi had connections. The problem of fading signal strength in long wires was resolved by Joseph Henry who built a relay to boost the impulses and made long-distance transmissions possible. In the 1850s the first lines linked San Francisco to Sacramento and other towns in California. By 1852, 23,000 miles of telegraph were singing across the land. Congress had no interest in public ownership of telegraphs; in France, by contrast, the government ruled in 1847 that the telegraph should be 'an instrument of politics and not of commerce'. In Britain the telegraph became state-owned under the Post Office in 1869.

The challenge for the British, an island people anxious to do business with the world, was to lay telegraph cables under the sea. Charles Wheatstone experimented with a cable waterproofed with tarred hemp, but it was unsuccessful. Wherever there was electricity, however, there was the inquiring genius of the physicist Michael Faraday. He recommended that the copper core of a cable should be coated with gutta-percha. First seen in London in 1843, the grey plastic-like sap of the Malayan gutta tree was similar to rubber, and manufacturers used it to make shoe soles and, eventually, golf balls. In 1850 the brothers John and Jacob Brett ordered a cable coated with gutta-percha, loaded it aboard a steamship and chugged for twenty-two miles across the English Channel to Calais, paying out the cable from a reel as they went and sinking it with weights. It worked briefly until a French fisherman hauled it up with his anchor and hacked it in two.

This set the scene for the development of undersea cables, a long story of failures, disappointments and dogged persistence. The Brett brothers laid a second Channel cable in November 1851, the world's first commercial submarine telegraph. It was welcomed as a link that would cement peace between two countries that had been enemies for centuries. A newspaper cartoon showed Britain and France as 'the new Siamese twins' linked by an electric wire which enabled the closing prices on the Paris stock exchange to be known in London the same day. Engineers laid an Irish Sea cable in 1853 to connect Dublin to London, Manchester and Liverpool. In that year the invention in Austria of a duplex system enabling messages to travel along a single wire in both directions simultaneously ended the laborious process of transmission in one direction at a time. In 1854 John Brett's Mediterranean Telegraph Company laid a cable from Genoa to Corsica, but it failed to make a link between Sardinia and Algeria because the sea was too deep. Nevertheless, during the Crimean War, British and French engineers laid the longest submarine cable in the world, across the Black Sea, to set up the first telegraph used in wartime.

The international telegraph transformed newspaper reporting. The German-born Israel Ben Josaphat Beer, who changed his name to Paul Julius Reuter, settled in London in 1851 and founded the first telegraphic news agency. The rapid dissemination of news was amazing. Tsar Nicholas I of Russia died at one o'clock on 2 March 1855 and his death was announced that same afternoon to the astonished British Parliament. *The Times* was sceptical of Reuter's service until he obtained a scoop, an advance copy of a momentous speech by Napoleon III in 1859, warning of war against Austria. It was telegraphed from Paris to London as the Emperor was delivering it, and as soon as it was translated Reuter gave it to *The Times* which published at once in a special midday edition, only an hour after the Emperor had spoken. That same year Reuter did a deal with two news agencies, in France and Germany, and the three divided the globe between them. Reuter's agency developed an international network of correspondents and emerged as the chief international news organization of the world.

Until the 1850s, communications in India were maintained by postal couriers and by two semaphore links in Calcutta, the British Indian capital. The semaphore masts were thirty feet high and erected twenty miles apart. In 1839 William Brooke O'Shaughnessy, a twenty-nine-year-old doctor in Calcutta, experimented with electromagnetism and sent a signal over twenty miles of wire. In 1851 the East India Company commissioned him to build India's first electric telegraph, a thirty-mile link from Calcutta to Diamond Harbour on the Bay of Bengal.

Lord Dalhousie, Governor-General of India from 1848 to 1856, called the telegraph 'an engine of power' and supported it as enthusiastically as he backed a range of grand civil engineering projects. He had a conqueror's mind and a robust sense of the British Raj's moral mission, and he was an energetic modernizer of India. He ordered the construction of canals and the completion of the Grand Trunk Road across northern India from Calcutta to Peshawar. In 1853 he began the railway age in India with a line from Bombay to Thana, the foundation of one of the world's largest rail networks.

As in Britain, the telegraph in India grew with the railways, but the web was cast much more widely. The pioneering Dr O'Shaughnessy built the first long-distance telegraph, 800 miles from Calcutta to Agra. By 1855 more than 4,500 miles of wire ran from Calcutta to Delhi, Bombay, Bangalore and Madras, and north-west to Lahore, Rawalpindi and Peshawar, the very edge of Afghanistan. A line ran north from Rangoon to Meeday in Burma. Construction parties in Assam armed themselves against tribesmen and tigers. Almost everywhere white ants ate the poles, porcupines and bandicoots undermined them, monsoon floods toppled them, villagers stole them for firewood and nimble monkeys crocheted the wires.

The oldest original telegrams surviving in India date from the beginning of the Indian Mutiny in May 1857. A telegram from Delhi to Ambala in Punjab reported: 'Cantonment in a state of siege. Mutineers from Meerut ... Several officers killed and wounded. Troops sent down but nothing certain yet.' Later there came another message from Delhi: 'We must leave office all the bungalows are being burnt down by the sepoys of Meerut. They came in this morning and C. Todd [Charles Todd, a telegraph officer] is dead we think. He went out this morning and has not returned yet. We heard that nine Europeans were killed. Good-bye.'

These messages alerted military commanders in Punjab who prudently disarmed their Indian troops to stop the fire spreading. 'The electric telegraph has saved us,' said a Punjab official. It certainly helped the British to keep control in the pivotal region of Punjab. Although mutineers destroyed some of the telegraph lines, much of the network survived and telegraphists followed the advancing British troops.

What they could not do, however, was to telegraph to London. There was no submarine cable. Dispatches were sent from Bombay by ship to Suez, then overland across the desert to Alexandria and thence by ship to Trieste, the nearest telegraph station. News of the Mutiny took forty days to reach Britain.

As early as 1856 businessmen in Britain and the United States thought seriously of laying a cable across the North Atlantic. Cyrus Field, a thirty-six-year-old New York paper merchant who knew little about telegraphy but had vision in abundance, saw the prospects of huge profits. Samuel Morse told him that a transatlantic cable was feasible, and an oceanographic survey showed that a plateau on the ocean floor was ideal for a cable from Ireland to Newfoundland. In the autumn of 1856 Field sailed for London, the world's financial capital, to found the Atlantic Telegraph Company and drum up money.

Although there were many who doubted that such a project was technically possible, 350 £1,000 shares were issued. The British and American governments backed the venture and promised warships to lay the cable. Charles Bright, aged twenty-four, was appointed chief engineer and the physicist William Thomson joined the board and developed the highly sensitive mirror galvanometer that enabled a Morse signal to be sent by the extremely low current in a long submarine wire. The cable, 2,500 miles long, was ordered from the Glass Elliott works in Greenwich. It consisted of seven copper wires wound together and armoured with three layers of gutta-percha. It was about the diameter of a medium cigar. No single ship in Britain or America had the capacity to carry such a cable and the load was divided between two vessels, USS *Niagara* and HMS *Agamemnon*. The plan was to splice the cable in the middle of the Atlantic.

In August 1857 the cable was secured at Valentia Bay in Ireland and slowly paid out from the stern of the *Niagara*. When 400 miles of it had been laid it snapped with a bang and fell to the seabed. This was a blow, but Cyrus Field, who was aboard the *Niagara*, was not a man to wring his hands and he at once made plans to try again. In June of the following year *Agamemnon* and *Niagara*, each carrying half the cable, rendezvoused in mid-ocean. The cable was spliced and the ships steamed away from each other, *Agamemnon* towards Ireland and *Niagara* towards Newfoundland. Again the cable parted. The chairman and vice-chairman of the company resigned, calling the venture hopeless, but Field, always spurred by calamity, went to London to rally the rest of the board. The ships sailed again in July, the crews joined the cables in mid-Atlantic and the *Niagara* and *Agamemnon* were soon lost from each other's view as they headed in opposite directions, lowering the cable from improved reels.

Niagara reached Newfoundland on 5 August and the elated Cyrus Field ran ashore to announce that the Atlantic telegraph cable had been laid at last. He and his associates had already constructed a cable from Newfoundland to New York, in itself an engineering and financial triumph.

At six o'clock in the morning the end of the cable was brought from *Niagara* into the Newfoundland telegraph station and contact was made with Valentia. It was a magnificent moment. The cable men were promoted to the status of gods of progress. The excited Queen Victoria at once awarded a knighthood to Charles Bright for his engineering prowess. *The Times* pronounced that the Americans and the British had 'become in reality as well as in wish one country . . . in spite of ourselves, one people'. On 16 August the Queen sent the first transatlantic message, her ninety-eight words to President Buchanan taking more than sixteen hours to tap out in Samuel Morse's code. A celebratory torchlight revel lit up New York and amid all the bonfires, gun salutes and fireworks the City Hall caught fire and was severely damaged.

American writers foresaw that the world would be rendered more civilized by 'free and unobstructed interchange' brought about by the power of the telegraph. 'The whole earth will be belted with electric current, palpitating with human thoughts and emotions.' In Britain, too, the Atlantic cable, speeding 'messages of love from kind Old England's breast', was welcomed as the enemy of tyranny.

The messages of love, however, began to fade. Some fragments of international news were transmitted but the signals grew weaker. The line was failing and despite all the attempts at resuscitation it subsided 'into entire silence' after four weeks. The following year a cable to India laid in the Red Sea failed without a word sent. The newspapers that had made heroes of the pioneers now jeered at them. This was a gloomy and sobering time for the ocean cable men. It was difficult to raise money among investors whose feet had turned cold, and in any case the American Civil War of 1861–5 created tensions between Britain and the United States which were bad for any kind of business. A British government inquiry concluded that there was still much to learn about the science of deep-sea telegraphy, something that was obvious to everyone.

In 1865, however, Cyrus Field and the Atlantic Telegraph Company gathered the money for another attempt and won the renewed backing of the British and American governments. They now had a heavier and better-insulated cable of higher tensile strength. This one was the diameter of a very fat Cuban cigar. They also had the right ship for the job, a marriage of new technologies. Brunel's *Great Eastern* was the largest ship in the world and had room enough to carry the 2,500 miles of cable in its entirety, as well as space for 500 men. Apart from all the machinery, she also had a farmyard of bullocks, pigs, sheep and chickens to feed the cableers.

The Times assigned its famous reporter, William Howard Russell, to travel aboard the *Great Eastern* to cover what was essentially a British engineering event. Cyrus Field was the only American on the ship. On 23 July the cable was secured at the telegraph station in Valentia and the *Great Eastern* headed for Heart's Content in Newfoundland, gently spinning out the curving filament of cable with all the ease of a spider casting its web. European news reached the ship through the cable as it was laid.

Eighty-four miles out from Valentia a fault was found in the cable. The engineers repaired it. Then another fault was found and mended. Twelve hundred miles out the cable snapped. The *Great Eastern* retreated to Britain. Russell had his story for *The Times*.

Yet Cyrus Field and his fellow directors saw the episode as a setback not a failure. They raised more money and ordered a new and improved cable. In July 1866 the *Great Eastern* sailed once more, and this time the rumble of the cable unreeling from its drum was constant and uninterrupted, a heartening rhythm for Field and the engineers. The ship kept up a speed of six knots and made a perfect crossing from Valentia to the flag-decked village of Heart's Content. All the way Field was able to receive news and stock prices from London. The cable engineers were the newspapers' heroes once more. *The Times* saluted them as 'the benefactors of their race', and Queen Victoria knighted four of them, including William Thomson, who would later become Lord Kelvin. A baronetcy was conferred on Daniel Gooch, the engineer in charge of the operation who was chairman of the Great Western Railway and head of the syndicate which owned the *Great Eastern*. The US Congress awarded Cyrus Field a gold medal, and the British newspapers called him Lord Cable.

The Morse signals began to flow. The cost was stunning, £20 a message at first, but at last money was coming in rather than disappearing to the bed of the ocean. Daniel Gooch used grapnels to recover the broken 1865 cable which lay two miles down and spliced the ends together to complete a second Atlantic link. Altogether Gooch and the *Great Eastern* laid four transatlantic cables. Copper wire, gutta-percha and an incurable persistence joined Europe and North America in the immediate exchange of information and revolutionized the business of news and the market in money.

John Pender, a Manchester cotton merchant, had put £10,000 into the Atlantic venture. An optimist of the same unshakeable quality as Field, he decided that his business future was in the international telegraph and he set out to lay a cable to India. An overland telegraph to India had already been opened in 1865 but the trouble was that messages took a week to arrive because of the delays in transmitting them across the territory of the Ottoman Empire. The signals frequently faded and messages had to be transcribed at relay stations and retransmitted by clerks who often knew no English and scrambled the words into unintelligible puzzles. The cost was £5 for twenty words. The alternative route, across Persia, Georgia and Russia, had similar problems and was no better.

The advantage of a submarine cable under one management,

safe from political troubles and the threat of damage by tribesmen, was obvious. John Pender formed a company to do the job and hired the *Great Eastern* to lay the cable. It ran from Porthcurno in western Cornwall, across the Bay of Biscay, the Mediterranean and the Arabian Sea to Bombay, and was completed in 1870. Pender then won concessions to extend the line across the Bay of Bengal to Singapore, Bangkok, Saigon, Hong Kong, Sumatra and on to Darwin. He was the boss of it all. His Eastern Telegraph Company gradually gobbled up smaller companies and became so large that it was known as 'Father Eastern'.

For all the skill of the engineers at sea and the technicians who built the terminals, the telegraph in the end depended on the expertise and concentration of the operators. It was hard, tedious and ill-rewarded work. The feeble signals of the early transatlantic cables were converted by a needle and a mirror into a flickering dot of light. An operator deciphered the Morse and dictated the letters one by one to a colleague who wrote them down. William Thomson's perfection in 1870 of his 'siphon recorder', which traced the Morse signal in ink on a paper tape, was an improvement and provided a permanent record of a signal; but it was expensive and demanded an exhausting level of concentration on the part of the operator, who had to be able to transmit at ten words a minute. Early recruits at the Porthcurno station in Cornwall earned no pay until they achieved proficiency and spent four months learning the craft before qualifying for a wage of £1 a week. Many quit, some because they could not cope, some because they were illiterate and some because they drank too much.

John Pender's cable to India speeded accurate messages halfway across the world in hours rather than days. Some old India hands, however, noted the disadvantages. In the pre-telegraph age of distance and delay the authorities in India took action first and told London later. Thanks to the telegraph the problems of managing India were directed for discussion and decision to the desk of the Secretary of State for India in London. Brought closer to head office, the men who ruled India saw some of their independence whittled away.

22

Pictures of silver

'This he perched upon a tripod –
Crouched beneath its dusky cover –
Stretched his hand, enforcing silence –
Said, "Be motionless, I beg you!"
Mystic, awful was the process.'
Lewis Carroll, 'Hiawatha's Photographing'

Here they pose before the camera's crafty eye, the people with the nerve to rule half of creation.

Engineers, for example, pause in their bolting of bridges and stand warmly clad and braising in the heat of an Indian noon. District officers, groomed and packaged by public schools to govern an empire, sit lonely at their dusty desks and marshal their paperweights. Cinched memsahibs sit tight. Men at the Club smoke cheroots. Lancers lounge in rattan ease against a frieze of anonymous servants. British Other Ranks rest on their sweaty slog up the Grand Trunk Road. Sleek Lord Curzon, viceregal gaze supercilious, poses victoriously by a slain tiger. The King-Emperor accepts the fealty of pearly rajas.

And who are these men, arranged for a rare team photograph in their puttees, bow-ties and sun helmets? They are magicians who can pin a shadow to a paper and snare the passing second on a square of glass, posing with their mahogany and rosewood cameras, symbols of power and the new surveillance.

'The curious tripod,' wrote one of these shuttersmiths, 'with its mysterious chamber and mouth of brass, taught the natives that their conquerors were the inventors of other instruments besides the formidable guns of their artillery, which, though as suspicious perhaps in appearance, attained their object with less noise and smoke.'

It was known early in the eighteenth century that light darkened salts of silver. Experiments in the 1820s and 1830s by Nicephore Niepce, William Henry Fox Talbot and Louis Daguerre led to the invention in 1839 of the process which the astronomer Sir John Herschel named photography. Herschel himself invented the chemical solution which fixed and made permanent the fleeting images. The daguerreotype, which captured sharp images on a copper plate coated with silver iodide, made a sensational début before a crowd in Paris in 1839. Samuel Morse, the inventor of the telegraph code, who knew Daguerre, took the process from France to America that same year. In 1840 Morse's colleague, John Draper, posed his sister Dorothy in her pretty bonnet and made what was probably the first photo portrait in America. In Calcutta in 1840 Dr William O'Shaughnessy produced portraits with the process; and in London Queen Victoria and Prince Albert bought their first daguerreotypes and became photography enthusiasts. In 1843 the *Illustrated London News*, then a year old, used daguerreotypes to produce images of London. The majority of pictures made with the process were portraits; and then, as now, sitters were sometimes disappointed with the outcome of their ordeal. For a pleasing portrait women were advised to frame their lips as if they were saying 'prunes'. 'Cheese' obviously came later.

Within ten years of the daguerreotype's début studios sprang up all over New York. New photographers loaded their equipment into wagons and took the trails of silver to the West. A Mrs Davis started a studio in Texas in 1843. A daguerreotypist began work in San Francisco in 1850. Meanwhile, studios opened in Sydney and Hobart in 1848. An exhibition of daguerreian views of California was a New York sensation in 1851. When they heard of Daguerre's death that year American photographers declared a month's mourning and wore black. Daguerre's process was marvellous, but it had its faults. It was expensive and produced a mirror image. In the early years it required an exposure of up to thirty minutes so that the subjects often sat rigid until tears streamed down their faces. Because there was no negative each picture was a one-off and could not be duplicated.

The process perfected by the English inventor Fox Talbot was an improvement. His Talbotype, also called a calotype, was a sheet of paper coated with silver iodide which produced a negative from which a print could be made. The exposure time was reduced to about five minutes. Talbot published the first book of photographs, *The Pencil of Nature*, a collection of his own soft and delicate pictures, in 1844. In 1851 Frederick Scott Archer made the decisive advance with his wet collodion process, coating a glass plate with silver salts to make a negative from which prints could be reproduced; and this became the standard photographic technology for the next thirty years. Scott Archer, however, failed to patent his process, did not profit from it and died in poverty in 1857.

A decade or so after its invention photography was, with the electric telegraph and the steamships, part of the excitement of an age of advancing science, another proof of progress. Improvements in optics and chemistry placed photography in the vanguard of technology, and just as the ever-faster steamships and locomotives were welcomed as shrinkers of time and distance, so the photograph, freezing moments of light and time, demonstrated a scientific mastery.

Like the artists who preceded them with pen and paint, the early photographers were witnesses who brought the empire home. Their 'sun pictures' of the peoples, costumes, gaudy fruits and plants, animals, architecture and landscapes of India, Australasia, Canada, Africa and the Pacific nourished an empire of the imagination, satisfying a craving for ever more illustration as well as stimulating it. The first photographers were in Palestine and Egypt in the 1840s; and in the 1850s photographs of the Nile, the Pyramids and the Holy Land were on sale throughout the world.

Photographs were regarded as true and trustworthy, giving the appearance and detail of places that were otherwise names on maps, casting light on the mysterious and making sense of complexity. The advance of the photograph coincided with the flowering of the Victorian empire, and the camera was there to record it. As adjuncts of geography and exploration, the images were part of the wonder of empire although, in time, they reinforced clichés and European ideas of the Orient. They were a novel and fascinating archive and an increasingly significant part of publishing and journalism, the mirror of the imperial spectacle. The authorities intended that they should impress and show lands and peoples tamed and controlled. As photographs accumulated they evolved into an immense colonial memory and museum, a repository of imperial attitudes, ideology and culture. They became powerful.

The photograph was a geographer's tool and an explorer's unembellished evidence that he had hacked through jungles and toiled over deserts to far places. The images were a gift to the future.

Camera frontiersmen endured endless frustrations in their struggle to get pictures. In India the humidity of the monsoon season rotted and rusted the cameras. Dust and sand clogged and scratched them. Flies stuck to the sticky wet plates. Insects chewed the camera bellows. (A London camera-builder offered leather bellows impregnated with a repellent.) Heat made wet plates dry and peeled the emulsion from them or cracked the negatives. And heat, of course, often sapped the photographers.

Supplies of the collodion needed for the wet-plate process were difficult to arrange. The stuff was a mixture of nitrocellulose, also known as gun cotton, suspended in alcohol and ether. Since it was explosive P&O would not freight it, and it had to be manufactured locally. Water was another problem. Photographers often despaired of finding a clean supply for their processing. Glass plates were heavy and easily smashed. Working in the field, the photographers had to take large darkroom tents and chests full of chemicals. They set up the tent, fixed the camera on its tripod, coated the glass plate with emulsion, made the exposure and then retreated to the tent to develop the plate before the coating dried.

The need for long exposures drew photographers to subjects that did not move. Early mornings were a good time. The light was ideal before the sun rose too high, the air was still and the leaves of trees did not stir. The Italian Felice Beato, who in 1858 took pictures in the aftermath of the Indian Mutiny, attended an execution and before opening the shutter stepped forward to steady the two bodies swaying on the gallows. Beato, described by *The Times* as 'a prince of photographers', also photographed Lucknow, one of the emotive places of the fighting. Here he found that the bones of 2,000 sepoys, slaughtered by the British in a furious avenging of the massacre of women and children at Cawnpore, had been buried. He insisted that skulls and bones should be disinterred and strewn about to make his photographic composition all the more compelling. British army officers, after all, were his chief customers. The British appetite for such pictures, 'the pictorial romance of this terrible war', was profound and lingering; and for many years afterwards photographers visited the ruins and battlefields to make pictures for their highly saleable portfolios.

The East India Company endorsed photography enthusiastically. From 1855 cadets bound for India were taught how to take and develop pictures, and engineers were advised to keep a photographic record of the construction of bridges, railways and canals, a process 'with the advantages of perfect accuracy, small expenditure of time and moderate cash'. John McCosh, an army surgeon in Bengal in the 1850s, urged his fellow officers to amass collections of photographs of people, animals and buildings, while a doctor in Calcutta recommended photography of murder scenes on the grounds that any suspect would be unable to face 'the ghastly faces, the sight which has haunted his brain since the act was done'; and would, presumably, confess.

Called the mirror of truth, the camera was compared with the 'untrustworthy pencil'. And the pencil was decisively snapped with the East India Company's order that artists who had been making drawings of the sculptures, forts and tombs of India should use a camera instead. An architectural and archaeological photographer was appointed in Bombay in 1855. Today the collection of Victorian photographs of temples and other monuments is a treasure in its own right. Many of the new photographers had a deep interest in Indian history and culture and produced work which, a century and a half on, remains astonishing in quality and composition. Working with large paper negatives, Dr John Murray, who took up photography in 1849, made beautiful pictures of the Taj Mahal, and other Mogul architecture in Delhi and Fatehpur Sikri, during his years as Civil Surgeon in Agra. He also photographed scenes during the Mutiny. Linnaeus Tripe, an army officer in Madras, took remarkable pictures in Burma in 1855 and compiled volumes of south Indian studies which are among the best of any nineteenth-century work. David Lyon, who worked in south India in the 1860s, ingeniously positioned servants, holding reflectors and standing hidden behind pillars, to illuminate interiors of buildings. His prints survive in excellent condition. For all our modern lenses, filters and film, the masters of light who brought India to Britain have much to teach.

Undoubtedly trying, Victorian India was nevertheless magnificent for photographers. From the ragged fringes of Afghanistan to the temples of Comorin and the road to Mandalay, they compiled a record of an evolving empire, revealing their fascination with India and their belief in themselves as its rulers.

'I mounted my yak,' wrote Samuel Bourne, 'the sturdy, strong, shaggy-haired buffalo of Tibet, and right nobly and steadily did he carry me up the steep ascent . . . we encamped at an elevation of 17,000 feet.' Bourne rewarded his frozen and ravenous porters with four sheep. 'This was at four o'clock, and by eight not a vestige of the unfortunate animals remained except their skins.'

Like the intrepid painters before him Bourne went to India to make his name and fortune with his pictures. He had studied photography in the 1850s while working as a clerk in Moore and Robinson's bank in Nottingham. In common with many artists and new photographers, he indulged a passion for the picturesque and travelled in Scotland, Wales and the Lake District to frame wild and romantic views. He was nineteen when he heard what he described as 'the call of the East', quit his safe job and sailed for India. It was a good time to go. The demand for Indian pictures was large and seemingly insatiable. He arrived in 1863 and although he wrote that 'there is scarcely a nook or corner of the globe which the penetrating eye of the camera has not searched', he clearly did not believe it himself and was determined to find landscapes that had never before darkened the salts of silver.

Bourne travelled in India for seven years. He started in Calcutta but, finding it pictorially dreary, and in search of something more dramatic, he went up to Simla, the Himalayan balcony on which the British sat out the summers. There he joined the photographer

Charles Shepherd whose name endures with his in the Calcutta firm of Bourne and Shepherd. Simla, however, was not pretty enough for Bourne's liking: it had no river scenery, no ruins to support the pleasures of melancholy; nothing that surpassed the scenes he had photographed in Britain.

For all its pictorial shortcomings, however, Simla was an excellent point of departure for expeditions into the Himalayas. Bourne made three long trips, which he described for readers of the *British Journal of Photography*. Determined to bring back home the splendour of hitherto unphotographed glaciers and passes, he saw himself as a soldier of the lens and planned his photographic forays as campaigns. He necessarily travelled heavy.

> 'My photographic requisites consisted of a pyramidal dark-room tent ten feet high by ten feet square at the base, opening and closing like an umbrella . . . in this country I could not work in one of those suffocating boxes without elbow room and without ventilation. I like to have plenty of both, as I am jealous of the bloom which I have hitherto maintained on my cheeks, and of the hale and robust constitution with which nature has blessed me.

In addition to his portable darkroom, he took 250 glass plates twelve inches by ten and 400 plates eight inches by four and a half. Then there were bottles of chemicals packed into padded chests, two pairs of bellows, cameras and lenses, two developing baths and a quantity of large containers of distilled water.

The photographic equipment alone made loads for twenty porters. He also took sleeping tents, bedding, camp furniture, cooking pots, hermetically sealed food, sporting guns and books and a good supply of Hennessy's brandy in place of Bass and Allsopp's ale, which was too heavy to carry. 'It is advisable to take as many luxuries as possible . . . I was sometimes two months in some solitary and remote district without ever seeing a European, talking nothing and listening to nothing but barbarous Hindustani.' As well as porters he had six servants to cook and care for him and six more to carry him in a dandy, a sort of palanquin swinging on a pole.

During his first expedition in 1863 Bourne headed towards the Chinese border. More than thirty men carried his gear and he drove them as he drove himself, ruthlessly. He reflected that one slip on the precarious paths above the great river gorges through which he marched would have sent him falling for nearly a mile before his body 'in scattered fragments found a resting-place in the river below'.

Much of India seemed splendid, savage and enchanting to British eyes; and also immense. Here in the Himalayas Bourne was overwhelmed by the massive mountainscapes. 'It is altogether too gigantic and stupendous', he said, 'to be brought within the limits imposed on photography.' He concluded that the artists with canvas and paints, 'our brethren of the brush', had the advantage. Although he had difficulty in making images of huge glaciers, monstrous chasms and awesome cascades, he persisted; and he once waited six days for the weather to change so that he could take just two pictures. Working in the blinding whiteness at more than 15,000 feet, where 'everything wore an air of the wildest solitude', he found that to compose a photograph 'required all the courage and resolution I was possessed of. Having no water I had to make a fire on the glacier and melt some snow. The hands of my assistants were so benumbed that they could render me no service in erecting the tent, and my own were nearly as bad.' It certainly needed a resolute mind to take a photograph with frozen hands, coat the plate with emulsion, make the exposure and then develop the plate. At last he made the first ever photograph attempted at such an altitude, 'but my hands were so devoid of feeling that I could not attempt another'.

What happened next was perhaps not surprising. With snow falling heavily, his coolies believed they might die from cold. They threw down their burdens and fled to a village. To prevent himself from freezing, Bourne lay in bed for three days and nights, 'without a particle of fire', before fresh coolies arrived to rescue him. 'All my enthusiasm in photography and my desire to see the mysteries of nature at high altitudes would not have urged me to this undertaking had I known what awaited me in that miserable interval.' After ten weeks away Bourne returned to Simla with 147 photographs of scenery that was 'amongst the boldest and most striking on the globe'. He was pleased to learn that in India photography had been designated a fine art, which was not the case in London. 'Are we more enlightened in this land of rising British enterprise', he puffed, 'than the would-be patrons of art in clique-ridden England?'

Bourne nearly perished at the start of his second expedition, through Kashmir, in 1864. He was swept away in a fast-running river and one of his servants dived in and hauled him to safety. He was soon his old self. Watching a Hindu temple ceremony he talked to a priest and 'tried to show him the absurdity of his devotions; that there was but one God who required all men to worship Him. He listened attentively, and said that He might be a very good sort of God in His way but was inferior to his own, so I left him only more confirmed in the grossness of his own belief.'

Bourne had more difficulties with his porters in the cold and snow of the high mountains and, in spite of his threats, they 'refused to stir a single step, saying that I might shoot them if I liked; it would only be dying an hour sooner, as they were certain to perish in the snow'.

Marching from Dharamsala to Dalhousie he sent a man ahead to a village to recruit porters and arrived to find fifty men imprisoned in a loft. His recruiting sergeant had removed the ladder to prevent their escape. Next day some of these unwilling wretches dropped their loads and ran away and Bourne chased after them. At the door of a house two women denied the presence of any porters, but Bourne barged in anyway and found the runaways hiding beneath a bed. 'Dragging them forth I made them feel the quality of my stick, amid the lamentations of the aforesaid females.'

The expedition floated across a river on the only available ferries, inflated buffalo skins, known as mussocks. Bourne's mussock-man lay across the ungainly balloon and paddled with his hands and feet while the bearded and burly Bourne knelt astride him, an unusual spectacle.

Once again Bourne wrestled with the difficulties of bringing mountain grandeur into the compass of a glass plate. 'How often have I lamented that the camera was powerless to cope with these almost ideal scenes . . .' He was always impressed by the spectacle of regal peaks and saw the hand of God in natural wonders. 'What a puny thing I felt, a mere atom, and scarcely that in so stupendous a world.' Like most who have seen it, he was enchanted by the Vale of Kashmir and found the picturesqueness he longed for along the lakes and waterways of Srinagar. In his delicate pictures he showed it as the Shangri-la of British imagining, and he enjoyed a blissful spell when the light was good, the scenes sublime, all his chemicals were working and he could manage six negatives in a day. But he had little success in persuading the local people to pose naturally for him. 'Their idea of giving life to a picture was to stand bolt upright, with their arms down as stiff as pokers, their chin turned up as if they were standing to have their throats cut.'

In the high mountains Bourne 'suffered very much from ... the acute pain which cyanide of potassium produces on hands chapped by frost'. One day he set up his developing tent on a narrow mountain track. He was working inside when a pack pony tried to skirt it, slipped and flattened the tent and smashed bottles of chemicals. Bourne wriggled free, unhurt. After ten months away, his photographic supplies almost exhausted, he returned from the mountains with 500 negatives.

He left Simla on his third and final Himalayan journey in 1866, heading by way of the Kulu Valley to the source of the Ganges, with sixty porters carrying his equipment on their heads and backs. Bourne was always ready to suffer for his art; and he saw to it that his unsung and sometimes reluctant porters suffered for it too. Once, when they started grumbling and seemed unlikely to continue humping their burdens, Bourne

> took a handy stick and laid it smartly about the shoulders of several of them until they lay whining on the ground. I gave them little time for this luxury but made them buckle to their loads in double quick time. This bit of seasonable sovereignty had a good effect ... they stuck to me through heat and cold, climbing the highest and most difficult passes, and carrying their loads bravely over glaciers and places so difficult and dangerous that I, empty-handed, only passed with fear and trembling.

In the Manirung Pass, at 18,600 feet, the silvery Spiti River gleaming far below, he arranged a line of porters to stand in the snow in the foreground to provide a sense of scale. They stood motionless for an exposure of seven seconds. Bourne was exultant. He had endured bitter cold to reach one of the world's rugged high places. No photographer had climbed higher. 'It seemed as though I stood on a solitary island in the middle of some vast polar ocean, whose rolling waves and billows had been suddenly seized in their mad career by some omnipotent power and commanded to perpetual rest.' His pictures revealed the dramatic beauty of high and distant places. They underlined Bourne's own idea of himself as a self-sufficient imperial explorer of daring and discipline, a man willing to go to the extremes for what he called his 'toilsome art'. He continued by way of vertiginous passes to the source of the Ganges and made his way back to Simla after six months away. He had achieved a photographic, physical and personal summit. With bloody-minded determination and glass plates and acid he had captured the mountains.

He could not help but exult. 'Ah you gentlemen! and you, careless public! who think that landscape photography is a pleasant and easy task, look at me toiling up that steep ascent in the grey dawn of a cold morning in fear that my labour would be all in vain! See me sitting for ten mortal hours, shivering in cold and mist, on the top of the bleak pass, waiting for a break which would not come!'

Map-makers and surveyors made sense of vast colonial landscapes, measuring and bringing them to order on sheets of paper. The photographers played their part, with pictures of highlands and lowlands, rivers, deserts and jungles, as well as railways, bridges, residences, law courts, offices and all the apparatus of the Raj. The Archaeological Survey, meanwhile, amassed a magnificent record of temples, mosques and palaces. By the 1870s, India in almost all of its aspects had been photographically recorded, catalogued and filed.

Increasingly, the record included images of people. There was a growing interest in anthropology. Photography obviously had a documentary role to play. Authorities in India encouraged administrators to take pictures of the ethnic groups in their territories, their costumes, weapons and homes. Army officers were given photographic assignments, making portraits in villages and tribal areas. Throughout the empire, images of racial types became part of the record and, published in books, introduced Europeans to the human variety of the world. The first significant and ambitious work of ethnographic documentation, a government-sanctioned project called *The People of India*, was published in eight volumes, each containing almost 500 pictures, between 1868 and 1875. The photographs classified people not only by their type and home district but also by the qualities that colonial administrators saw in them. Thus they were labelled hard-working or soldierly or reliable or lawless.

Out of such pictures emerged the stereotypes of the snake charmers, water carriers, dancing girls, fierce tribesmen, loyal and stalwart soldiers, maharajas and their retinues; and also the cliché of the white man with his shorts and bush shirt, his rifle, his topi, his foot on a tiger or hand upon a rhino horn. Photographers responded to the public appetite for pictures of exotic people, the enigmatic faces of the world. Looking at them today we see the difficulties encountered in making natural and relaxed images. Like Samuel Bourne, many photographers complained of the shyness and stiffness of their subjects. Joseph Thomson, taking pictures during an expedition in East Africa, said that people could be persuaded to pose with 'soothing words aided by sundry pinches and chuckings under the chin'; but as soon as the photographer ducked under his black cloth to focus his lens, his subjects 'would fly in terror to the woods'. Many photographers found that the mere appearance of a camera on its tripod intimidated and frightened people; and some of them relished that very power.

Maurice Vidal Portman, an assistant superintendent at Port Blair in the Andaman Islands, began a detailed study of the islands' tribes and made hundreds of photographs. He posed his subjects against chequerboards to measure their height, and fixed their heads and arms in clamps to keep them rigid during long exposures. What lay behind the drive to make such human portraits was the belief that in many cases the people being photographed were doomed to be extinguished by the rush of civilization.

23

Rivers in the sky

'Back to his bounds their subject sea command
And roll obedient river through the land.
These honours peace to happy Britain brings.'
Alexander Pope, *Moral Essays*

In the winter of 1826 a survey team of eleven well-weathered men, two of them English, five Irish, one Scottish, one French Canadian and two American, struggled through the snowy landscape of Ontario. In the fading light of each bitter day they cut down spruce branches to make the floor of a rough deerskin wigwam. They crept inside, lit a fire and stoked their comforting pipes. Then, as John Mactaggart, their leader, recalled, they huddled in their piney fug and tried to sleep. 'Having lain an hour or so on one side, someone would cry Spoon! – the order to turn to the other. Reclining thus, like a parcel of spoons, our feet to the fire, we found the hair of our heads often frozen to the place where we lay.'

In the days that followed these necessarily intimate nights, the surveyors charted a route for the Rideau Canal to connect Montreal, on the St Lawrence River, to Lake Ontario, and thereby open a water highway into the west of Canada. It began as a military project, a product of the uncertainty that followed the war of 1812–14 between Britain and the United States. The Duke of Wellington thought the canal would secure supplies should the Americans attack British Canada. It was an ambitious undertaking. The canal was 124 miles long. To surmount the high ground on the route a great ladder of locks had to be built to lift barges up and down. John By, a Royal Engineers colonel from Lambeth in

London, directed the work and commanded an army of more than 4,500 craftsmen and labourers. Without benefit of machinery, using oxen to haul cartloads of stone from quarries, they built forty-seven masonry locks and fifty-two dams. One of the dams rose to sixty feet and for many years was the highest in North America.

A construction headquarters grew into a settlement called Bytown, notorious for drunkenness, brawls and fever. Over the years the malaria that struck every summer killed hundreds of labourers in Bytown and its satellite camps along the Ottawa River. Colonel By himself was so sick and his stout frame so much reduced that it was feared he would die. For all the hardships, and the impossibility of working in the depth of winter, the engineers and navvies finished the canal in 1832. Colonel By sailed to London expecting honour for his leadership but, typically, the government accused him of spending beyond his budget. A parliamentary inquiry cleared him but the injustice of it all broke his heart and he died at Frant in Sussex, undecorated. Canada, however, remembered and honoured him. His remarkable canal endures and you can admire the symmetry and grandeur of the stone locks he built in the heart of Bytown which, in 1855, was proclaimed by Queen Victoria as the capital of Canada and given the name of the local Indian people, Ottawa.

The 1820s were wonderful years for the civil engineers whose daring constructions formed the framework of the future in the early nineteenth century. Thomas Telford was one, a visionary and prolific builder of roads, canals, harbours and bridges. Nothing shows more clearly his genius for fusing beauty and strength in the landscape than the Pontcysylltau aqueduct he constructed between 1794 and 1805 to carry the Shropshire Union Canal 120 feet above the Vale of Llangollen. The 'stream in the sky' it was called, a pioneering work of soaring grace in cast iron.

Telford demonstrated his feeling for appearance again with the revolutionary bridge he designed for the Menai Strait, the vital link in the London–Holyhead coaching road. He chose to build it in iron, a material not well understood, because he admired the shapely Iron Bridge, built over the Severn at Coalbrookdale in 1779, the first important iron structure. Telford conducted numerous tests to ensure that the tough form of the metal, wrought iron, was strong enough for his Menai bridge. He started in 1819 and completed the tall masonry towers in 1824. Sixteen iron chains were suspended from tower to tower, high above the water, creating the curve which supported the 579-foot centre span. These chains were nine inches wide. As soon as they were in position two of Telford's agile and fearless men celebrated by walking across the strait on one of them. The next step was to hang 444 vertical rods from the chains to support the roadway. The Menai bridge set the standard for suspension bridges. It was the ancestor of them all. Telford built it in the years before any steam engine could have assisted the work and he relied on the muscle of men and horses and the lifting power of the tides. He was almost seventy when he completed the bridge in 1826.

Twenty years later, a mile west of this masterpiece, Robert Stephenson, the son of George, began to build his Britannia railway bridge over the strait, using the tides and hydraulic lifts to raise two wrought iron tunnels side by side on three limestone towers. The first train crossed the bridge in 1850.

President Thomas Jefferson laughed when he heard that New Yorkers planned to build a canal to connect their city to the whole American West. 'A splendid project,' he said, 'and it may be executed a century hence. It is little short of madness to think of it at this day.' Governor De Witt Clinton of New York was undeterred, turned the first spade of earth in 1817 and initiated perhaps the greatest American engineering feat of the century.

New York muscled its way into the heart of America. Armies of labourers, many of them Irishmen recently disembarked from immigrant ships, joined crowds of tree fellers and stonemasons to drive a canal 363 miles from New York to Buffalo on Lake Erie, linking the Hudson River to the Great Lakes. With picks and shovels they dug a channel forty feet wide and four feet deep and laid out a ten-foot towpath for mules and horses. They built eighty-three locks to raise barges nearly 700 feet over ridges and cliffs, and erected eighteen aqueducts on stone towers to cross valleys. Many men fell sick and died of malaria in the swamps. The project was a hard college for the engineers and craftsmen. They learnt as they worked and emerged as the experienced specialists who went on to father canals, bridges and roads throughout America.

When the Erie Canal was completed in 1825, at a cost of $7 million, Governor Clinton cruised in the celebratory flotilla of boats that left Buffalo and reached New York nine days later. There he poured a barrel of Lake Erie water into the harbour to symbolize the union of the fresh water of the lakes and the salt of the Atlantic.

The canal brought about dramatic changes. Until its completion the Appalachian mountains had formed a barrier between the prairies of the Middle West and the coast. Farmers had to ship their grain down the Mississippi to New Orleans. The Erie Canal, 'Clinton's Ditch' as it was called, cut through the mountains and gave access to Lakes Ontario, Erie, Huron, Michigan and Superior. It opened a route for westbound settlers; and many of the men who built it stayed on and settled the new canal towns of Rochester, Syracuse and Utica. Transport charges fell sharply. It had cost a hundred dollars to haul a ton of freight by wagon from Albany to Buffalo. It now cost $10 in a barge pulled along the canal at a steady mile and a half an hour. In the first year 7,000 barges went up and down the canal and the following year there were 19,000, laden with wheat, flour, beef, pork, timber and whiskey, heading east; and textiles and iron goods heading west. The immediate success of the waterway, the volume of freight, the large income earned from tolls and the rise of towns along the way, set off a surge of canal building. By 1840 more than 3,000 miles had been constructed.

Clinton's Ditch settled conclusively the competition for economic supremacy between New York, Boston and Philadelphia. George Washington had predicted in 1784 that New York would emerge as 'the seat of empire', and thanks to the Erie Canal it grew rich and famous as the chief port, Atlantic headquarters, entrepôt and financial centre of the United States. The canal transformed New York into America's imperial city.

It also started a migratory swarm to the small village of Chicago on the muddy shore of Lake Michigan, strategically sited as a gateway to the Mississippi, to the open West and the Great Lakes. In the 1830s a count of girls of marriageable age in the village revealed only two. The wedding of a postal clerk and an Indian chief's daughter was attended by warriors who frightened the white guests by sporting tomahawks and scalps in their belts and whooping during the church service. In 1837 a journalist wrote of the shortage of women and described the scene when steamers arrived from Buffalo. The bachelors of Chicago, he said, 'flock to the pier and stand ready to catch the girls as they land'. The Erie Canal and a rush of speculators ensured that in a few years there was no shortage of women, or anything else, in the booming city of Chicago.

24

Mount Hopeless

> '"I expect you will consider it imprudent, Mr Voss, if I ask whether you have studied the map?"
> "The map? I will first make it."'
>
> Patrick White, *Voss*

In 1853 the Society for the Diffusion of Useful Knowledge, a British philanthropic body founded in 1826 to feed the hunger for information about the world and exploration, published a map of Australia. It lies beside me as I write. It shows the interior, a territory larger than Europe, as a featureless blank. Three words are printed across it: Tropic of Capricorn. Otherwise there is nothing. The heart of Australia is as silent as the plains of Mars.

At the time the map was published people from the British Isles had been settling the Australian colonies for sixty-five years. The permanent pattern of habitation was being laid down, the great majority of Australians becoming a coastal people, making their homes within twenty miles or so of the thundering surf of the Pacific and Indian oceans.

A smaller number were looking for water and pasture inland. In New South Wales pioneers had pushed out from Sydney along the valley of the Hunter River and had pierced the high sandstone wall of the Blue Mountains to settle the plain of lovely woods and grasslands beyond. In the colony of South Australia squatters had planted wheatfields and orchards and established sheep and cattle runs up to 150 miles north of Adelaide. Similarly, in Victoria, a rural society of hard-working farmers and small towns was taking root around Melbourne. The discovery of gold in New South Wales and Victoria in the 1850s promised prosperity. In Western

Australia, meanwhile, settlers were making tentative footholds beyond Perth.

To those who pondered it from the very edges of the settled lands, the outback in mid-century appeared devoid of history and chronology, empty of remnants and ruins, the vacant space where the Aboriginal people seemed to be fleeting shadows, their steps so light upon the scorched surface that they barely left a print.

Not only had no one ever crossed this infinite and menacing desert, there was a strong opinion that it was impassable. It had already swallowed adventurers who had ridden towards the fateful shimmer of the distant sand ridges and were never seen again. Like all such secret lands down the centuries the centre of Australia was attended by myths and wistful longings. The mystery fed a notion that beyond the dunes and hot pebbles and saltbush of the deserts there stretched a blue inland sea, a fish-filled antipodean Caspian watering a subtropical paradise of pasture and woodlands. It grew in men's dreams as a place where cities might rise and conquer the emptiness.

Edward John Eyre, son of a Yorkshire vicar, sailed for Australia in 1832 when he was seventeen. By his twenty-fifth birthday he had travelled more of the outback than any other man. He set up as an overlander, a sheep and cattle drover, taking flocks and herds from New South Wales to South Australia, a journey of three and a half months. He made good money and used his profits to finance his dream of discovering what lay in the interior, making horseback expeditions into the territory north of Adelaide. He travelled through a creek-veined land of stark beauty, red and blue and umber, and penetrated the Flinders Ranges. In the 1920s the compelling gold and chocolate majesty of these mountains would inspire the paintings of Sir Hans Heysen, the Australian artist, who was fascinated by them all his life.

Looking out from the higher ridges Eyre saw how the light played tricks and distorted distances. Staring at hypnotic mirages he found it hard to decide 'what to believe on the evidence of vision . . . the whole scene partook more of enchantment than reality'. He came upon the immense salt lake that now bears his name and concluded that it was part of a great horseshoe of marshes that barred any access to the centre. He observed little in the landscape that was not 'desolate and forbidding'.

At the northern end of the Flinders he climbed a hill to survey the land. He was so sure that he would discover only an expanse of arid salty crust that even before he reached the summit he had named it Mount Hopeless. If anyone had told him that within twenty years there would be sheep and cattle stations in and around this range he would surely have laughed. He could not know that the saltbush, the principal plant of the arid regions and to his mind a symbol of a desiccated land, would nourish and fatten cattle. Its virtues would be a significant discovery in Australia's pastoral history.

In October 1840 Eyre began the most famous and dangerous journey of his life, a walk of more than a thousand miles along the scimitar sweep of the Great Australian Bight. He was looking for a sheep and cattle route between Port Lincoln in South Australia and Albany in the far west. His track lay across the flat expanse of the Nullarbor, Latin for treeless, a desert extremely hot by day and bitterly chill after sunset. For companions he chose his white overseer James Baxter and three Aborigines, Wylie, Joey and Yarry. They took some horses and a walking larder of sheep. Each day Eyre scouted ahead of the party to seek water. When it was hard to find the men moistened their lips by fashioning brushes of grass to mop the morning dew from the leaves of bushes.

In six months they covered half the distance to Albany and by then it was horribly clear that there was no stock route and that their situation was desperate. During a night at the end of April 1841 Eyre walked away from the camp to round up the horses and heard a gunshot. He returned to find that two of the Aborigines had murdered Baxter and fled with most of the food and guns. Eyre was left with Wylie and the suspicion that he, too, had been part of the plot.

> In the wildest and most inhospitable wastes of Australia, with the fierce wind raging in unison with the scene of violence before me, I was left with a single native whose fidelity I could not rely upon ... Three days had passed since we left the last water ... Six hundred miles of

> country had to be traversed before I could hope to find any more.

The only weapon Eyre had was a rifle jammed with a bullet in the barrel. He held it over a fire until the gunpowder exploded and the ball whizzed past close to his head. Now at least he could hunt and protect himself. He and Wylie walked on through scrub and prickly desert spinifex. They fed on one of their horses and made a feast of a kangaroo that Eyre shot. They descended the cliffs and Wylie found a dead penguin on the beach and ate it, all but the beak and feathers. Three hundred miles from Albany the tall and skinny Eyre saw a whaler offshore and made a smoky fire to attract its crew's attention. Soon he and Wylie were being generously fed on board and Eyre luxuriated in deliverance, drinking wine and reading newspapers. The pair stayed aboard for ten days and then resumed their trek, loaded with food. In July 1841, nine months after starting and long given up for dead, they reached the small settlement of Albany. They were the first men to journey across the country to the far west, having discovered little of any use except the harshness of the land and the depth of their own determination.

Eyre's courage and fanatical edge had won him his place in history. Later he would win one in infamy. As Governor of Jamaica in 1865 he suppressed with great brutality a black rebellion in which seventeen whites were killed. He ordered the hanging and shooting of 440 people, the flogging of 600 more and the burning of a thousand homes. The government recalled him. His harshness was popular with the white plantocracy and defended in Britain by Dickens and Tennyson and hundreds of Anglican clergymen. Other clergy, however, were roused to furious attack.

The advance of the electric telegraph in the 1850s and 1860s filled Australian imaginations with the idea that the end of their isolation was at hand. The fastest sailing ships of the time took at least three months to sail from England to Australia. When they arrived in Adelaide, Melbourne and Sydney, reporters boarded them to pick up the English newspapers and copy out the latest news, months old but to them fresh and avidly read.

The first successful international cable, linking London and Paris, had been laid in 1851; and Australian hopes were raised three years later by the stringing of a ten-mile telegraph line between Melbourne and Williamstown. By the end of the 1850s a wire connected Sydney, Melbourne and Adelaide. As the international telegraph from London snaked across Europe, towards India, Ceylon and Singapore, Australians yearned to see the south coast of their continent stitched to the north by wire. If an expedition could travel across the central void and chart a route, the telegraph builders would surely follow, picking up the thread from Asia and taking it almost 2,000 miles across Australia to bring the chief cities within hours of London. It would make a great difference to commerce. Within hours, not months, Australia would have the latest news of trade, commodity prices and imperial wars. It would know, as quickly as people in the English countryside, of Royal doings and the winner of the Derby.

There was also the matter of security. Many Australians believed that should the Russians or the Chinese threaten an invasion they would be able to summon the Royal Navy by telegraph, a 'hotline' to the big guns.

Yet the underlying impetus for expeditions into the centre was curiosity about 'the ghastly blank', the natural longing to enter the eerie landscape and raise the veil. The old questions still excited speculation: was there a mediterranean sea; were there streams and pastures fit for sheep; what did the heart of the continent look like; and could Europeans survive there?

Curiosity nurtured a desire for glory. There would be a crown for the conqueror of the fearful desert. More than anything else, Captain Charles Sturt, a handsome India-born British soldier, ached for that reward. He had already discovered much of what was known about the hinterland in his expeditions of 1828 and 1829 which led him to the great southern rivers, the Darling and the Murray, and their abundantly watered grasslands. In the years that followed he became obsessed with reaching the geographic heart. 'Let any man', he wrote, 'lay the map of Australia before him, and regard the blank upon its surface, and then let me ask him if it would not be an honourable achievement to be the first to place foot in its centre.'

In August 1844, aged forty-nine and partly blind, Sturt led fifteen men out of Adelaide determined that the first foot to step on Australia's moon should be his. Indeed, he felt that in the matter of exploring the centre 'no man had a greater claim than myself'. Such was his belief in the inland sea that he carried a boat to sail upon it. It was not long before the sun made fritters of his hopes. He was 500 miles from Adelaide when the heat trapped his expedition at a waterhole in a terrible labyrinth of sand hills. Sturt and his companions must have felt that demons had stoked the furnaces for them. In six months of 'ruinous detention' temperatures soared to 130 degrees Fahrenheit and beyond, agony for men and horses. And as if the roasting wind and scorching days were not punishment enough the explorers were tortured by the dazzle of the moon at night. The graphite fell from Sturt's dried-out pencils, heat burst the thermometer and the party's supplies shrivelled. The air thickened with red dust and the men's skins blackened with the onset of scurvy. One of them died from it.

When at last rain freed the explorers from the trap Sturt was determined not to retreat and pressed onwards and northwards into what he described as 'a region that almost appears to be forbidden ground'.

Months later, months of dust and torment from thirst and flies, he and his men had crossed almost half the continent and had penetrated to latitude twenty-five degrees, farther than any other explorer. They entered the region later named Sturt's Stony Desert and also found a stream they called Cooper's Creek. Sturt acknowledged at last that the desert had beaten him and turned back. The return journey was a version of fire-walking. 'The ground was so heated', he observed, 'that our matches, falling on it, ignited.' The sun was blood-red and 'the silence of the grave reigned around'. Sturt collapsed, weak from disease. Unable to stay in the saddle, he was carried in a cart.

At midnight on 19 January 1846 Charlotte Sturt answered a knock at her door in Adelaide. Swaying on the threshold was a tall and skeletal figure. It was the wreck of her husband. She had had no report of him during the eighteen months he had been away and feared that he had perished. She fell in a faint at his feet. A month later the civic and business figures of Adelaide gave a dinner to honour the gallant pathfinder who had sought to raise the British flag in Australia's navel. Sturt rose to speak and the diners fell silent and expectant. As he began to describe his ordeal he faltered and then broke down in tears.

Two years later the explorer Ludwig Leichhardt, an immigrant from Prussia who had developed a mystic commitment to the outback, led a party of seven men, with a mob of horses, bullocks and goats, into the uncharted bush west of Brisbane with the aim of crossing to the Indian Ocean coast. The Australian Nobel laureate Patrick White, haunted and inspired from childhood by the strange landscape of the outback, made Leichhardt the model for the hero of his 1957 novel *Voss*. Leichhardt's expedition vanished, as a drop of water in the sand, and nothing was ever found of it, not a hat nor a bone.

Australians noted ruefully that they had more knowledge of remote parts of Africa than they had of the centre of their own country. But there was confidence that modern science, the sextant and theodolite and breech-loading rifle, as well as organization, would at last make man the master of the mystery.

Shared hopes for a telegraph line linking Australia to London and mutual curiosity concerning the wilderness never persuaded the young Australian states to join in a united expedition across their continent. As rivals they wished neither to divide the glory nor to pool the expense.

In 1859 South Australia offered a £2,000 prize to any man who could traverse the continent and thereby show the way for a telegraph line from the north coast to Adelaide. At the end of that year the Philosophical Institute, which was soon to be the Royal Society of Victoria, decided to mount its own expedition to reach the north and bring the wire and its commercial rewards to Melbourne.

So now there was a contest. Adelaide's man was John McDouall Stuart, a native of Fife, who had hungered for years for the unknown red world and, like Captain Sturt, longed to be the first to reach the core. Stuart knew very well that any Australian explorer joined a brotherhood, if not a cult, of suffering. He had himself been a member of Captain Sturt's expedition, the draughtsman who sketched the landscape and drew the charts. Sturt had commended the 'zeal and spirit' that made Stuart such a dogged traveller and enabled him to endure the gruelling travel in spite of the poor health he invariably suffered.

The other thing that kept Stuart going in the bush was the prospect of whisky at the end of the long trail. 'He drank as much of it as he could,' wrote one of those who knew him, adding that 'he was not particular about a bottled natural history specimen or two. But he had the pluck of a giant in his puny frame, coupled with prudence and good judgment.' By 1860 Stuart, a bushy bearded bantam in a blue coat, had accumulated years of experience of outback surveying and prospecting and knew how to survive on the anvil of the desert. In March of that year he and two companions, and thirteen horses, set a course for the centre. Stuart harboured hopes of glory but no illusions.

When I followed part of his route, trailing a plume of red-brown dust in a four-wheel-drive vehicle, I could only marvel at what he and men like him achieved. As I walked on the cindery plain I felt the weight of heat and silence and the brooding temper of the desert. Standing at the centre of an immense rusty disc I thought I could be persuaded by a plausible rogue that the world was flat. When I stopped near Lake Eyre I walked through saltbush and spinifex and screwed my eyes against the brilliance of the salt crust. A mirage of islands floated on the white haze. The lake barely supports life; but about twice every century it is inundated and becomes the inland sea of legend; and in the meantime, about every eight years, it floods enough to be a breeding ground for great flocks of water birds. It is dramatic and magical and can be dangerous. Near here, not long before I passed, a young woman left the protection of her sand-bogged vehicle to seek help. She wandered for eighteen miles, scribbling despairing notes, until the sun sapped her of life. A rough cross, erected where her body was found, bore the old word always used for those who died in the desert: Perished.

For Stuart, survival depended on his ability to find water. His horses would live only two days without it. From Marree to

Oodnadatta he followed the strange mound springs, the chain of artesian wells like miniature volcanoes that are still bubbling and which formed Stuart's stepping stones through the salt lakes towards the centre. By now he was suffering from ophthalmia, a painful inflammation of the eyeball, and was almost blind in his right eye. On 22 April, seven weeks after he started, and 120 miles north of the waterhole that would later be called Alice Springs, he wrote in his journal that he had reached the 'Centre of Australia'. He underlined the words twice. It is a place of red sandstone hills, pleasing grasslands and eucalyptus trees. A brick monument to the explorer stands beside the bitumen ribbon of the Stuart Highway.

By this point in his journey Stuart's limbs ached with scurvy and were growing black. His mouth and gums were so tender he could eat only a pap of flour and water. Yet he rode on. Astonishingly, he and his companions persisted for another 150 miles until disease, thirst, failing horses and an attack by Aborigines, who showered them with boomerangs, forced them to retreat. Recuperating in Adelaide in October 1860, after the return journey of 1,200 miles, Stuart noted laconically in his journal that his emaciated companions looked a hundred years old.

As Stuart was making his way south from this failed first foray, Melbourne's man was setting off in style at the head of the Victorian Exploring Expedition. Robert O'Hara Burke, aged thirty-nine, was a police superintendent stationed in Castlemaine, forty miles from Melbourne. Born in Galway, he had matured into a slightly eccentric, slightly reckless bachelor, known for the warmth of his character and his dash. He had been a soldier in Austria and had a duelling scar on his cheek. He was brown-haired, blue-eyed and handsome; and like many men of the time he had taken cover behind a huge dark beard.

Curiously, Burke was no explorer. He had, one might say, no track record: no reputation for judgement, resolve or tenacity, and no particular bush skill. Perhaps the Melbourne organizers appointed him leader because of his apparent charm and because they esteemed him as a country gentleman, a rank confirmed by his membership of the exclusive Melbourne Club. He was full of vigour, excited by the prospect of being first across the continent. Do or die, he had promised. Meanwhile, he had his eye on a pretty actress of sixteen who, he thought, would not be able to resist him on his triumphal return.

His surveyor was William Wills, a twenty-six-year-old Devonian, who had 'a longing desire' to explore the Australian interior and shared his chief's hopes of fame. Exploration was in his family: his cousin Harry had perished in the Arctic with Sir John Franklin's expedition in the 1840s. Carefully, Wills packed his sextant, telescope and other instruments into boxes. He was a navigator; his boss was not. Burke was known to lose his way frequently, even on well-beaten paths.

Businessmen in the state of Victoria raised £9,000 to fund the exploration. No expedition in Australian history had been better equipped. The nineteen explorers had twenty-three horses, six wagons, lime juice to prevent scurvy, the latest firearms, twenty-one tons of provisions expected to last for eighteen months and charcoal water purifiers to ensure that no one would go thirsty.

They also had the expedition's guarantee of success, twenty-seven camels. The sponsors had sent George Landells to Peshawar on the North West Frontier of India to buy them and to hire four Frontier men to drive them. During his trip Landells met John King, a young English soldier, and invited him to join the Australian venture. King was a quiet man who had witnessed some of the horrors in the orgy of British vengeance that marked the aftermath of the Indian Mutiny. He had seen Indian prisoners taken to the mouths of cannons and blown to pieces in a shower of blood and limbs.

A stone monument in a park in Melbourne marks the place where, on 20 August 1860, the explorers harnessed their animals and packed their wagons. Sixty gallons of rum were loaded on the carts, for the camels not the men. Landells, who cut an exotic figure in his Frontier dress of pantaloons, long smock and turban, believed that the beasts marched better on a daily ration of grog.

At last, restless and anxious, Burke's men formed a long and straggling line to start their journey north. Melbourne took the day off, a crowd cheered and a band played. Burke looked magnificent, a gallant figure mounted on a grey horse.

Five hundred miles away, in Adelaide, the Homeric John McDouall Stuart rested only a month after his long desert punishment. He learned that Burke and Wills were on their way and announced that he was ready to have another go. South Australia's government put up the money and the wiry, dried-up Stuart set off again in November 1860 with eleven men and forty-nine horses. The delighted *Melbourne Punch* of 8 November 1860 encapsulated the rivalry, publishing a cartoon showing the heavy-bearded Burke atop a camel looking down at Stuart cantering beside him on a horse. 'The Great Australian Exploration Race' read the caption, which was just how the public saw it, and beneath it the doggerel declared:

> A race! A race! across this land
> From south to northern shore.
> A race between two colonies!
> Each has a stalwart band
> Sent out beyond the settled bounds,
> Into the unknown land.
> The horseman hails from Adelaide
> The camel rider's ours –
> Now let the steed maintain his speed
> Against the camel's powers.

When it left Melbourne Burke's expedition was a proud cavalcade. Quite soon, as it moved slowly northward, it began to unravel like a woollen snagged on a thorn. Quarrels and resentments rumbled. Burke's impatience soured the air and brought about a virtual mutiny. Six of the party quit in disgust. Others were dismissed. Burke fired the camel master who at once denounced him as mad. A Melbourne newspaper reported that he was disagreeable to the men and bullied them 'soldier style'. The editorial opinion was that 'he is not the right man for the work he has undertaken'.

As a sideshow to the drama of human bickering the male camels fought each other for the females' favours.

All the animals and wagons were plainly overloaded. Burke ordered the dumping of some of the supplies, including sugar and lime juice. Meanwhile his backers in Melbourne put him under pressure with messages which both urged him on and told him to cut costs. Burke forced the pace, feeling that the honour of Victoria was in his hands. But he had made a bad start. The only cheering news was that Stuart had failed in his first expedition.

After two months of toil Burke's party had covered only 450 miles. It ventured into the wilds with its morale in tatters and the remaining men confused. In November 1860 the expedition reached Cooper's Creek, which Captain Sturt had found and charted fifteen years before. Burke was in luck. The temperature was more than 100 degrees Fahrenheit but water flowed in the Cooper and the billabongs mirrored a luxuriance of green vegetation, eucalyptus and coolibahs.

This was the truth and marvel of the desert, that for all its furnace heat, aridity and the glare that made eyes ache, there were thin and often elusive veins of moisture and rare and unpredictable rains that gave startling birth to an abundance of plants of the brightest colours. At Cooper's Creek, when Burke arrived, Aborigines fished and feasted, kangaroos came to drink and pelicans hunted. The stream was an oasis teeming, buzzing and cackling with life. Here the expedition had reached its critical point; and here Burke, whose judgements and orders had rarely been clear, or the product of measured thinking, made his fateful decision.

He would go for glory. He, William Wills, Charley Gray and John King, a small and lightly laden team, would make a daring thrust to the Gulf of Carpentaria, a race-winning commando raid. Burke would come back as a man of renown. He ordered four men to stay at the Cooper, to build a stockade, keep a cache of food and wait three months for his return from the north.

Any idea that the expedition had a scientific aspect was now cast aside. Burke's poor management and irascibility had brought all his men to a dangerous corner; and he had planned no sensible escape route.

He and the three chosen for the attempt to reach the ocean loaded six camels and a horse with a store of dried beef and pork, flour, rice, sugar, salt and tea: iron rations for ninety days. To save weight they took no tents, just bedrolls. On the morning of Sunday, 16 December 1860, they shook hands with the Cooper's Creek party and set off.

Wills, the second-in-command, noted the time, 6.40. Without his assiduous note-keeping we should know very much less about what happened. His boss, no proper explorer, kept no proper journal, and his notebook remained almost a blank. Wills was also the only one who could use a sextant and knew where he was. That day, as on every other, he pointed the way. To reach the coast they would have to travel about 900 miles.

At first they enjoyed pleasant countryside, passing through creek-fed greenery, and made good going. Trudging for eleven, twelve, even sixteen hours a day, they sometimes exceeded thirty miles in twenty-four hours. They rose early to march before the sun grew too hot and then stopped to breakfast on tea, dried meat and a damper, an unleavened loaf of flour and water baked in the ashes of a fire. They had much the same meal for their supper and sometimes walked at night, silvered by the moon. One of Burke's scanty notes recorded a meeting with amiable Aborigines who gave them fish. The four men awarded themselves a creekside holiday on Christmas Eve, an idyll but for the flies and the mosquitoes. They trekked across part of Sturt's Stony Desert to what is now Birdsville, their bodies scourged by thorns and steadily eroded by heat, dust and dehydration. Fortunately they found a branch of the Diamantina River to take them north. In the middle of January 1861, however, they and the camels had a terrible ordeal getting through the Selwyn Range and passed close to what is now Cloncurry.

We may imagine that Burke felt his hopes rising, that he was within sniff of the sea. But he was in the tropics at the wrong time and had ignored advice that it was dangerous to strike out for the Gulf of Carpentaria during the rainy season. He and his hapless men soon discovered all the wretchedness of swamp travel. Pestered by mosquitoes they floundered in the mud, their strength sapped by high humidity. The camels hated it. They sank to their knees in the oozing mud. Each step was agony, for their big feet were like sink-plungers, hard to withdraw from the slime. One of them fell, exhausted, and was abandoned. It was evident, too, that the others had reached their limit. Burke told Charley Gray and John King to look after them while he and Wills set off for the gulf in hosing rain. Wills walked and Burke rode his horse, but no animal could get far. On foot they staggered through a mangrove swamp, camped and then struck out for the sea.

At last, on 11 February 1861, when they were fifty-seven days out of Cooper's Creek, their gaze wandered over a stretch of drab tidal swamps. No doubt they had dreamed of the explorer's moment, the triumphant and inspiring view of their goal. But this was as far as they would get, about a dozen miles from the sea. They scooped up some water and tasted salt. It was not a glorious moment, but it would have to do.

In their march north they had shown remarkable endurance. Now they had to get back. They rejoined the others and braced themselves for the suffocating humidity, incessant rains and quagmires; and then for the kiln of the desert. They started on 13 February and, since they had already eaten the bulk of their supplies, they ran the risk of starving. They moved slowly, thrashed by storms, and Burke became very ill. They ate a snake, but otherwise did not seek out much bush food. They shot and ate a camel and, later, consumed the horse. Still their strength drained away. Charley Gray was found cramming flour into his mouth to assuage his hunger and Burke gave him a beating. Three weeks after that, on 17 April, Gray died near Lake Massacre. His companions spent a day hacking out a grave to give him a respectful burial.

Four days later, and sixty-six days' march from the gulf, Burke, Wills and King stumbled into the camp at Cooper's Creek. We can only guess that as they covered the last mile or two they would have dreamed of food and looked forward to the glow of glory as they told their waiting friends of their adventure. Burke would surely have envisaged his vindication and have relished the idea of seeing all his critics confounded, all his follies buried in Australia's acclaim.

Here beside the creek was one of the great tragic scenes of nineteenth-century exploration. Burke called out. There was no answer. The camp was empty, the ashes of the fire still warm. Nine hours earlier, that very morning, the Cooper's Creek team had given up on Burke, had packed up and headed south to civilization with six camels and a dozen horses. They had waited four months and then agonized for a further week. Just before they left they cut a message in the bark of a coolibah tree: 'Dig Under 3ft N.W.' and, on a branch, the date, 'Apr.21 1861'.

Burke slumped to the ground in shock. Wills and King dug as instructed and found a month's supply of meat, rice and sugar and a bottle containing a message explaining that the Cooper's Creek men had left because one of them was seriously ill. The note did not say that all of them were growing weaker and showing signs of scurvy. Their situation was desperate. Burke had staked everything on his dash to the north and had made no secure provision for the survival of his expedition after the gulf had been reached.

Fish, birds and water were abundant at Cooper's Creek. By hunting and gathering the explorers could have lived there, restoring their strength and awaiting rescue, or, perhaps, making themselves fit enough to journey more than 360 miles south to civilization. Burke, however, weak and malnourished as he was, decided they should walk about 150 miles to the settled district of South Australia and in particular to the police station near Mount Hopeless.

Within fifteen days of setting off their two remaining camels were dead, they had lost hope of reaching Hopeless and they were back on the Cooper, a few miles from where they had started, and very hungry. Aborigines fed them fish and a paste made from seeds pounded into flour, but Burke had no wish to become friendly with them and repaid their hospitality with pistol shots over their heads. Not surprisingly, the food supply ceased. The three men made their own seed flour. It gave them indigestion and caused them to pass huge stools, an exhausting process.

Slowly they starved. Towards the end of June Wills wrote a letter to his father, concluding 'My spirits are excellent.' In his last diary entry he noted that he was skin and bone and was waiting, 'like Mr Micawber, for "something to turn up"'. He died a few days later. Burke and King struggled on for a few miles and then Burke lay down in his ragged clothes, with his revolver in his right hand, and died. King wandered off along the creek.

Around this time, in July 1861, John McDouall Stuart, on his second expedition from Adelaide, was facing defeat again. He had reached the springs he named Newcastle Waters, about 400 miles south of Darwin, and tried several times to find a way out through the scrub. But his men were faltering and the horses were sagging. They would all die if they continued. Once again, Stuart retreated and reached Adelaide in September.

At the end of June 1861, while Burke and Wills were dying beside Cooper's Creek, the Exploration Committee in Melbourne asked Alfred Howitt, an experienced and efficient bush man, to travel north to find them. On 15 September one of Howitt's horsemen, searching the bank of the creek, came upon a group of Aborigines and among them found the skinny and ragged figure of twenty-two-year-old John King. He had been looked after by the kindly Aborigines for two and a half months.

King carried a pouch containing the last letters of Burke and Wills and also their watches. He led Howitt to the water holes where his companions had died and the searchers buried the bodies. Howitt and his men mounted up and took the distraught John King slowly back to Melbourne.

On the way, on 2 November, Howitt stopped in Bendigo to telegraph the news of the deaths of Burke and Wills. In Melbourne a theatre manager went on stage to break the sensational report to the audience. The city shuddered and next day there was barely a face that did not shine with tears. The hitherto unknown and lowly ex-soldier John King came back, pale and frightened, to a mobbing by thousands of Melburnians.

The government in Adelaide persuaded the exhausted John McDouall Stuart to try again for the prize of a continental crossing. In October 1861, while John King was being brought home, Stuart rushed through his preparations and started for the north with eleven men and seventy-one horses. On the way out of Adelaide one of the animals reared, kicked Stuart unconscious and trampled on his right hand. This injury was so serious that there was a risk of amputation. Stuart had to rest for six weeks before riding to catch up with the expedition.

As he convalesced he heard the news from Melbourne of Burke's triumph and disaster. It made no difference. He believed he could come back alive with a feasible telegraph route. He caught up with his men and fell into the discipline of his overland routine. He took a compass bearing each morning, lit his pipe and urged his horse forward on the bearing. The pace was steady and relentless and no man was allowed to drop out of line or to drink water without permission. Behind his calm front Stuart confided his private fears to the journal he kept so meticulously. 'I feel my capability of endurance beginning to give way,' he wrote.

Short of time and food he and his men fought their way north through the confusing landscape until, on 24 July 1862, Stuart reached the coast near Darwin and savoured the moment which had been the focus of his longing for nearly twenty years: 'I was gratified to behold the water of the Indian Ocean . . . I dipped my feet and washed my face and hands in the sea . . .'

Sickness quickly corroded the elation. Heading south, Stuart recorded that 'I get no rest day or night from this terrible gnawing pain.' His legs were blackened by scurvy. His men's eyes, he noted, were empty of any expression. The horses were sacks of bones. Men and animals were dangerously short of water. Stuart was paralysed and hardly able to see when his companions lifted him from his mount and put him in a stretcher slung between two skinny horses. 'I am doubtful', he scribbled in September, 'of being able to reach the settled districts.'

One of his men said: 'Mr Stuart, you are not fit to travel today.' Stuart mumbled: 'I must go on . . . if I stop all will be lost.'

In 1862 the Melbourne Exploration Committee sent Alfred Howitt back to Cooper's Creek to retrieve the bones of Burke and Wills from their shallow graves. Dingoes had scrabbled and gnawed at the skeletons. Howitt put everything he could find into boxes and headed home. He found it more convenient to ride by way of

Mount Hopeless. In December he arrived in Adelaide and the remains were carried through the streets and then shipped to Melbourne. On the last day of the year, in a bizarre ceremony, the bones were placed in coffins. A number of people bent to kiss Burke's skull, as if it were a saintly relic. On 21 January 1863, eighteen months after their deaths, Burke and Wills, Victoria's martyrs, were accorded all the solemn magnificence of a state funeral and borne in a carriage drawn by six black horses through streets draped in black and crowded with 40,000 mourning citizens.

That same day John McDouall Stuart had recovered enough of his strength to appear in a triumphant procession that filled the streets of Adelaide. Willpower had carried him home. He was half-blind again and his right hand was ruined. His own diagnosis was that he was 'a sad, sad wreck of former days', but he thanked God for the strength 'to overcome the grim and hoary-headed king of terrors'.

The Advertiser in Adelaide was exultant.

> The feat is unparalleled; many have attempted it, but none except Stuart and his party have achieved it. Burke nearly effected the object, though not quite; but Stuart succeeded completely . . . he alone wrested from the interior its long hidden secret. What was the map of Australia in our school days? What was it ten years ago? It was a vast blank . . .

The South Australia government awarded Stuart £2,000 for his exploit. The glow soon faded, as it always does. The following year, his health deteriorating, this most remarkable of explorers decided to return to Britain. His departure from Adelaide went unnoticed. In 1865, after some pressing, South Australia grudgingly paid him another £1,000. In 1866, only three years after the crowds had mobbed him in Adelaide, Stuart died in London, all but forgotten.

25

The bush staccato

'Faintly as from a star
Voices come o'er the line;
Voices of ghosts afar,
Not in this world of mine.'
Robert Service, 'The Telegraph Operator'

Eight years after Stuart's expedition hundreds of men followed his tracks to build the Overland Telegraph Line clear across Australia. The surveyors navigated by his map, 'so correct', said one of them, 'that we used to put a protractor and scale on it, get the bearing and distance and ride on with confidence'.

There was no difficulty at all in recruiting workers. Adventurous men flocked to the hiring offices anxious to wield pick and shovel to make history; or, at least, money. Seven of them died, of thirst, scurvy, fever, tuberculosis and drowning, a remarkably small number considering the hard work, the broiling heat and the rains and floods of the northern wet season. The line was to stretch for 1,975 miles from Darwin to Adelaide and Charles Todd, Postmaster-General of South Australia, directed the construction and divided it into three sections. The north and south went to private contractors and the isolated and hazardous central section to the government. A team of men assigned to the central section took eight months to ride to the place where they would begin their work.

The telegraph cable from Europe curved across land and sea to Banjoewangie in Java and a submarine line was laid beneath the Timor Sea to come ashore in Darwin. The first telegraph pole was erected there on 15 September 1870 and the northern construction team headed south while the southern team headed north. Hundreds of bullocks, horses and camels carried reels of wire, loads of building stone, insulators, signal keys and sounders, glass jar batteries nine inches tall, chemicals and 36,000 twenty-foot poles weighing 5,000 tons. To feed themselves the men drove thousands of sheep and cattle along the trail. As well as setting up the poles and stringing the wire they built twelve substantial repeater stations of stone or timber and dug wells and made water tanks for them. The stations were necessary because the Morse signal faded after about 180 miles and had to be repeated by operators sitting day and lantern-lit night at their benches.

In March 1871, looking for a likely site for a station at the midway point of the line, a thousand miles from Darwin and a thousand from Adelaide, the surveyor William Mills discovered a waterhole he called Alice Springs, after Charles Todd's wife. The man appointed as the first station-master at Alice Springs, C.W. Kraegen, was among those who died during the telegraph's construction.

As Charles Todd himself reported, Kraegen and two other men, Mueller and Watson, were riding up to the newly built station at Alice Springs to open communications.

> The second day out they camped without water, and then they lost their way and failed to discover water. On the following day, Kraegen, being the strongest of the three, pushed on for water but did not return. Mueller and Watson wandered that day and the next in all directions under a burning sun, vainly looking for water, and finally, after killing one of their horses and drinking its blood, succeeded in recovering the track and retraced their steps. Search was made for Kraegen who was found lying dead at the foot of a telegraph pole. His remains were interred on the spot . . .

As the two ends of the line snaked towards each other Todd introduced a pony express to bridge the message gap; but it was not needed for long because the entire line was completed in twenty-three months. An operator tapped the first message on the 12,500-mile London-to-Adelaide link on 23 June 1872. Next day the Banjoewangie–Darwin cable broke down and the fault was not repaired until 21 October. This was regarded as the official opening date and Charles Todd felt confident enough to announce that the Australian colonies were now 'connected with the grand electric chain which unites all the nations of the earth'. At first it was an expensive connection. To send four words to the earth's ends cost the equivalent of a week's pay for a working man.

Within two years termites had eaten many of the telegraph poles. Iron replacements were shipped from Britain. The line often broke down, struck by lightning or broken by Aborigines who took the wire for fish hooks and smashed the porcelain insulators to make sharp spearheads and scraping tools. Every time there was a breakdown men rode from the repeater stations along the line of poles until they found the damage. The journey often took them several days.

In their lonely outposts, more than a hundred miles from their nearest neighbours, the operators were permitted eavesdroppers, the first to hear the news from London, passing it on like a baton. At Barrow Creek station on a hot February evening in 1874 seven men and a boy sat out on the veranda, smoking and doubtless discussing the international news to which they were privy.

Out of the indigo night fell a shower of spears. As the men bolted for the shelter of the courtyard one of them was pierced through the heart. James Stapleton, the station-master, tried to close the courtyard gates and was seriously wounded by four spears as he did so. Two other men were also hit, but they and their colleagues slammed the gates shut and drove off the attackers with rifle fire. Like most telegraph stations Barrow Creek was built to be defended and had loopholes for guns. One of the wounded men tapped out a message to Alice Springs which was sent on to Adelaide. There Charles Todd dispatched his carriage to bring Mrs Stapleton to the telegraph office. Twelve hundred miles north, the dying station-master insisted on being helped to the Morse key to send a message to his wife in Adelaide. Todd himself sat at the receiving desk and took down the words, 'God bless you and the children.' Without a word he passed them to Mrs Stapleton who sat beside him.

It was a sensational story, all the more so because it was reported swiftly by telegraph. It emerged later that the Aborigines may have been aggrieved because Barrow Creek station was built near their waterhole. Whatever the reason the outraged Adelaide press demanded retribution. A police officer was assigned to lead a punitive expedition of mounted constables and telegraph workers which was specifically exempted from any legal restraint. Its job was to teach the natives a brutal lesson. For some months the posse ranged over thousands of square miles of territory and shot every Aborigine it could find, perhaps fifty, including women and children. A hill near Barrow Creek, called Blackfellows' Bones, and a watercourse near Tea Tree, called Skull Creek, are a reminder of the bloody revenge. These murders set the pattern for dealing with 'outrages' for many years and established the outback policeman as an officer free to enforce the law as he saw fit.

In its first year the Australian Overland Telegraph charged a minimum of £10 for a telegram. It nevertheless handled 9,000 messages. The service gradually became cheaper. Farmers quickly discovered that it enabled them to find profitable markets for their wheat.

Shipowners sent telegrams to the captains of their vessels in Adelaide, Melbourne and Sydney instructing them where to find a cargo. Telegraphed orders provided British manufacturers and investors with clear and rapid information about the extent of the Australian market: they no longer had to guess. Australia was small, with just over a million people, but it was growing in commercial strength.

Meanwhile, there was the vital matter of cricket. In 1873–4 English cricketers toured Australia and for the first time the scores were sent to newspapers by telegraph. Returning from the tour W.G. Grace was gratified to find that, thanks to the telegraph, 'our doings in Australia had been followed with the keenest possible interest'.

The telegraph line formed a bridge into the centre of Australia. The repeater stations, built beside water sources, were the launching places for explorers heading into the unmapped outback, and oases for pioneer farmers looking for pastoral land. They also

e stupendous progeny of Mr Brunel': launched at Bristol in 1843 the steamship *Great tain* was the largest and most powerful ship in the world

nner: bridge builder Robert Stephenson)3–59) was, with Brunel and Thomas 'ord, one of the engineering geniuses of :ain's industrial advance, and lies beside 'ord in Westminster Abbey

Another anxiously chewed cigar as Isambard Kingdom Brunel (1806–59) watches an attempt to launch the *Great Eastern*, his third shipbuilding masterpiece, in London. Problems with the vessel fatally undermined his health

Working on the railway: engineers camp in the Rocky Mountains, British Columbia, dur
the epic building of the Canadian Pacific line

A loop in the ribbon of iron: the 2ft-gauge Darjeeling railway, built in 1879, took British
officials to their cool summer quarters in the Himalayas

ht Canada's big feat: railway through the of high mountains, npleted in 1885, ined Canada itself. re a train passes below ›unt Stephen in the ckies

'ow On the fringe of Raj: Gulistan railway :ion, near the Afghan ntier

On the stage: Wells Fargo office, Virginia City, Nevada, 1866

Twilight: Plains Indians depicted by George Catlin in his self-imposed mission to record their vigour before the West was settled

end of Africa, famously found, though lost: David Livingstone (1813–73)

Wales, America, Africa: Sir Henry Morton Stanley (1841–1904), reporter and explorer

oding: scholar, swordsman, linguist, entalist, master of disguise, African lorer and translator of the *Kama Sutra*, Richard Burton (1821–90)

The Nile is mine: John Hanning Speke (1827–64) ran to claim the prize and faced a public showdown with Burton and perhaps humiliation

Above left Desperate in t
outback: Robert O'Hara
Burke (1821–61) gambled
life on a dash for glory.
Portrait by William Stru

Above right 'We wear th
Queen's scarlet': more th
a police force, the Moun
built an image as nation-
builders and Canadian ic

Left 'I must go on': Joh
McDouall Stuart (1815–6
conquered the 'vast blan
at the heart of Australia

er pitch: disease ended the first attempt to build the Panama Canal in the 1880s. ıedical campaign brought success in 1914

llion-murdering Death': Sir Ronald Ross (1857–1932) fought the microscope war against aria

Past meets future: Emperor Napoleon III visits the new electric telegraph office in Paris

Radio days: Guglielmo Marconi (1874–1937), the determined pioneer of wireless telegraphy, was helped by news of the Prince of Wales's knee and Dr Crippen

Just the ticket: Thomas Cook (1808–92), entrepreneur, fixer and marvel, made dream travel come true. Very sound on lunch

provided a tenuous link with medical assistance: the station-masters telegraphed to Adelaide with details of symptoms and injuries and a doctor sent a message back with recommendations for treatment.

Alice Springs station survives today as a museum, an evocative place among gum trees and rock-strewn hills, where it is easy to imagine that it was a link in the global chain from London to Adelaide, its handful of people the guardians of the slender thread. The compound had a stone building of eight rooms, the walls of which were very thick to keep out the heat. There was a kitchen, a blacksmith's shop, a harness room, a battery store with hundreds of wet-cell jars, a well and a stock of sixty horses. The telegraph office itself was manned day and night in four-hour shifts, the operators fluently translating the Morse they heard through the amplifier and transmitting it on their repeater keys. The news flowed through them as if their ears and fingers were brass and copper and their nerves were electric wire. Here in this outback junction, men listened intently, scribbling the telegrams, speeding the news of great events, the mundane traffic of business and the personal information of births and deaths.

Beside the Todd River, two miles south of the Alice Springs telegraph station, the settlement of Stuart grew into the first permanent town in the heart of Australia and in the 1930s adopted the name of Alice Springs.

Ernest Giles was the more thoughtful sort of explorer. His journeys through the deserts of Australia took him close to death several times. He might easily have died of thirst or from the thrust of an Aborigine's spear, and during all of his expeditions he experienced much pain and protracted suffering. Exploring a particularly harsh stretch of country he summarized it in three words: 'Dreadful, gloomy, desolate'. He was no stranger to failure. It was no wonder he considered that 'the exploration of a thousand miles in Australia is equal to ten thousand miles in any other part of the earth's surface'. Half-dead upon his failing horse he began to ask himself what was the meaning of it all.

Giles was a ponderer. He carried poetry books in his saddlebag and peppered his journals with verse. He wrote a definition of a good explorer. 'He should be able to make a pie, shoe himself and his horse, make an observation, mend a watch, kill or cure a horse, make a pack-saddle and understand something of astronomy, surveying, geography, geology and mineralogy.'

As a seasoned bushman, he knew how to cook a horse and described how it should be done. The delicacies of the tongue and brains were eaten first and the flesh was cut into strips and smoked over a fire. The hooves and bones were meanwhile boiled to produce 'a jelly that stank like rotten glue'. This was nevertheless nutritious 'and at breakfast when this disgusting stuff was coagulated, we would request one another with the greatest politeness to pass the glue pot'.

Giles was one of the explorers who used the new telegraph line as a starting point for expeditions into the still unexplored regions of central and western Australia. No European had followed the sunset and crossed the desert from Alice Springs to the Indian Ocean. Giles decided to give it a go.

He needed an achievement. A Bristolian educated at Christ's Hospital in London, he had migrated to Australia and found failure as a digger in the goldfields of Victoria, was made redundant by the Post Office in Melbourne and started working as a surveyor. In 1872 he led two other men on a horseback expedition from Alice Springs, aiming to reach the coast. To his surprise he found a beautiful flowery and palmy glen amid the stony desolation but otherwise the journey was unpleasant and frustrating. Flies settled thickly in the nostrils, ears and eyes of the men and their mounts, dense thorny scrub made ribbons of the animals' legs and furious Aborigines sprang from rocks and yelled at them. Giles never forgot the terrifying sound. Horses often angered Aborigines because they drank so much of the meagre supply of desert water. In the end it was a lack of water that defeated Giles and his companions and they turned back to the telegraph line.

But Giles was a questing man and always willing to risk his life. He returned the following year in another attempt to reach the west coast. In April 1874 he and one of his party, Alfred Gibson, left their main camp for a foray into what Giles called 'the worst desert upon the face of the earth'. They were dangerously short of water when Gibson's horse collapsed and died. For a while they took it in turns to ride Giles's mare, Fair Maid of Perth, but Giles knew they would die of exhaustion and thirst if he did not take action. He ordered Gibson to take the horse and ride thirty miles to a place where they had earlier cached kegs of water. Gibson was then to make for their camp, about sixty miles farther on, get help and fresh horses and return to rescue him.

As Gibson rode off Giles started on the long walk to the water cache. When he reached it he found that Gibson had been there and had left him two gallons in a keg and a few dirty strips of smoked and dried horsemeat. Giles reckoned it 'highly probable that I should be dead of hunger and thirst long before anybody could possibly arrive'. There was nothing he could do but strike out for a distant creek.

The keg of water weighed thirty-five pounds. He hoisted it to his shoulder and bent double under its weight. He longed to drink the water but knew he had to ration it. 'It was the elixir of death I was burdened with, and to drink it was to die.'

He walked by night, rested in any shade he could find when the sun rose high, and covered about five miles a day, eking out the water from the burden on his back. By the third or fourth day he had consumed the scraps of horsemeat. Thorns punctured his arms and legs, and the holes festered into ugly pus-filled sores. He lost track of the passage of time and often drifted into unconsciousness.

He had been travelling for about five days when he drank the last of his water. He hobbled on 'in April's ivory moonlight' and at last staggered to the creek. He could hardly stop drinking. 'Oh, how I drank! how I reeled! How hungry I was! How thankful that I had escaped from the howling wilderness.' He stayed, drinking and drinking, and then heard a squeak and saw a tiny wallaby, thrown from its mother's pouch and still just alive. 'Like an eagle I pounced upon it and ate it, living, raw, dying – fur, skin, bones, skull and all. The delicious taste of that creature I shall never forget. I only wished I had its mother and father in the same way.'

He was barely alive when he reached the safety of his base camp and woke one of his party. The hapless Gibson had never been seen and had evidently perished. Giles named the desert after him and recalled the conversation they had had when they first rode out together, into the hammering heat. Gibson had told him that his brother had died with Franklin in the Arctic. 'How is it', he asked Giles, 'that in all these exploring expeditions a lot of people go out and die?' Giles replied that apart from accidents and attacks by natives a want of judgement or knowledge often brought about death. 'Well,' Gibson had said, 'I shouldn't like to die in this part of the country.'

Giles confronted the desert a third time, in 1875, and led a party from Port Augusta, north of Adelaide, into the Great Victoria Desert and across the emptiness of Western Australia. A hundred spear-throwing Aborigines attacked his party at Ularring and were driven off by gunfire. 'The spears were long and barbed,' Giles wrote, 'and I could not help thinking how much I liked them on my outside rather than my in.' Like many of his contemporaries he thought that Aborigines would inevitably become extinct as white men settled the country.

> The Great Designer of the Universe permitted a fiat to be recorded, that the beings whom it was His pleasure to place amidst these lovely scenes must eventually be swept from the face of the earth by others more intellectual, more dearly beloved and gifted than they. Progressive improvement is undoubtedly the order of creation, and we perhaps in our turn, may be ruthlessly driven from the earth by another race of yet unknown beings of an order infinitely higher and more beloved than we.

He reached Perth after six months and, like a pilgrim, bathed in the ocean. Shortly afterwards, the local people shook their heads as they watched the indefatigable Giles commence a return journey. This time he suffered from ophthalmia and, temporarily blinded, his eyes badly irritated and a magnet for flies, he had to be led by one of his companions. In honour of his agony he named a group of mountains the Ophthalmia Range. Other explorers had named their places of torment Mount Dreadful, Mount Despair and Mount Misery. At one stage Giles noted ruefully that 'the pale spectre of despair comes to our lonely tent'. He crossed his old foe, the Gibson Desert, and reached the telegraph line nine months after he had set out from the coast. The Royal Geographical Society awarded him a gold medal for his assiduous and courageous gathering of knowledge of unknown territory.

Giles asked the panjandrums of Adelaide for the reward of some official position but, prigs that they were, they refused on the grounds that he was a man known to enjoy a drink, that he was 'not always strictly sober'. He died at Coolgardie, an obscure clerk, in 1897.

A telegraph cable was laid from Sydney to New Zealand in 1876; and the following year a telegraph from Perth to Port Augusta connected the West to the rest of Australia. In 1902, a cable running from Queensland to Vancouver by way of Norfolk Island, Fiji and Hawaii linked Australia to North America. That year, too, a cable was laid from Durban to Perth. Telephone wires were added to the Australian Overland Telegraph poles in the 1940s.

26

The cameleers

'The beautiful horse sinks to the insignificance of a pigmy when compared to his majestic rival.'

Ernest Giles, 1872

Scorched on the bare brown plain, Marree is these days a township of fifty people, a couple of stores and petrol stations, a police post and a three-storey hotel which was built in 1885. Most afternoons a noisy crowd of travellers and white and Aboriginal labourers fills the smoky bar.

At the end of a track, a mile or so out of town across the desert, is a cemetery fenced with drooping rusted wire, a desolate and stony patch for bones. Some European pioneers lie here but many of the graves are those of Afghans, the mounds surmounted by white stones and crumbling scraps of wood bearing the names of Khan, Rasheed and Wahub. The forebears of these men came to Australia in the nineteenth century from Kashmir, Punjab, Baluchistan, Egypt, Iran and Afghanistan; but whatever their origin they were known everywhere as Afghans, or just 'Ghans. They brought with them no wealth, only the skill and experience that made them valuable in the exploration and settlement of the Australian interior. They were the cameleers. For more than sixty years, from the 1860s to the 1920s, they drove the camel trains that carried into the distant parts of half the continent everything that settlers and farmers needed: mail and medicines, food and furniture, trousers and timber, pots and parlour pianos. Horses and bullock-teams and railways played their part, but for many years the principal carriers of the outback were the long-suffering, unhurryable and supercilious camels.

The first suggestion that camels should be used to explore the Australian desert was made in 1839 by a writer in the *Sydney Herald.* A few years later, John Horrocks, a South Australian farmer, bought Australia's first camel, which had been shipped from Tenerife, and used it to explore the brown and purple hills of the Flinders Ranges. During such a trip in 1846 the beast jerked its head, and the gun that Horrocks was holding discharged, causing a wound from which he died a few weeks later, aged twenty-eight.

Camels made their proper Australian début with the Burke and Wills expedition in 1860. A few years later, in 1865, Sir Thomas Elder, a Scottish immigrant and landowner, imported 124 camels and thirty-one handlers and started a breeding station at Beltana in South Australia. It became a base for expedition camels. They quickly proved their value, travelling easily over the sand ridges that defeated horses and bullocks, and coping with the deserts of small sharp stones known as gibber plains. They survived on wizened vegetation, nibbling at saltbush and the leaves of prickly acacias and thorn, and drank foul water that horses could not stomach. Left to forage, Australian camels grew larger, fitter and more fecund than their Asian ancestors.

In the 1880s Marree was a grand junction for camel trains, a bustling green oasis and the heart of Afghan activity in the outback. It had herds of hundreds of beasts. The Afghans planted date palms, which survive to this day, and built a mosque, the first in Australia. They drove strings of camels to the desert. Between thirty and seventy animals walked in single file, the nose rope of one linked to the tail of the beast in front. They hauled freight along the Birdsville Track, which ran for 300 miles into Queensland, and walked the 500-mile Oodnadatta Track to Alice Springs and beyond, serving areas where horses could not work and railways did not reach.

Across the desert, William Creek, population eleven, is little more than a roadside pub and store. A memorial here salutes the men who built the Overland Telegraph, and the camels that hauled the poles, wire and insulators. Camels supplied the gold rushes in the Kimberleys, Coolgardie and Kalgoorlie. Famously, a camel that strayed in the Kalgoorlie district in 1894 was tracked by its handler who found it browsing. One look told him that the beast was nibbling the meagre vegetation on a reef of gold and he went at once to tell his boss. The camel's picnic spot became the Wealth of Nations mine, the richest in Western Australia.

A young man working in the diggings at Coolgardie in 1894 described the arrival of 'a seething mass' of gold-hungry men at the railhead: 'Most had never seen a camel, except in the zoo. Then suddenly they found themselves in the midst of thousands of the hideous brutes, roaring and bubbling, tangling up everything. Afghans were yelling and thrashing them . . . then hoisting incredible weights on their backs.' Adults warned their children that camels were fierce and unpredictable and frightened them with a story that a camel had once killed its Afghan handler by biting off his head.

In many parts of Australia there was open hostility between camel-train owners and the white carters. Carters in Queensland and New South Wales petitioned their governments to ban camel trains. In the goldfields of Western Australia dislike of the cameleers led to the founding of the Anti-Asiatic League of Coolgardie. 'The presence of the saddle-coloured aliens in our own midst is a matter of deep regret,' said the *Coolgardie Miner* in June 1897. 'They must be permitted no miners' rights.' Another newspaper described the camel drivers as 'abhorrent Mongolians'. Writers railed against white women who married 'unclean and unspeakable' Afghans. Every time horses bolted at the sight or smell of a camel, perhaps upsetting wagons and injuring passers-by, the event was reported in the newspapers.

For all that, the camels were a welcome sight to those at the end of the long supply lines. Doris Bradshaw, daughter of the telegraph station-master at Alice Springs at the turn of the century, recalled in the engaging memoir of her childhood that before the railway reached Alice Springs in 1929 the arrival of the camel train was a festival.

> No besieged legion getting its first glimpse of relief troops shouted more enthusiastically than we did on the day our annual stores arrived. Everyone stopped work. We deserted

the schoolroom and our governess; Aborigines came running from the creek; the staff stood around while my father ceremoniously took delivery of the station's goods. Slowly, silently, the great packing cases creaking as they swayed, the long string of camels came padding into the compound. The animals knelt, grunting and sometimes squealing, and squatted on the ground. The packs, weighing as much as 500 pounds, were unlashed. I have no idea what we and the whole of the Inland would have done without the camels. They were stand-offish, unlovable, smelly in the extreme and yet indispensable.

In many mining communities the arrival of the camels caused particular celebration because they brought twenty-five-gallon casks of beer. Suppliers preferred to send alcohol by camel train rather than by horse-drawn cart because the cameleers followed Islamic teaching and did not drink. White carters, on the other hand, often returned to their employers reporting mysterious losses from their cargoes of beer and whisky.

The camels opened up Australia. They transported surveyors and explorers forty to sixty miles a day, hauled construction equipment for water stations along the stock routes, and the food, posts and wire for the teams that built the long-distance fences protecting pastoral land from dingoes, kangaroos and rabbits. They covered an eighty-mile postal round in a ten-hour day and carried water into arid districts where creeks and wells might be dry. They were the only suitable transport for police officers patrolling the immense desert plains of South Australia, Western Australia and the Northern Territory. They took machinery to the new mines and sheets of corrugated iron six feet long to roof the new houses. On their backs were loaded piles of ore from the mines and great bales of wool from the sheep stations.

The spreading network of railways and the increasing reliability of cars and lorries put an end to the camel trains. Redundant camels ran wild in the bush and became a pest. The Camel Destruction Act of 1925 empowered police to shoot the strays. Camel rides are now a tourist attraction, and Alice Springs salutes the Afghan cameleers who for years were its lifeline with a street named Mahomed.

From the 1870s to the 1890s Major Arthur Glyn Leonard was a British Army transport officer and baggage master charged with keeping troops and their equipment on the move. As an old campaigner in Afghanistan and Sudan, he knew the freight capacity of elephants, bullocks, horses, mules and donkeys, knew how far they could walk in a day and how many hours it took to load them in the morning. He, more than anyone in uniform, understood the beast on which the British so often depended during military operations. He wrote the military textbook on the camel, and sketched out its history. Camel-mounted Persians defeated Croesus at Sardis in 557 BC, he noted, partly because horses could not bear the smell of camels; similarly, Roman cavalry horses were confused by the camel's 'uncouth figure and odious smell'.

Although he was their sympathetic champion and enthusiast, Major Leonard had no illusions about camels. He concluded, after twenty years of experience, that they were gloomy and undeniably stupid.

> The normal condition of the camel is one of moral torpor and insensibility. His intense Medusa-like composure is repellent to the ordinary observer. He declines your advance, refuses to become your friend. His eye never lights up with love at the approach of his master. No endearment is of any avail. The horse is a gentleman and the camel is a boor, wrapped entirely in himself and his grievances.

Leonard reflected Kipling's view of the creature known to British soldiers by its Hindustani name of oont:

> O the oont, O the oont, O the Gawd-forsaken oont!

Many soldiers, elevated to the hump to form a camel corps in Egypt and the Sudan in the 1880s, no doubt expressed the surprise of Sergeant Eagle of the Royal Marines who found himself

> . . . riding on an animal
> A Marine never rode before,
> Rigged up in spurs and pantaloons,
> As one of the Camel Corps.

The ancients knew the camel well, said Major Leonard, 'but in this enlightened nineteenth century we know less about him than any animal under the sun'. He complained that many British soldiers believed 'the fallacy of indefatigability', that camels stored fluid in their humps and could travel long distances without water. 'The camel can go for days without water, to his detriment, but not days and days. Old hands know they work better with daily water. I have seen thousands of camels die from exhaustion brought on by ill-treatment.'

British soldiers often overloaded their animals, but Major Leonard said no camel would carry more than 600 pounds and 'in our little wars 300–350 pounds should be the limit'. Even in skilled hands, the loading of such burdens took a long time. 'When I was Brigade Baggage Master from Kabul to Peshawar, and had the arranging of several thousand camels, we never got off the camp ground in a morning in under four hours. In the Khyber, cramped for space, it took nearly seven hours.' Baggage camels could walk twenty miles a day, twenty-five at a push, at two and a half miles an hour. A riding camel could cover up to sixty miles a day; and General Charles Gordon, the hero of Khartoum, once rode 250 miles in the Sudan in fifty-two hours.

Although they were certainly not tireless, camels endured gruelling punishment and continued plodding even when close to exhaustion. Major Leonard once saw a camel, shot through the head, hump and hindquarters during a skirmish, shrug off the injuries and keep walking. He warned that it was not easy to put down a sick camel. 'I have seen six bullets put into the poor creature's head without the desired effect.' He recommended a single shot behind the ear.

In his opinion most hired camel drivers were 'worthless scamps', and he offered this prescription for their treatment: 'Don't bully, be fair, punish promptly. If necessary, the only way to appeal to them is by a flogging.' But he also had scant regard for the British soldiers put in charge of camel drivers. 'For supervision over natives the British soldier is useless because he himself, besides ignorance or inexperience, requires constant supervision.'

Leonard called the camel the 'living automaton' and compared its performance with that of other transport animals. The elephant, for example, could haul 8,000 pounds but was not well adapted to transport work, needed good roads, ate a lot, had tender feet, was costly to maintain, could not stand cold or gunfire and required mollycoddling. 'In 1878, marching from Jacobobad to Kandahar, my company escorted a heavy battery drawn by eleven elephants and 300 oxen. We only lost one elephant but this was due to the . . . attention lavished by the officer in command feeding them chappatis, rice, sugar and two bottles of rum each evening.'

In India bullocks carried only 160–200 pounds, were good only in favourable conditions, were unfitted to stony country and were obstinate. On the march to Kandahar, Major Leonard said, 'the road was dotted with the carcases of bullocks and camels'.

From Kabul to Kandahar in 1879, 10,000 men under General Roberts with mules, donkeys and ponies marched for nineteen days at nearly fifteen miles a day. Major Leonard admired the toughness of ponies and donkeys and said that even the smallest donkey would carry two boxes of ammunition at two and a half miles an hour. The mule was faster, capable of up to three and a half miles an hour, and was overall the handiest and hardiest of pack animals. 'In India I knew mules which had worked twenty-two years. The mule is accused of being obstinate but this arises from ill-treatment as well as from the woeful ignorance of the animal's ways that generally prevails among Britishers. He strongly objects to being hit over the head as I have frequently caught Tommy Atkins doing.'

In 1884 the House of Commons learnt that in the Afghan campaigns 60,000 camels and 30,000 other baggage animals had died. Major Leonard did not think much of the Army's senior management and said that during the Afghan war and Nile expeditions baggage animals were bought by purchasing officers who had never seen a camel outside a zoo.

> The Army cannot make war without good transport. Being ready for war is the best way of preserving peace, but

> British-like we never act up to the precept. In all of our recent expeditions the transport has been nothing more than a mere collection of animals, bought not for fitness, but by the gross, looked after by men totally inexperienced and ignorant of the work. Here we are, at the end of this very advanced nineteenth century, in as rotten a state of unpreparedness as ever we were.

The care of animals was no better when Britain forced the outbreak of war in South Africa in 1899. The story of the British military horses was a tragedy and a scandal. The year before the war the budget omitted any kind of care for sick animals. 'The veterinary department was destroyed as an active service unit,' the history of the Royal Army Veterinary Corps recorded. The army was no better equipped to look after its horses than it had been in the Crimea, the Peninsular War or in Flanders in 1799.

It was not surprising therefore that in South Africa the army lost 67 per cent of its horses, 326,067 animals in all, and 35 per cent of its mules, 51,399. As with the troops, the majority of the casualties fell to disease rather than to gunfire. The RAVC history said that 'never in the history of any British war has there been such a deliberate sacrifice of animal life and of public money. For this no one was ever called to account.' Indeed some officers believed that the war was prolonged by the immobility the Army brought upon itself by its mismanagement of horses and its neglect of veterinary care.

With little veterinary knowledge, the British authorities in South Africa managed the horses with methods that were 'not only archaic but opposed to all common sense and experience. Typical of the ignorance, an officer travelling in a steamer bringing horses on a troopship in 1901 insisted on shooting 251 horses for the throat disease glanders in defiance of a civilian vet's diagnosis that they were merely suffering from colds.'

Horses landed after long voyages were allowed no time to acclimatize and were sent into action at once. The remount system, the supply of fresh horses, was a fiasco and thousands of horses were left to starve.

The Boers knew the value of horses and how to treat them. They provided their own animals and tended them carefully, often walking rather than riding. The British policy of converting infantrymen to mounted infantry, and teaching them only the rudiments of riding, meant that few men knew anything about the care of the animals they rode.

27

The driver cracked the whip

'The first thing we did on that glad evening at Saint Joseph was to hunt up the stage-office and pay a hundred and fifty dollars apiece for tickets per overland coach to Carson City, Nevada.'

Mark Twain, *Roughing It*

As the stagecoaches vanished into the British sunset, they burst into the dawn of their legendary age in America and Australia and South Africa. For the frontier generations they were supremely romantic, the reality as well as the symbol of speed and progress. Probably, the romance lay mostly in recollection. But, as Hollywood has always known, the stagecoach hurrying across the open plain is a powerful image; and, into the twenty-first century, the Wells Fargo Company values it still as an icon.

The forging of the long-distance stagecoach routes across the American West followed closely on the discovery of gold in California in 1848. 'California's made of gold,' went the song, 'we'll get as rich as Lima.' More than a quarter of a million men poured into the territory from many parts of the world and, as in all gold rushes, created a makeshift and turbulent society with very few services, amenities, courtesies or principles. California was 3,000 miles from the east coast of the United States, and there was no transcontinental railway or road, only rough trails. It was easier to get there by sea. The voyage from New York to San Francisco by way of Cape Horn, however, was frequently punishing and took four or five months. It was faster to sail to San Francisco from Sydney than from New York. An alternative was to go by ship to Panama and cross the isthmus by mule to join another vessel on the Pacific side.

The California miners ached for letters and parcels from home and a means of exchanging, storing and transporting the gold they found. Henry Wells and William Fargo, two former horseback postmen, saw their opportunity. They opened their first office in San Francisco in 1852 and started shipping gold to New York. Within a few years they dominated the collection and transport of letters, parcels and gold in California, with fifty-five offices in towns from the Rocky Mountains to the Pacific. They grew into an institution, a framework of stability in rollercoaster times. It was said that the first facility set up in a mining camp was a saloon with a poker table; and the second was a Wells Fargo office. Always conscious of the need for a good image, Wells and Fargo saw to it that their employees were sober and respectably dressed. They shrewdly transported newspapers free of charge so that other papers copied the news with a printed acknowledgement that Wells Fargo was the bearer of it. 'Wells Fargo', said the American journalist Samuel Bowles, 'is the omnipresent universal friend and agent of the miner, his errand man, his banker, his post office.'

Wells was the financial brains in the partnership. Fargo, who first delivered letters in New York state at the age of thirteen, was the visionary organizer. He and Wells combined with other businessmen in 1850 to found what eventually became the American Express Company, taking mail from New York to Buffalo, Chicago and St Louis. In 1857 they joined other firms to form the Overland Mail Company which won a $600,000 government contract to operate a twice-weekly stagecoach service from the Mississippi, where the West began, to the Pacific. It ran from St Louis and Memphis to San Francisco, taking a wide loop close to the Mexican border by way of El Paso, Tucson, Yuma and Los Angeles, the longest stagecoach journey in the world at almost 2,800 miles.

The inaugural trip in 1858 took three weeks and the coaches called at staging stations ten to fifteen miles apart to change horses. The fare was $200 and each passenger was allowed forty pounds of baggage. Wells and Fargo won control of the Overland in 1860 and six years later bought out their big rival, Ben Holladay, whose network of routes, 20,000 wagons and coaches and 150,000 horses, mules and oxen, earned him the label of 'Napoleon of the West'. Holladay sold out partly because of the financial losses he had suffered through Indian attacks but also because he foresaw the inexorable march of the transcontinental railroad and its threat to the stagecoaches. Wells and Fargo, however, felt that there was plenty of life and profit in the stagecoach and, with Holladay's empire in their ownership, they commanded the stage and mail business in the immense region west of the Missouri.

For performance and reliability on the frontier trails one type of coach stood out above all the others. It was expensive but it was the one that drivers and serious operators like Wells Fargo wanted. Its builders, J. Stephens Abbot and Lewis Downing, constructed the first model in 1826 at their workshop in New Hampshire and named it after the town where they lived: Concord.

The Concord weighed over a ton and was usually drawn by four or six horses or, when the going was muddy or sandy, by a team of mules which coped better on bad ground. It could cover more than a hundred miles a day, about the same as a ship with a good breeze on the ocean. The wheels were notably strong and so skilfully jointed that they needed no screws or bolts. The body was ovaloid rather than a box and was suspended on thoroughbraces, shock-absorbing straps of ox-leather, so that it swayed like a cradle or hammock, backwards and forwards and side to side – 'a great swinging stage of the most sumptuous description', wrote Mark Twain, who made a twenty-day trip in one. The body was light and the undercarriage heavy and stable so that the coach's centre of gravity was low. The body was painted a shining red, the undercarriage yellow, and artists decorated the doors with landscape scenes and gold-leaf lettering.

In the most popular model nine passengers sat inside on three upholstered seats with six or more outside; indeed it sometimes squeezed twenty-one people aboard. The driver, known also as the whip or reinsman, was described by flowery writers as the Jehu, from the Old Testament description of the son of Nimshi, 'for he driveth furiously'; and certainly coachmen were reminded by their employers that 'every minute is valuable'. They were the skilled captains of their craft, awesome walnut-faced figures with three pairs of reins in their hands and a whip with which they could flick a fly from a horse's flank. They roasted in summer and froze in winter, even in their swaddle of buffalo robes. In the nature of things, many were characters simply waiting for writers to come along and describe their exploits and their picturesque, spitting, swearing, wisecracking, ornery ways. Charlie Parkhurst was a famous Concord whip in California for thirty years, tobacco-chewing, cigar-puffing and properly laconic, and also one-eyed after being kicked by a horse. Charlie died in 1879 and while being prepared for burial was discovered to be a woman.

Sitting beside the driver, guarding the iron-bound treasure box filled with gold or silver, was the messenger, armed with a Colt revolver, a rifle and, most effective of all, a sawn-off double-barrelled shotgun. Wyatt Earp, who rode as a messenger for Wells Fargo in 1879, described the shotgun pellets as 'forty-two leaden messengers, each fit to take a man's life'. The first hold-up of a Wells Fargo stagecoach took place in 1855. Although bandits killed and injured drivers, guards and passengers, they generally came off worse because of the stagecoach crew's firepower and Wells Fargo's persistence in hunting them down. In fourteen years from 1870 bandits robbed 313 stagecoaches, killing four drivers, two messengers and four passengers, and stole more than $400,000, the equivalent of about $9 million today. Sixteen robbers were killed and seven hanged.

Apart from attack by bandits or Indians, travellers also risked being swept away in river torrents, injured in an overturned coach, and frozen or boiled according to the season. For some the journey west was an unforgettable adventure, and the story of it was inflicted on countless grandchildren. The views in the Rocky Mountains were spectacular and the sheer immensity of the plains overwhelming. But no stagecoach, not even a Concord, could be truly comfortable. It provided an ordeal of guaranteed pain, motion sickness, lack of sleep, foul and expensive food and the probability of disagreeable fellow-passengers. Jammed thigh by thigh with only fifteen inches of buttock room, sleepy heads banging together, teeth gritty with dust, desperate for proper horizontal rest, passengers often lost their tempers. We can imagine too the torture of bowels and bladders and understand their constant complaints about primitive lavatory facilities at stopping places.

An article in the *Omaha Herald* in 1877 told passengers to expect annoyance, discomfort and some hardship. They were advised to take two thick blankets and to wear shoes and gloves two or three sizes too large, not tight boots, because extremities swelled when people sat for a long time. They were to avoid alcohol in cold weather because drink would make them freeze twice as quickly. They were not to discuss politics or religion or point to places where there had been murders. And they were not to grumble when asked to get out and walk.

> Don't growl at food at stations, don't smoke a strong pipe inside especially early in the morning, spit on the leeward side. If you have anything to take in a bottle pass it around, a man who drinks by himself in such a case is lost to all human feeling. Don't swear, nor lop over on your neighbour when sleeping. Don't ask how far it is to the next station until you get there. Never attempt to fire a gun while on the road. Don't keep the stage waiting; many a virtuous man has lost his character by so doing.

The stagecoaches were fast and modern but not quick enough for some. To speed the mail a freighting company started the Pony Express from St Joseph, Missouri, to Sacramento, California, guaranteeing delivery over the 1,966-mile distance in ten days, less than half the time of any stagecoach service. Although the charges were very high, $5 for half an ounce at first, the equivalent of more than $100 today, there was never any hope of the promoters meeting their costs, let alone making a profit; but it was marvellous frontier theatre. It was set up with 190 stations, between ten and fifteen miles apart, 500 first-class horses and a team of young and daring riders, dressed as lightly as jockeys, who swore 'before the great and living God' that they would not use foul language or fight with fellow employees. Their number included sixteen-year-old William Frederick Cody, who would mature into Buffalo Bill, scout, hunter, dime novelist and the impresario who took his Wild West show to Queen Victoria's golden jubilee in London in 1887.

The first Pony Express dash was in April 1860. Thereafter, every day, eighty riders, forty in each direction, galloped the mail from station to station, pausing for just two minutes to change horses. At the end of a seventy-mile stretch they passed the mail to another rider. Mark Twain described the thrill of watching a Pony Express rider streaking across the prairie – 'man and horse burst past our excited faces and go swinging away like a belated fragment of a storm!' The letters, a maximum of twenty pounds in weight, were wrapped in waterproofed silk inside the four leather pockets of a mochila, a lightweight saddlebag invented for the Pony Express and thrown over the saddle to be secured on the pommel and cantle. The mail carriers, unarmed for the sake of lightness, were sometimes attacked by Indians and a few were wounded. The short career of Charlie Miller, a rider at the age of eleven, was ended by an arrow wound; but only one messenger was ever killed. The staff who manned the stations were much more vulnerable and several were murdered by Indians and the buildings wrecked.

In April 1861 Wells Fargo took over the western end of the route from Salt Lake City to Sacramento and cut the prices; but the service had only a few months to run. The new electric telegraph showed that a thin wire beat a fast horse. In October 1861 the Western Union Telegraph Company completed the first transcontinental line, and an inaugural message was tapped out from California to President Lincoln in Washington telling him that East and West were now united. The construction teams worked westward from Omaha and eastward from San Francisco and moved towards each other to join the wires in Salt Lake City. Poles, cables and insulators for the western end of the line were shipped from the east coast around Cape Horn and landed in San Francisco, while tens of thousands of poles were delivered by wagon to workers who planted them across the plains.

Two days after the telegraph line was completed the Pony Express closed. The last breathless rider galloped in with the last of the 35,000 letters carried in the mochila. Dramatic, legendary and unviable, the Pony Express ran for nineteen months and cost its backers more than $100,000. The electric telegraph, however, took time to spread across the West, and more modest and local pony expresses operated until 1870.

A photograph of 1868 shows thirty stagecoaches loaded on flatbed train wagons heading west to serve the growing Wells Fargo network. After 1869, when the transcontinental railway was completed, train services superseded the coaches on the trunk routes. But there was still plenty of mail and passenger business for stagecoaches in the mining towns and the new settlements on the frontier; and there were more stagecoach robberies in the 1870s and 1880s than in any other period. Wells Fargo ran Concord stages to the remoter areas until 1918.

Seventeen-year-old Freeman Cobb travelled from Massachusetts to the California goldfields and made his money not from gold but by driving a Concord stagecoach. In 1852 he heard that gold had been found in Australia and he crossed the Pacific with the hope of making a fortune. Melbourne swarmed with people anxious to get to the diggings at Bendigo, about a hundred miles north. A man wrote home to *The Cambrian* in Swansea that 'six married couples are crowded into one sleeping room in one of the most respectable boarding-houses'. Cobb went up to Bendigo and saw the diggers' canvas settlements. They were full of dishevelled men with clay-stained whiskers, English, Irish, Scottish, Welsh, Italian, French and American, dizzied by grog and fantasies of gold. They dreamed of women too, and were so short of them that they formed clodhopping male couples to polka the evenings away at their dances.

The bush roads from Melbourne were rough and dusty, or treacly and almost impassable in wet weather, jammed with teams of bullocks drawing wagons at perhaps two miles an hour. Most hopefuls walked to the diggings, the price of a horse and its feed being far beyond their pockets. The cost of freighting goods into the goldfield made every man's jaw drop. It was more expensive to cart clothes and picks and shovels from Melbourne to the goldfield than to ship them around the world from Britain to Australia. Anyone with a transport skill, a wagon driver, a farrier or blacksmith, was in short supply and demanded good wages.

It was clear to Cobb that prosperity lay not in gold but in a stagecoach line. He raised money with friends and in 1856 had two Concord coaches shipped out from Boston. Cobb and Company began a regular service from Melbourne to Port Melbourne at five shillings each way and opened routes to Bendigo and other parts of Victoria, employing American coachmen who promised not to drink while working. A Cobb coach left Melbourne at six in the morning and covered the seventy-seven miles to the gold town of Castlemaine by 2.30 in the afternoon.

After seven years Cobb left Australia with a fortune and returned to the United States; but Cobb and Company prospered and expanded, becoming as renowned a name in Australia as Wells Fargo was in America. It was reliable and punctual and extended its network throughout Victoria, South Australia, New South Wales and into Queensland. In the years from 1890 to 1903 it had 6,000 horses working each day and its longest route covered 2,500 miles. The stagecoaches gave way to the railways but the horse remained an important power unit and the horse population peaked in 1918. Cobb ran coaches in rural Australia until 1923; and the poet Henry Lawson remembered the excitement of them:

> Behind six foaming horses,
> And lit by flashing lamps,
> Old Cobb and Co, in royal state,
> Went dashing past the camps.

'Possibly no country worshipped the horse with the same fierce veneration as Australia in the nineteenth century,' wrote the historian Geoffrey Blainey. But one of the problems with horses was the cost of feeding them. Rural growth depended on the farmer's ability to sell his crops and that in turn depended on his having the horses or bullocks to reach the market and the fodder to fuel them. The boom in gold and wool made agricultural progress possible. They were so valuable that they easily covered the costs of cartage to the ports and onward transport to Europe. 'Gold and wool created far more employment and wealth than any other commodities,' said Blainey, 'and provided more incentive for money and men to cross the world to Australia.' It was Australia's good fortune that the gold of Victoria and New South Wales lay relatively close to the ports and to the farmland and forests that provided miners with meat, corn and timber. There was also an efficient network of carts, stagecoaches and, eventually, of railways.

At our distance the horse age may look romantic, but Blainey reminds us that the manure left in city streets was as detrimental to public health as the exhaust fumes which replaced it and that kitchen windows swarmed with flies. Moreover, 'horses were tyrants and helped to keep alive the cult of early rising because thousands of Australians had to wake to muster horses, feed them and harness them for the day's work'.

The library in Kimberley, South Africa, erected in 1887, was one of the first substantial buildings in town. The diamond rush of the 1870s created a raucous shanty sprawl where larky ladies bathed in champagne at the expense of some of the shovellers. But in the 1880s the town started putting down roots, and the library was a mark of the Victorian respectability many citizens craved.

Here I read two letters written by David Livingstone. Describing journeys during the 1860s, he apologized for his scrawl and joked that an inky spider had roamed over his notepaper. He wrote with wry exasperation of the struggle he had in urging oxen to drag a heavy wagon. 'All this blessed morning hauling the beastly thing up the hill . . . the oxen not behaving like gentlemen at all.'

From the 1650s and into the twentieth century oxen were the heavy-duty transport of South Africa and bore the greatest burden in the European colonization. 'Staring and worn and parched with thirst,' as the song 'The Cry of the Ox' put it, they toiled into South African culture. The lurching ox-drawn wagon was both the vehicle and the symbol of the migrating Boers, the African farmers of Dutch descent. In the 1830s, to escape British rule and the strictures against the ownership of slaves, they began their Great Trek out of the Cape Colony, the defining Biblical exodus of their history.

Given the nature of the beast, the landscape and the climate, travel by ox-wagon was a slow punishment for animals and men.

Wagons took five weeks or more to cover the 400 miles from Durban to Pretoria and usually made only two trips a year because the tracks were poor. Edward Sandeman, an Englishman, landed in Cape Town in 1878 determined to experience the romance of a long trek. In Pietermaritzburg in Natal he fitted himself out for an eight-month haul, hiring a staff of four men and buying two horses, a wagon and a team of fourteen oxen to haul it. The wagon was eighteen feet long and four feet six inches wide and carried more than two tons. Sandeman was shown that there was an art in loading it, so that equal weight bore on each wheel. He stowed Australian tinned meat, coffee, potatoes, candles, rice, a dozen bottles each of brandy and gin, and three gallons of rum. His medicine chest included castor oil, Epsom salts, chloroform and rhubarb pills. He packed tools to repair the wagon and plenty of ammunition for his guns. He took a supply of knives, beads, salt and blankets for bartering. For the Boer farmers he would meet along the way he carried a store of sweets. He was issued with a permit for the trek after he had sworn that his behaviour would be orderly.

As in a ship at sea the wagon's enemy was chafe, the rubbing and grinding on axles and harness, and for all its apparent strength it needed constant maintenance. The rutted and rocky tracks wore it down. Wheelwrights, blacksmiths and wagon builders were always scarce and their bills invariably stunning; sometimes they were slow and incompetent and always they had to be gently coaxed and flattered. Finding that the iron tyre on one of his wheels needed tightening, Sandeman waited two days for his turn in a queue at a blacksmith's forge. Later, he waited a week for a broken wheel to be repaired.

The animals, too, needed close attention. Disease killed tens of thousands of oxen. They died from lung sickness and, when pressed too hard, from exhaustion. In wet weather the yoke excavated deep sores on their necks. They went hungry when plagues of caterpillars ate the grass. Sandeman met wagons hauled by skinny oxen. Horses were vulnerable to what was simply called horse sickness and which was frequently fatal. When looking for horses to buy, experienced men sought animals that were 'salted', meaning they had had horse sickness and had recovered. The only medicine for 'unsalted' horses, a mixture of yellow soap and tobacco water, was not much help. Both men and animals were also menaced by the tsetse fly, the cause of sleeping sickness. Sandeman was annoyed to find that the flies bit him through his thick moleskin trousers.

In the rainy season the wagon sank to its axles on muddy tracks and had to be unloaded before the oxen could haul it out, a task that took hours. Sandeman found it was wise to cross streams whenever the chance presented itself because sudden rainstorms made them unfordable. Storms raged over the veld with frightening ferocity and lightning sometimes surged through the harness chains and struck dead a whole team of horses and oxen in an instant. During winter on the treeless high veld Sandeman and his men awoke in stiff and frozen blankets; and at noon they sweltered.

Sandeman's wagon was typical of the long-distance transport on the South African frontier. It was in such vehicles that men who could afford to do so travelled to the new diamond diggings at Kimberley. The discovery of diamonds in this inhospitable landscape in the late 1860s, and the rush that started in 1870, began the transformation of South Africa from a farming country to an industrial one and created an insistent demand for faster transport.

Crowds of urgent and anxious men hurried and jostled down the gangways of steamers at Cape Town looking for the speediest means of getting to the diggings which lay 750 miles distant. In May 1871 David McIntyre went to Ceres, north of Cape Town, and bought a seat in a Kimberley-bound mule wagon. He was 'tightly jammed and bitterly cold'. The road was horrible and the wagon rolled and plunged and 'even the laughter of the three ladies subsided to an occasional squeak'. The wagon stuck in the mud of the Braak River and the driver hoisted his passengers on his back and carried them ashore before walking to a farm to rent a team of oxen to haul his vehicle out. It was four o'clock in the morning when the party arrived in Hopetown, and the hotel was full. 'Four of us slept on the billiard table,' recalled McIntyre, 'one at each corner.'

A few months later Frederick Boyle arrived at the Cape from

England and found that every wagon heading out was full. The Inland Transport Company ran a weekly wagon to Kimberley and charged a stiff £12. This was too expensive for many would-be prospectors who simply shouldered their possessions in a sack and set off on foot. Some pushed wheelbarrows. The better-off purchased a mule or a donkey.

Boyle went by train to Wellington, sixty miles from Cape Town. There his ticket secured him a place on a wagon drawn by eight horses and fitted with eleven seats, nine inside the canvas cover and two outside. Luggage was squashed in with the passengers and there was barely any leg room. Food was stuffed into canvas bags swinging from the roof. Two days out of Wellington the wagon embarked on its crossing of the formidable Karoo, the semi-arid plateau which encompasses a third of the land of South Africa. Boyle looked with increasing dismay at the desolate stony flatness. 'An abomination,' he decided.

Tortured by flies, floured by dust and desperate for sleep, he and his companions crumpled. Cramped for hours on their seats they suffered severely swollen legs and ankles. When at last they were allowed to get out they tottered on aching and discoloured limbs; and some could not walk at all. The wagon took a week to cross the Karoo and on the way it passed knots of men, walking towards the diggings and dazed by thirst, who begged for water. 'There are', Boyle noted, 'continual discoveries of unknown dead in this wilderness.' I drove across the Karoo in an air-conditioned car and only when I got out and walked for a while in the heat and felt the sweat prickle under my shirt was it possible to imagine how harsh and frightening the burnt land must have appeared to those who crossed it in wagons or attempted it on foot. I had the luxury of enjoying the drama of the immense plain with its small flat-topped hills, windpumps and red-roofed farmsteads shaded by cypresses. From a veranda I watched the swarming stars and the silver scratches of meteorites. Boyle and thousands like him had the stars too, but not much food or rest.

Boyle reached Kimberley after eleven days in the wagon and found a small hotel where, being a young English gentleman, he was relieved to find that his fellow guests at dinner were decent and mannered, that Boers did not come into the hotel and that 'the poorer sort take luncheon in their tents'.

The grand sight of Kimberley was, and remains, the Big Hole, the largest pit excavated by man. Thousands of miners dug ever deeper and hauled up stones and slurry on cobwebs of rope. By the time it was closed in 1914 the Hole had yielded 28 million tons of gravel and mud and three tons of diamonds.

At home in America in 1870 after his Australian adventure, Freeman Cobb heard the news of the Kimberley diamonds and, restless for another frontier, sailed for South Africa. The following year he decided to run stagecoaches from Port Elizabeth to Kimberley, 428 miles to be covered in five days with passengers quartered overnight in hotels. He imported six Concords and started a regular service in 1872. The roads, however, were particularly rough, the cost of fodder high and many of the horses died. Although the Port Elizabeth route was shorter than the 700-mile road from Cape Town many diamond hunters preferred to get going as soon as they landed and not make the extra sea journey to Port Elizabeth. The company failed in 1875 and Cobb himself died in 1878 aged forty-six.

John Gibson left Chichester in Sussex for South Africa, became station-master at Wellington, the end of the line from Cape Town, and saw that instead of flagging men off to the diamond diggings he could profitably take them there himself by stagecoach. With his brothers James and Fred he started the Red Star Line in 1879 and within a year was running four coaches a week each way. As it grew, Red Star became familiar throughout South Africa, linking the advancing railheads to the mining towns. When gold was discovered on the Witwatersrand in Transvaal, at what became Johannesburg, the Gibsons promptly started a coach service. The decision by Paul Kruger, the Transvaal president, to stop British railroads entering the Transvaal helped Red Star and other coach companies to prosper.

Gold mining in Transvaal was funded at first by the diamonds dug at Kimberley. One evening a diamond trader hurried into Gibson's Kimberley office and ordered a coach to rush him 200 miles north to the Witwatersrand to stake a gold claim. It was a desperate drive. 'Three times was our coach completely overturned and righted,' wrote the driver. 'We escaped with a good shaking. We showed no mercy to ourselves and very little pity to the horses, five of which were killed through being overdriven.' By the time it reached the goldfield the half-wrecked coach was held together by wire and rawhide. The single passenger tottered out and staked his claim, two hours ahead of a rival. Red Star was soon operating fourteen coaches a week from Kimberley to Johannesburg. The volume of mail was so heavy that a thousand horses and mules were needed to keep the coaches running.

When gold was found in Barberton, in Transvaal, in 1884, a town was rapidly conjured up and many men died in the rush to reach it, killed by wild animals, tribesmen and malaria. So many horses and oxen died on the track into the district that it was said to be paved with bones, the worst road in Africa.

Stagecoach journeys in America and Australia were rough and dangerous enough, but at least the passengers were not menaced by lions. In South Africa the roaring of lions terrified horses and mules and set them bolting. On the Pietermaritzburg–Leydsdorp road lions sprang at a team of mules and rolled the coach over. Two gold prospectors drove off the lions with gunfire and found three passengers in the wreck, one of them dead. The stagecoach proprietor persuaded the government to send a hunter to clear the road and he did so, killing two lions a day. The South African stagecoach experience also included attacks by elephants. Enraged tuskers butted a coach off the road near Naboomspruit and stamped it to matchwood as the passengers fled. A driver of another coach, seeing an elephant about to charge, shouted at his passengers to jump for it. They leapt out and returned later to find their hero dead in the smashed coach. On the road out of Mafeking a stage driver saw the sky ahead darkened by a swarm of locusts and feared that they would engulf his coach. He unhitched the mules and ordered his frightened English passengers to mount up and ride as hard as they could to find shelter in Mafeking. They reached it just in time. Apart from lions, elephants and locusts there were highwaymen. Fred Griffiths, who drove coaches until 1913, was held up seven times in his career.

The wagons gave way to the railways. As soon as the first train reached Johannesburg the business of carting coal by ox-wagon was over and the freight price fell at once from twenty-six shillings a ton to six shillings. To the end the stagecoach operators sought ways of beating horse sickness. In 1893 Doel Zeederberg hitched a team of zebras to a coach running from Pietermaritzburg to Pretoria and the town turned out to see the show as the coach with its ten passengers set off. But the zebras tired easily and had to rest every few miles. President Kruger watched the pantomime from his veranda in Pretoria. 'It is an evil day', he pronounced, 'when wild animals are made to do work which the Almighty has ordained should be performed by domestic animals.'

28

That art of masts

'With tallow casks all dunnaged, with tiers on tiers of bales,
With cargo crammed from hatch to hatch, she's racing for the sales;
A clipper barque, a model ship, a "flyer" through and through,
O skipper bluff! O skipper brave! O would I went with you!'

E.J. Brady, 'The Ways of Many Waters'

Gold fever launched fast ships. The nuggets dug in Australia in 1851 created an impatient demand for ocean expresses that would race the flying fish. The shout sped round the world, spuds of gold at the turn of a spade, and hundreds of thousands of frantic men clamoured and scrabbled.

The fossickers of Ballarat and Bendigo sent Australia crazy. The stampede to the diggings emptied offices, shops and ships. Policemen abandoned their posts and shearers and parsons their flocks. In many houses there was not a servant to be had and the Governor of Victoria had to groom his own horse. Champagne-swigging diggers swaggered into Melbourne with gold in their moleskin trousers and bankrolls in their wide-brimmed cabbage tree hats. The first Australian gold reached Britain in 1852 and the lust for it spread like flu. Men quit their jobs and homes and sought the fastest passage to Australia. Eighty-six thousand migrants sailed that year from Britain for Melbourne and Sydney, a number not exceeded until 1913. The population of Australia was 400,000 before the gold rush and three times greater within twelve years.

Australian gold inspired the last great age of long-distance sailing when ships driven by lofty pyramids of canvas were the supreme expression, as John Masefield had it, of 'beauty in hardest action', of 'that art of masts, sail-crowded, fit to break'.

Until prospectors stumbled on the golden veins of Victoria and New South Wales the passenger traffic from Britain to Australia was small. The country seemed less inviting than America and far more distant. Emigrants faced a five-month voyage, and the doubtful benefit of the longest holiday they would ever have, alternately frightening and dreary. Few travelled in any comfort. There were certainly good ships commanded by professional masters, and for those who could afford such luxury there were decent cabins and dinner with the captain. But most passengers sailed in small and poor vessels, lived dismally and, compared with convicts, seemed to lack only chains.

'These colonial wagons', said Basil Lubbock, a chronicler of sail, 'were navigated by rum-soaked, illiterate, bear-like officers who could not work out the ordinary meridian observation with any degree of accuracy, whilst they were worked by the typically slow-footed, ever-grousing Merchant Jack of the past two centuries.'

James Turner, who sailed with his family to Australia in 1829, a year after the first law was passed to improve ship safety and comfort, reported that one of the horses on board was 'a spirited stallion who has been stamping to our great annoyance, he being opposite our cabin which is nearly in total darkness, and filth in every corner. I really now find advantage in being a very little fellow with a very little wife and children.' Apart from the stamping stallion and the thumping sea there was the frequent uproar of argument in the herd of economy passengers. 'A general quarrel in the cabin between the Scotch and Irish,' Turner recorded one day, 'Captain Byrne and Mr Green quarrelled. Brown drunk.'

Although the British Passenger Acts led to some improvements on migrant ships, the ocean passage was for most an ordeal of unmitigated wretchedness. The first class had their cabins and the second class their cramped cubicles, but nine-tenths of those who sailed for Australia endured months in steerage. In this dark, congested and smelly space men, women and children were racked in tiered shelves with three feet of headroom. The berths were six feet by six feet, each one separated from its neighbour by a twelve-inch-high plank. Four people were allotted to each berth, there to undress, bicker, sleep, keep their children in order and endure sickness, all in the public gaze. The lack of privacy was eventually ameliorated by a regulation insisting on the installation of a two-foot-high board dividing each berth into two and preventing any immoral commingling after the compulsory bedtime at ten.

A ration of rice, flour, sugar and tea, with molasses to spread on the bread, was doled out to each passenger; and children qualified for a full ration at fourteen. People organized themselves into messes of a dozen or two dozen, fetching their porridge and stew from the cook in the galley and eating from the plates they themselves provided. They washed their clothes in cold water, often sea water, and dried them by tying them on the rigging.

A report to the British Parliament on conditions in emigrant ships in 1844 noted that it was 'scarcely possible to induce the passengers to sweep the decks after their meals or to be decent in respect to the common wants of nature; in many cases, in bad weather, they would not go on deck, their health suffered so much that their strength was gone and they had not the power to help themselves. Hence the between decks were like a loathsome dungeon.' So cramped, ratty and squalid were conditions below decks, so foul the accumulation of vomit, excrement, wet and rotted straw, bedding and food, that when the deck hatches were lifted they released a disgusting miasma, a stench worse than any pigsty. Many passengers died or half starved and, as the report to Parliament said, 'the strongest maintained the upper hand over the weakest'.

Many of those who sailed to Australia from the late 1830s did so on subsidized tickets. This was the solution proposed by Edward Gibbon Wakefield to the problem of peopling such a vast and distant land. He set it out in a pamphlet in 1829 in which he purported to be a Sydney landowner. In fact he had never been to Australia and wrote his manifesto while serving three years in Newgate prison for abducting Ellen Turner, a schoolgirl and heiress, and marrying her at Gretna Green.

But from his cell Wakefield saw Australia's problem all too clearly: a colony brutalized by convict labour which he regarded as a version of slavery, a land where children's minds and property were laid to waste and drunken beasts of men resorted to prostitutes. He urged an end to the transportation of convicts and, in its place, the recruitment of free migrants of all trades, complete with plenty of young single women, to found a civilization in the southern sun. The way to do it was not to give away Australian public land but to sell it at fairly high prices and to use the revenue to fund assisted passages. Wakefield's ideas were implemented. Free migrants settled in their thousands. And while many of those who rushed to the goldfields paid their own way, half of them travelled cut price or free.

The unmarried women who sailed in the steerage of migrant ships were confined, as if in a harem, to a walled-off compartment in the stern, as far as possible from the seething testosterone of the single men's quarters in the bow. On deck the girls were segregated from other passengers by a rail and forbidden to speak across it, their virtue guarded by unsleeping matrons. The regular landings of nubile cargoes on subsidized fares gradually righted the imbalance of the sexes, and helped to civilize the country, but men and women did not draw level until 1916. As Geoffrey Blainey has pointed out, the shortage of women affected Australian society in many ways, in the male camaraderie enshrined in the idea of mateship and the masculine passion for sport. Australia paid an economic price for balancing the sexes and broadening its society. Its land was as much as four or five times dearer than that in the United States and Canada; and high prices kept farmers confined to a particular piece of land and brought many to ruin.

When the Australian gold rush erupted there were too few ships to meet the demand. The Atlantic trade in delivering European migrants to America was in full flow. Some owners diverted vessels from the India trade to Melbourne but not enough of them to carry the thousands desperate to travel. Shipowners wanted fast ships but most British vessels were not as fast as those built in the north-east of the United States and Canada, and which had already broken records carrying gold-seekers around Cape Horn to California. In 1849 alone, 90,000 passengers arrived in California from the Atlantic ports. British owners turned to Boston and Nova Scotia to charter and purchase fast ships for the Australian run. Many of them were Liverpool men; and the word spread that if you sailed from Liverpool, you sailed fast. The city quickly became the main European springboard for Australia, and its wharves were thick with masts.

In a daring stroke, James Baines, the thirty-year-old founder of the Black Ball Line, not related to the New York line of the same name, ordered four ships from Donald Mackay of Boston. Mackay was a Nova Scotian, a leading builder of the clippers whose lines gave them the speed that clipped hours and days off passage times. Mackay's clippers were large as well as fast, the jumbos of their time, with berths for more than 400 passengers. Mackay designed them to be sailed by a crew of twenty rather than the thirty or more needed to handle a British vessel of a smaller size. In 1854 Baines took delivery of four ships from Donald Mackay, *Champion of the Seas*, *Lightning*, *James Baines* and *Donald Mackay*, all of more than 2,000 tons and capable of high speeds.

The shrewd and genial Baines became a Mersey legend, typical of the new breed of Liverpool owners who looked dashingly modern compared with their staider counterparts in London. Liverpool set the pace in ocean transport.

Ships with faster hulls, more efficient gear and taller masts able to spread more sail were only part of the revolution. In the 1850s captains learnt from John Towson, an Englishman, and Matthew Maury, an American, that they could save more than a thousand miles on voyages to Australia and find winds and currents that would give them more speed.

Hitherto, captains had followed Admiralty instructions, steering clear of the feared shoulder of Brazil, staying close to the Cape of Good Hope and holding a course to the north of the Roaring Forties to reach the ports of Australia. Towson instructed seamen that since the earth was a sphere the shortest route across any ocean was a Great Circle and that ships would save distance by heading south into the Forties and Fifties. Here they would find the westerlies that blow hard and consistently for thousands of miles all the way to south Australia. Towson was an examiner of captains and mates in Liverpool and disseminated his ideas to receptive men for whom speed meant more money and glory.

Matthew Maury was an influential naval officer, a pioneer meteorologist and oceanographer, who became the United States Navy's superintendent of charts and instruments. From the 1840s he published maps showing ocean currents, prevailing winds and compass variations. These, with his sailing directions and his 1855 book on oceanography, became sacred texts for seafarers. He showed captains how to cut more than a month off the voyage from New York to San Francisco.

He, too, advocated a passage to Australia along 'the highest degree of south latitude which it may be prudent to touch': in other words, a route only for the finest ships and the best masters. Before Towson and Maury, a voyage of four months to Australia was considered a good time. In 1850 a ship steering the new course sailed from Plymouth to Adelaide in seventy-seven days and became the talk of every waterfront and shipping boardroom.

The ships of James Baines's Black Ball Line set the records. In 1852 the young, hard-driving skipper James 'Bully' Forbes sailed the 2,500-ton *Marco Polo*, with 930 British migrants aboard, to Melbourne in seventy-four days. He also claimed a best day's run of 364 miles. Incredulous Melbourne officials refused to believe him until he showed them the Liverpool newspapers he had bought the day he sailed. To stop his crew running off to the goldfields he had them jailed for trumped-up offences so that he could easily retrieve them from the cells and sail home on time. He reached Liverpool in seventy-six days and tied a banner in his rigging proclaiming *Marco Polo* as 'The Fastest Ship in the World'.

In 1854 Captain Charles McDonald sailed the *James Baines* from Liverpool out to Melbourne in a record sixty-three days and claimed a best day's run of 423 miles. He delivered 700 passengers and 180,000 letters, having contracted with the government to land the mail in sixty-five days or pay a penalty of £100 a day. In the same year 'Bully' Forbes sailed the *Lightning* back from Melbourne in sixty-four days, a time never bettered, and his cargo included a million pounds' worth of gold. Ships returned from Australia by heading south of New Zealand and racing in the high Forties and Fifties of latitude, their decks awash in gales, all the way to Cape Horn. Some were wrecked on the rocks of the Auckland Islands lying at the earth's ends between New Zealand and the Antarctic. Ten survivors of the wrecked clipper *General Grant* waited there for eighteen months before a passing ship rescued them.

There had never been such exhilarating and terrifying power in sail. Captains and owners boasted of remarkable speeds, always good for business. Certainly the best vessels could roar along at fifteen knots and more in bursts. Captain James Learmount, a clipper master, cautioned that it was easy to overestimate a sailing ship's speed and believed that no ship ever covered 400 miles in twenty-four hours. Nevertheless, captains bent their ships and crews to the challenge of fast passages, driving so hard that it became impossible for them to heave to or bring the vessel up into the wind. There was nothing for it but to keep going with several men gripping the wheel and struggling to keep the ship straight before the wind. The steersmen knew that if they lost control the ship would lurch broadside to the seas and be overwhelmed. All this was the stuff of wild tales. Sailors' yarns had it that 'Bully' Forbes once stood guard with a pistol in each hand to stop his crew shortening sail in a gale.

British legislation in the 1850s imposed examinations of competence on masters and mates and, with the growth of emigration, higher standards of safety and comfort for passengers. Nevertheless a passenger to Australia in 1852 recorded: 'Weevils in the biscuit, maggots in the plums, garlic in flour, rice rotten, everything very bad . . . The cook pleads guilty to a charge of abusive language.'

The *Eagle*, sailing to Australia in 1853, seemed to be well-run. Passengers travelled in four classes and, as in every ship, single men and women were quartered as far from each other as possible, with a buffer of married people between them. First- and second-class passengers had only to provide their bedding and napery. Steerage passengers had to bring their own utensils as well. There were three meals a day: breakfast at eight, dinner at one and tea at six.

Breakfast for most passengers was the same every day: biscuits, butter, cheese and coffee or tea, with porridge and molasses on Saturday. A typical dinner was pea soup, pork and biscuits, or salt beef and plum duff.

Some ships were lavishly furnished with stained glass, mirrors, ornamented pilasters and crimson velvet seats. Aboard the Black Ball Line's *Lightning* in the mid-1850s first-class passengers were served sumptuous dinners of soup, cod with oyster sauce, roast beef, goose, mutton, veal and ham pie, plum pudding, orange fritters, almonds and raisins, and champagne and hock. Captain Anthony Enright posted a list of fifteen rules for the 495 passengers. Bedding was to be aired on deck twice a week. Steerage passengers were told to appoint constables to keep order between decks and their reward was a morning cup of coffee or a glass of grog. Second-class passengers were not allowed to promenade on the windward side and third-class passengers were only permitted on the quarterdeck from seven until nine in the evening.

The most popular pastime on board was dancing to a band on the poop. Even in rough weather the quadrille went on. Whether first, second or third, each class held its own concerts and there were debating and choral groups. On Valentine's Day the captain's jolly steward delivered the cards and left a trail of fluttering eyelashes. With a star captain like Enright in command passengers expected fast voyages, but luck was still a factor. One ship took 196 days to make Australia, having spent the first dreary fortnight limping across the Irish Sea from Liverpool to Cork.

At the end of the glorious 1850s traffic to the gold diggings declined. Gradually the American clippers were replaced. Built of soft wood they became waterlogged in two or three seasons and their performance fell away. British shipyards took the initiative and led the world in design and construction, building strong composite vessels with teak and oak planking laid on iron frames.

Speed remained the heart of it all. Daring masters ran their risks and kept their topsails and topgallants set when more cautious men would have reefed them. Captain Richard Angel was such a man. With his ship pressed hard in a gale off Cape Horn his mate asked if he should shorten sail. 'Get the royals on her,' retorted the captain, 'and if you can't find anything else to set, go below and ask Mrs Angel to lend you her petticoat.'

In September 1874 the sailing ship *Cospatrick* left London with a crew of forty-five and its cabins filled with 429 men, women and children, emigrants bound for New Zealand.

During the night of 17 November, as the ship sailed in a light breeze south of the Cape of Good Hope, fire broke out in the forepeak, raced through the decks and up the tarred rigging to consume the masts and spars. Terrified passengers swarmed on to the deck, shouting for the boats to be lowered, and there was such a stampede that the first boat capsized, spilling people into the sea. Only two boats got away, one with forty-two aboard, the other with thirty-nine. As the burning ship slowly sank and the flaming masts fell, the horrified spectators in the boats saw Captain Elmslie throw his wife over the taffrail and jump after her.

The survivors wore only their nightclothes. The boats had no masts or sails and no food or water, and one of them, commanded by Henry Macdonald, the second mate, had only one oar. The boats kept company for two days but were separated during bad weather. 'Thirst began to tell severely on all of us,' Macdonald recalled, giving his account at the subsequent inquiry.

> A man named Bentley fell overboard while steering and drowned. Three men became mad that day and died. We threw the bodies overboard. On the twenty-third the wind was blowing hard and a high sea running. Four men died, and we were so hungry and thirsty that we drank the blood and ate the livers of two of them. We lost our only oar. On the twenty-fourth there was a strong gale . . . six more deaths that day. On the twenty-fifth there was a light breeze and it was awful hot.

Maddened by thirst, some of the passengers drank sea water and made their suffering worse.

> Early in the morning of the twenty-sixth, not being daylight, a boat passed close to us. We hailed but got no answer. She

was not more than fifty yards off. She was a foreigner. I think she must have heard us.

One more died that day. We kept on sucking the blood of those who died. The twenty-seventh was squally all round, but we never caught a drop of water. Two more died. We threw one overboard, but were too weak to lift the other. There were then five left, two able seamen, one ordinary, myself and one passenger. The passenger was out of his mind. All had drunk sea water. We were all dozing when the madman bit my foot and I woke up. We saw a ship bearing down on us, the *British Sceptre* from Calcutta to Dundee. We were taken on board and treated very kindly.

Of the 474 who had sailed in the *Cospatrick* only three survived.

29

Posh

'You have it, Madam.'
Benjamin Disraeli, 1874

Along the route pioneered by the visionary Thomas Waghorn in the 1830s, British travellers bound for India crossed the Mediterranean to the Egyptian port of Alexandria, swayed in camel caravan over the sands of the isthmus of Suez and joined a steamer bound down the Red Sea for Bombay. After the Peninsular and Oriental Steam Navigation Company took over the Waghorn trail in 1847 around 3,000 people traversed the desert every year. To keep and transport them in comfort P&O maintained hotels, vegetable and fruit gardens, carriages, more than 400 horses and 3,500 camels. These Suez overlanders were an elite. Heavier traffic, like troops and bulk supplies, sailed by the longer traditional route around the Cape of Good Hope.

A canal linking the Nile to the Red Sea was an old dream. The Egyptians began to dig one about 4,000 years ago. Napoleon saw Egypt as the strategic key to India and planned a canal, but Nelson smashed all his Egyptian schemes at the Battle of the Nile in 1798. In April 1859 the French diplomat Ferdinand de Lesseps, backed by French money, began work on a canal. The British declared it a hopeless project, saying that even if the ditch were dug the hot wind would quickly fill it with sand. Lord Palmerston, the Prime Minister of the time, dismissed it as both impossible and 'a device for French interference in the East'.

De Lesseps, however, had a rock-like belief in himself. Under his direction swarms of labourers hacked with picks and shovels, and dredgers scooped up thousands of tons of silt. In 1869, after ten years of work, de Lesseps had his golden day. The lovely Empress Eugènie came to open his 100-mile thread of waterway and the first ship steamed from the Mediterranean to the blue Red Sea. Thomas Cook escorted a party to the ceremony, sixty clients paying 50 guineas a head, and installed them in a luxurious tented camp in the desert, carpeting the sand beneath the palms. He set the pattern for Middle Eastern tourism. After this initial success Cook's odysseys in Egypt and the Holy Land, lasting seventy or a hundred days, became popular experiences for better-off Victorian tourists. Cook's Nile steamers were famously comfortable, built in Scotland and shipped out in kit form to be floating palaces for elegant guests who appreciated surroundings that put them at a little distance from Egypt.

Confronted with the reality of the canal, the British still argued complacently that the Cape route would be more important than Suez. P&O remained dubious about it and wondered what would come first, bankruptcy or total silting. What experts had not foreseen soon became clear: any mariner could have told them that from the perspective of the steamship companies and the naval strategists the canal changed the chart of the world. The steamer journey from Britain to India was reduced from three months to three weeks. If necessary, troops could now be rushed to India. Having dismissed the canal as improbable, the Foreign Office was galled to observe that four out of five of the hundreds of ships using it sailed under the British flag. Seeing the steamers floating slowly over the desert P&O ended its camel service on the old Waghorn trail and ceased investing in the dockyards at the Cape.

The realization that the Suez Canal was now the virtual threshold of India grew in British minds. There was always the Russian threat to the Indian empire to consider. Benjamin Disraeli, who became Prime Minister for the second time in 1874, saw a permanent danger in the French owning a waterway that Bismarck, the Chancellor of Germany, called the spinal cord of empire. Hearing that the squandering Khedive, the ruler of Egypt, was bankrupt and desperate to sell his canal shares, Disraeli coolly borrowed

£4 million from the banker Edmund de Rothschild, bought the shares and, as if presenting the canal to Queen Victoria as a gift, told her: 'You have it, Madam.' She did not really have it. The Khedive's shares amounted to a 40 per cent holding. But it was a significant stake; and, whatever the reality, Britain certainly seemed to own the canal. *Punch*'s cartoon showed Disraeli handing a swagbag of money to the Khedive while the British lion rested, smugly couchant, near the Pyramids with the key to India in its paws. East of Suez now meant the Indian empire; and Britain's paramountcy at sea was complete.

Increasingly, that dominance depended on the development of steam power. The Suez Canal, after all, was a channel for steamships. It would have been prohibitively expensive to tow sailing vessels along it, and in any event the way the winds blew in the Red Sea was no help to sail.

On reflection, the Suez Canal seems to have drawn a line in history, marking the supremacy of steam over sail; but the reality was that the sailing ship, and especially the British sailing ship, remained significant for the next twenty years. In 1864 British shipyards launched the largest tonnage of sailing vessels ever built in one year, 272,500 tons; and in the same year completed 159,400 tons of steamers. The following year the total tonnage of British sailing vessels afloat amounted to 4.9 million tons compared with a steam tonnage of 800,000 tons. In the year after the Suez Canal opened ships of the British empire amounted to almost 6 million tons under sail and 1.2 million tons under steam.

Through constant experiment British engineers made marine engines increasingly efficient and powerful. By 1860 condensers supplied constant fresh water to engines, improving performance and eliminating the old bugbear of salt-water encrustation. The difficulty of making a watertight sleeve for propeller shafts was solved. John Elder's development of the compound engine in the 1850s boosted horsepower and fuel efficiency and, thereby, the range of steam vessels. Between the 1850s and the late 1870s the consumption of marine coal was halved. The opening of the Suez Canal greatly shortened distances and brought much high-value trade, such as tea, within the steamship's reach. It cut the China voyage, for example, by 3,000 miles, saving ten to twelve days.

The heavy tolls charged by the Canal authorities, however, helped the sailing-ship owners. They made the transporting of bulk cargoes like jute and rice uneconomical for steamships. In 1876 a shipowner remarked that on some routes steam simply could not compete with well-run sailing vessels.

The limits of engine power and coal-bunker capacity also excluded steamships from long-distance non-stop voyages. Steamships relied on a network of coaling stations, islands and coastal bases where they could take on food, water and, above all, coal. They were especially important in the thinking of the Royal Navy's planners.

According to Admiral Sir John Fisher, the great Victorian and Edwardian sailor, the world was locked up by five strategic keys, Dover, Gibraltar, Alexandria, the Cape of Good Hope and Singapore, and 'these five keys belong to England'. Other outposts were keys, too, more than 150 coaling stations where the Royal Navy could be certain of fuel and supplies. Malta, Aden, Ascension, Port Stanley, Bermuda, Fiji, Esquimault, Trincomalee and Broome were all vital red dots on the map. Many of the coal bases, especially the remoter ones, were supplied by sailing ships. P&O employed 170 sailing ships to stock its bunkers in the Mediterranean, India and the Far East. Every naval captain kept in mind the location of stations and the limited capacity of warships which, when they steamed at full speed, had a monstrous hunger for coal. Every ship's company hated refuelling, a dirty and often dangerous job which coated ships and men with dust and glittering grit. Tons of coal were poured through the chutes into the bunkers where it was stowed by sweating sailors as blackened and glistening as the miners who had dug it out, deep in the pits of Wales.

In 1828 the First Lord of the Admiralty stated that the navy would discourage the use of steam vessels because steam was calculated 'to strike a fatal blow at the supremacy of the Empire'. Admirals worried that the paddles of steamers could be smashed by gunfire and that the coaling stations necessary to support them would be expensive to supply and defend.

Many senior officers also clung to the beauty and romance of sail. In the 1850s big ships were still painted in black-and-white chequerboard style and armed with broadside guns. In 1859 the author of a midshipmen's manual wrote: 'Although we are living in the steam era and ours is a steam navy, I have in this work wholly excluded the consideration of steam power, as, owing to the great cost of coal and the impossibility of providing stowage for it, the application of steam power must be strictly auxiliary and its employment on general service the exception rather than the rule.' The navy's instruction book of 1862 said: 'There is no greater fallacy than to suppose that ships can be navigated on long voyages without masts and sails.' Some captains used their engines with distaste, and perhaps the grimy business of loading coal added to the prejudice that some felt for steam. In the transition, when hybrid ships had sails and auxiliary engines, officers frowned at the smuts on sails, uniforms and scrubbed decks.

For all the doubts, however, the Admiralty advanced with steam as fast as possible and was quick to seize its advantages. The demands of the Crimean War hastened the process. The withdrawal from sail was long and difficult. The navy could hardly embrace, immediately and wholeheartedly, a decidedly unreliable and unproved new technology. Its needs were quite different from those of the shipping companies. Its commitments covered the world. Its ships had necessarily to accommodate large crews, heavy guns, ammunition and supplies, and be able to keep the sea for long periods without resupply. For a number of years only sailing ships were suitable. Until the 1840s the only powered vessels were paddle steamers, known as flappers, which were limited in size and performance and which were far from ideal as gun platforms. For both the navy and the merchant marine the unreliability and limits of engines dictated the era of sail-assisted steam. In 1853 there were almost 10,000 ships registered at Lloyds of which under 2 per cent were steamers.

Engines became more efficient during the 1840s, however, and ambitious young men saw that experience in steam was the way to advancement in the navy. It was a reflection of trends in society where professional skills were growing in science, engineering and medicine. Like Kipling's McAndrew, the engineers were a different breed, many of them men from the north pushing their way through southern snobbery. The invention of screw propulsion was the great advance: it did away with the paddle problem and prompted thoughts of iron ships and giant guns. The last pure sailing vessel for the Royal Navy was built in 1849. In 1852 the first successful screw-driven warship was launched.

As steamships evolved in the 1870s, they shed their masts and sails and clipper bows and hybrid looks and adopted the unmistakeable steamer shape. Passengers, too, were in danger of taking on a steamer shape. Aboard P&O ships bound for the east there was an emphasis on plentiful food and abundant wine, especially in first class, which was the larger class, since there was no emigrant class to India. Until 1874, when P&O started charging for it, passengers drank as much claret, champagne and spirits as they liked because the fares included drinks. Menus offered a choice of turkey, pork, mutton, goose, duck, beef and chicken. The appropriate beasts had to be carried in the onboard farmyard and supplied with fodder and straw, so that the sounds of the wind and sea were augmented by moos, grunts, clucks and squawks. The ships steered east or south-east on the voyage to India so that the north-facing side of the ship was cooler; and the privileged booked to travel port side out and starboard home: the acronym POSH for this arrangement was not, however, the origin of the word. Passengers leaving London boarded special P&O trains, so that they could stretch a point and believe that Victoria station was the gateway to India.

In the year that the Suez Canal was opened, a beautiful new ship splashed into the water at Dumbarton in Scotland. She was a clipper, a three-masted square-rigged vessel of fine lines built for long-distance racing. New and advanced as she was, she was already out of date; but she captured the last of the limelight as one of the fastest ships in the world.

The Suez Canal sounded a knell for sail, and many sensible owners put their money into power, investing in the steamship's commercial virtues of speed and predictability. There were more than a few owners, though, who thought that sailing ships still had a future. Sail owners reckoned that steamships were not yet the masters of the long-distance routes to China, Australia, New Zealand and South America. The distance from Britain to Sydney was 12,340 miles by way of the Cape of Good Hope, 11,200 miles by way of Suez; and sailing ships were faster. There was no Panama Canal, there was work for Cape Horners to do, and perhaps the Suez Canal would fill with sand.

Certainly John Willis hoped so. An owner who had been a captain in the China tea trade, he was committed to sail. He saw a niche and commissioned the Scottish designer Hercules Linton to draw him a ship of 963 tons, 212 feet long, thirty-six in the beam, her planks fastened to an iron frame. Owners gave the clippers heroic and fancy racehorse names: *Spirit of the Age*, *Coral Nymph*, *Dawn of Hope*, *Chariot of Fame*, *Neptune's Car*, *Seaman's Bride*, *Blackadder* and *Glory of the Seas*. Willis named his ship *Cutty Sark* and embellished her with a figurehead of the witch in Robert Burns's 'Tam O'Shanter' wearing the skimpy blouse called a cutty sark.

In that year of 1869 the clipper *Thermopylae* completed her maiden voyage to Melbourne in sixty days, a feat never bettered. That was the sort of record Willis was after. He sent his ship to Shanghai for a cargo of tea. There were still tea merchants who believed tea was better carried in a wooden ship than in an iron steamer where it might be contaminated. *Cutty Sark* brought it back in 109 days and went out again the next year. In 1872 she and *Thermopylae* loaded tea together at Shanghai and raced to London. *Cutty Sark* lost her rudder in a storm and soon lost the replacement, too. With immense difficulty her crew made and fitted a third; but she lost the race. She carried tea from China eight years running from 1870 to 1877 and was outstripped by *Thermopylae* every year but one. It was always dangerous and exacting to keep a ship racing at high speed. Part of the price of China tea was damage, shipwreck and drowned sailors.

By 1873 it was clear to owners and captains that, with the Suez Canal not yet filled with sand, the steamers were winning the tea trade. The clippers could make only one China voyage a year, the steamers two. *Cutty Sark* and her sisters fired the imagination with riproaring voyages of a hundred days, but the dogged steamers came home in less than half that time. After 1877 the tea clipper age was over and like other vessels *Cutty Sark* had to go tramping for cargoes, loading anything she could get: jute, rice, molasses and kerosene. Now she performed the dreary chore of humping coal from Australia to China, 1,200 tons at a time.

But from 1883 she had her long Indian summer, twelve years of fame racing Australian wool around Cape Horn to London mostly under the command of Captain Richard Woodget. He typically sailed her out to Sydney in seventy-five days and back to the English Channel in eighty, and once came back in sixty-seven. They were years of adventure and danger and consummate seamanship in the great gales of the southern ocean, of frightening confrontations with icebergs on the way to the Horn. 'We found ourselves surrounded by icebergs', Woodget wrote in his log in February 1893, 'cracking like thunder, yet we could not see them.' A few weeks later he recorded that 'John Doyle and John Clifton, whilst making fast the outer jib, were washed from the boom and drowned. I heard crying from forward and saw the two men struggling. There was too much sea to lower a boat. What a gloom it cast over the ship!'

Cutty Sark, the champion of the wool clippers, brought home her last Australian cargo in 1895. John Willis had been right. *Cutty Sark* had had her glorious time and had earned her keep for twenty-five years; and her name endures. 'It was thrilling on the old *Cutty*,' said Captain Woodget. 'She sailed like the witch she really was.' But her sun had set and the boys who trained aboard her and learned to sail around Cape Horn were going off to careers in the P&O steamers. The Suez Canal, constantly dredged, never did fill with sand.

30

Laying the ghost

> 'How can you put so much devilry
> Into that translucent phantom shed
> Of a frail corpus?'
> D. H. Lawrence, 'Mosquito'

Quinine was a magic potion in the nineteenth century. No one knew how it worked, but no one knew either how malaria was spread. Quinine was made by stewing the bark of the South American cinchona tree. Apothecaries made a powder of it with pestle and mortar and mixed it with water or wine. It did not stop the malarial parasite taking hold but it could reduce the dangerous effects of the disease. It was an erratic remedy because the bark was of variable quality, its potency deteriorated with time, dealers adulterated it with bark from other trees and there was no guarantee of a regular supply.

In the 1850s Clements Markham persuaded the British government to send him to Peru to gather cinchona plants and seeds which could be transplanted in India as a reliable source for the quinine that was badly needed to fight malaria. Markham had joined the Royal Navy at the age of fourteen in 1844, and took part in an expedition which searched for Sir John Franklin in the Arctic. He retired as a lieutenant at twenty-two and became an explorer.

In March 1860 he and John Weir, a gardener, set out on mules into the Andean forests. They travelled through the jungles beyond Lake Titicaca, scaled steep cliffs and fought their way through torrents to reach the cinchona trees. To ease the hardships of travel Markham adopted the Peruvian custom and chewed coca leaves.

> Besides the agreeable soothing feeling it produced I found that I could endure long abstinence from food and it enabled me to ascend precipitous mountain sides with a feeling of lightness and elasticity, without losing breath. This latter quality ought to recommend its use to members of the Alpine Club and to walking tourists in general.

Markham and Weir put the plants into Wardian cases. These were miniature greenhouses, sealed glass boxes more than three feet long and weighing more than 300 pounds, invented by Dr Nathaniel Ward to protect plants in London and developed to keep seedlings safe on long journeys. John Weir paid a high price for gathering the cinchona seeds. He was crippled by fever. Another collector was similarly ruined and the government, typically, refused him a pension until Lord Palmerston intervened and got him £50 a year. Shipped from Panama the cinchona eventually reached the Nilgiri Hills of southern India where they were planted. Many of the plants were rested at Kew and sent out to India when the heat in the Red Sea had abated.

The cinchona planted in India and later in Ceylon earned the government large profits. Markham wrote that his work brought great blessings to the human race, rendering the supply of quinine 'more certain, cheaper and more abundant'. At least half of the 2 million deaths a year in India would be prevented by putting 'some cheap form of quinine into every shop in the country at one rupee an ounce or less. This measure is as important as any other question connected with Indian administration, for it involves the annual saving of many lives, of large sums of money and of an immensity of misery.'

In the 1870s doctors understood very little about any of the tropical diseases. But something needed to be done. India was the heart of the empire and a regular flow of people went there: civil servants, planters, missionaries, businessmen, engineers, factory managers, railway builders, doctors and thousands of soldiers. They all needed protection. In many other parts of the empire, too, malaria was a clear obstruction to progress. The killer had to be found.

In the late 1870s and 1880s a number of doctors began to guess or suspect that the mosquito played some part in the spread of the disease. Patrick Manson, working in China in 1876, found that mosquitoes were somehow transmitting the parasites of elephantiasis. Charles Laveran, a French army doctor examining sick Foreign Legion soldiers in Algeria, where malaria was rife, saw parasites in blood he was examining under a microscope. Malaria became a medical frontier in which the latest microscopes, able to magnify a thousand times, were the vital weapons in new wars against enemies too small for the eye to see.

Scientific researchers like Manson had very little to go on and had to teach themselves. They needed enthusiasm and Manson had it in abundance. In London in April 1894 he shared it with Ronald Ross, a thirty-seven-year-old Indian Army doctor. Ross had never seen the malaria parasite. Manson set up a blood slide on his microscope and showed it to him. It was he who put Ross on the malaria trail.

Ross was born in India in 1857, a few days before the Mutiny broke out. The eldest of ten children of an Indian Army general, he trained as a doctor at St Bartholomew's Hospital in London. It was an exciting time to be in medicine. For most of the nineteenth century there was little that doctors could do, most of the medicines they prescribed were useless and diseases raged more or less unchecked. The French chemist Louis Pasteur showed the way ahead. His brilliant work on bacteria was changing the understanding of the way that infections were transmitted. His book on germ theory was published while Ross was completing his medical training. At the same time Charles Darwin's thinking on evolution influenced the way biologists approached their work. Medicine and science were bubbling with new ideas.

After his meeting with Manson in London, the encounter that altered the course of his life, Ross returned to India in 1895 determined to find out how the mosquito spread malaria. He could barely wait to begin work. 'Wild with excitement', he kept in touch with Manson through a stream of letters. Manson encouraged him every step of the way. 'You are doing most valuable work . . . a light will dawn on you soon such as will astonish you and the rest of mankind.' As well as being Ross's mentor and adviser, Manson acted as his publicist in London, ensuring that the results of his experiments and trials were known in the right places and that his articles were published in the medical journals. He also kept him up to date with the work being done by others. Ross was well aware that he was competing with other intellects in an international race for knowledge.

In India Ross worked doggedly on, spending hours over his microscope looking for parasites in the mosquitoes he dissected. There were many failures and frustrations. In spite of his own daily dose of quinine he caught malaria himself and was left very weak by it. The crucial step forward came when he began to look for the malaria parasite in birds; and in 1898 he demonstrated convincingly the mosquito's part in what he called 'the great malaria problem'. Other researchers, including Giovanni Grassi in Italy, claimed similar discoveries, but it was Ross who was awarded the Nobel Prize in 1902 for discovering how malaria was transmitted from man to man by way of the mosquito.

Ross and Grassi, caught up in the jealousy of discovery, disputed the matter for the rest of their lives; and Ross even distanced himself from his great supporter, Patrick Manson.

In the Spanish-American war of 1898 United States troops invaded Spanish colonial territories in Cuba and the Philippines. To support the actions in Cuba the American battleship *Oregon* sailed from San Francisco, rounded South America and reached the Caribbean after a spectacular voyage of sixty-eight days. To both the American navy and the public the lesson was plain: there was a vital strategic need for a canal through the isthmus of Panama, the junction of the two American continents, to shorten the passage from the Atlantic to the Pacific and to strengthen the imperial reach of the United States.

During the war 379 American soldiers were killed in action and 5,083 died of disease, most of them from yellow fever contracted during the occupation of Cuba. Indeed four-fifths of the troops suffered from the disease. The Americans sent an army doctor,

Major Walter Reed, to fight the disease in Cuba although he had no idea of the cause of it. Many doctors at that time thought the fever was a contagion and could be spread by the clothing and bedding of sick men. A Cuban doctor, Carlos Finlay, suspected that the mosquito was the cause, but his work was largely ignored and Reed himself was doubtful.

Three young doctors, however, believed Finlay was right and as part of their research two of them submitted to mosquito bites and suffered the delirium and other horrors of yellow fever. One recovered, one died. Walter Reed reviewed their work and was persuaded that the *Aedes aegypti* mosquito was the culprit. To provide conclusive evidence the contagion theory was tested by putting volunteers into the beds of men who had died of yellow fever and swaddling them with blankets stained with the victims' characteristic black vomit. Other men were confined to special tents where doctors introduced mosquitoes which had fed on the blood of yellow fever victims. The men who slept in the soiled beds suffered no harm. Those bitten by the mosquitoes fell sick with yellow fever.

Now that they had identified the enemy the Americans could wage war. In 1901, Major William Gorgas, an army physician who himself had suffered yellow fever while serving in Texas in 1880, turned sanitary engineer and set out to clean up Havana. He isolated sick men in hospitals screened with wire gauze to disrupt the chain of infection and ordered military patrols to search rigorously for ponds and puddles where mosquitoes bred. They poured petroleum on breeding grounds to kill the larvae, drained ditches and imposed fines on people with uncovered water tanks and jugs. The patrols made house-to-house searches and inspected every water receptacle in the city. Such official toughness angered many Cubans but by 1902 there was not a single case of yellow fever in Havana. It seemed like a miracle. Gorgas was a hero, the conqueror of the scourge.

In the early years of the twentieth century the United States took up the challenge of building the Panama Canal, what the historian James Bryce called 'the greatest liberty man has ever taken with nature'. The ebullient Theodore Roosevelt, President from 1901 to 1909, believed strongly that America's manifest destiny was to control the Caribbean and the western hemisphere. 'Do things!' was one of his slogans. 'Speak softly and carry a big stick,' summarized his foreign policy.

Roosevelt had no doubt at all that the Panama Canal would be the greatest work of the American empire, not only a permanent monument but also a practical and progressive demonstration of imperial power.

It was an old idea. The Spanish conquistadors had suggested a canal in the sixteenth century. In 1848 fighting between Mexico and the United States ended with Mexico ceding to the Americans the territories of California, Texas, New Mexico, Arizona and parts of Colorado, Wyoming, Nevada and Utah. The Americans thus acquired a Pacific frontier and a need for better communication. The gold rush to California underlined the point. Prospectors sailed from New York to Nicaragua, transferred to smaller craft to cross Lake Nicaragua and completed the journey by land. It took a month.

In 1881 a French company headed by the energetic Ferdinand de Lesseps, the hero of Suez and a man 'born to pierce isthmuses', as an admirer said, began to cut a forty-seven-mile canal through Panama. At Suez de Lesseps confronted the desert; but in Panama there were all the horrors of swamps, fast-growing jungles and an eight-month rainy season. More than 20,000 men died of malaria and yellow fever. Some of the construction bosses said that liquor and loose living caused the fevers.

The excavations slowed, debts mounted, the investors' money poured into the quagmire and the digging ended in scandal with de Lesseps, his son and others tried for fraud in Paris. Poor de Lesseps, much more an enthuser than an engineer, had been caught up in a hopeless scheme. He was sentenced to five years, a fallen god, but was spared prison. Just before his death in 1894 at the age of eighty-nine, the courts cleared his name.

Roosevelt saw his chance. Panama was a province of Colombia. The Americans encouraged its secession and sent warships to stop any Colombian action. A treaty gave the jubilant Roosevelt everything he wanted, control of Panama and the right to build a canal. As soon as the Americans took up the task in 1904 yellow fever struck. In 1905 hundreds of skilled white workers fled. A large number of coffins were piled up on wharves and in railway stations to accommodate the expected fever victims and one newly appointed American engineer and his wife even brought their own coffins with them.

William Gorgas was sent to Panama and, against the initial opposition of administrators who did not believe mosquitoes had anything to do with disease, began the kind of clean-up he had carried out in Cuba. His men installed a good water supply, drained swamps, cleared jungle to destroy mosquito breeding places, spread oil on ponds, screened the hospitals, cleared gutters and built pavements to prevent puddling.

His critics, who saw the workforce as mere helots, complained about the high cost of keeping the men healthy, fed and in reasonable accommodation, about one-twentieth of the cost of the canal. As it was, about 6,000 workers died in nearly ten years of work, crushed in frequent landfalls, drowned or blown up by dynamite. For all the preventive work that Gorgas carried out in draining and screening, men still fell to fevers because there was not enough protected accommodation for a huge labour force. Still, in 1912, of the 40,000 men employed on the project, only 7,000 went down with malaria and only thirty-two died, a huge improvement on the early years.

Without Gorgas's campaign against yellow fever and the subsequent drives against malaria and typhoid, the canal could hardly have been built. Forty miles long, with huge locks that raised and lowered ships eighty-five feet, it was a magnificent engineering feat. The first ship went through from the Atlantic to the Pacific in nine hours and the Americans exulted. They had shown the world what Americans could do; and they had removed Cape Horn from the seafarers' map.

31

Halt in the name of the Queen

> 'This Mounted Policeman was instantly part of me, as ineradicable as a scar. I can see him yet in the brilliant uniform, stooping to close the shack door, as if in symbolic relinquishment of the West he had made tame.'
>
> Wallace Stegner, 1955

Under the Convention of London in 1818 a line was drawn across North America along the forty-ninth parallel of latitude from the Lake of the Woods to the Rocky Mountains. It was acknowledged on both sides as the border between the United States and the territory of His Britannic Majesty, otherwise Canada.

No fence or wall marked the line and for the traders, trappers and Indians who roamed across it at will it was hardly a boundary at all. Yet it slowly emerged as the frontier between two cultures, two bodies of law, two interests, two different ways of looking at human activity.

On the immense rolling ocean of prairie it also indicated the beginning of history, the arrival of civilization and order in the wilderness. On either side of the forty-ninth parallel there were different experiences of the phenomenon of the frontier, two different Wests. To the vigorously pushing Americans of the nineteenth century the expanding frontier was a world of adventure, opportunity and profit, and its very size was a protection from any possible enemies. The politician and soldier William Gilpin spoke for all the destinarian Americans who believed that the settling of the West fulfilled God's purpose. 'Divine task!' he proclaimed. 'Immortal mission!'

To many Canadians the American West really did look wild. It seemed an unstable and undefined territory where land was for the grabbing, only the strong prevailed and the law had at best only a tenuous hold among the cowboys and swarms of miners and chancers. After the Civil War, too, it seemed that the sabre-waving United States Cavalry was pursuing a policy of massacre as the remedy for the Indian problem.

The unruliness and naked materialism of the American frontier suggested to Canadians that the 'divine task' might include the American annexation of Canadian territory. In 1846 the magnificent lands of Oregon, in the north-west, which were claimed by both Canadians and Americans, became part of the United States after Americans poured in and settled the issue by weight of numbers. Then, in 1867, the United States bought Alaska from Russia, for $7 million. In that year, the semi-independent Canadian provinces of Ontario, Quebec, Nova Scotia, New Brunswick and Manitoba formed the self-governing Dominion of Canada. One of the purposes of this confederation was to keep the almost empty hinterland of the country safe from American expansion. In 1869 the new government in Ottawa bought Rupert's Land from the grasping Hudson's Bay Company for £300,000 and a handsome grant of territory. The Company relinquished its royal charter and its trading monopoly in the huge and hardly known North West Territory. The purchase imposed an unwelcome burden of responsibility on Sir John Macdonald, the Prime Minister. 'I would be quite willing to leave that whole country a wilderness for the next half-century,' he said, 'but I fear if Englishmen do not go there Yankees will.'

Ottawa had no wish to see Canada's West become the kind of turbulent society flourishing below the forty-ninth parallel. There were already alarming symptoms of it in the operations of the American whiskey smugglers who made a lot of money brewing and selling 'Bug Juice', a mixture of molasses, ginger, red pepper, tobacco and whiskey. From their bases forty miles inside Canada, their log-built settlements jokily named Fort Whoop Up, Fort Slide Out and Fort Stand Off, they bartered hooch and guns with Indians in exchange for furs and horses.

The smugglers paid two cups of firewater for a buffalo robe

and four gallons for a prime pony. To buy a rifle an Indian had to bring in a stack of pelts as high as the gun. The Indians were unaccustomed to alcohol and after drinking Whoop Up whiskey they rampaged across the prairie, stealing horses and looting trading posts to fund their addiction. The cynical smugglers pushed northwards into Canada. As they themselves pointed out, there was no one to stop them.

Sir John Macdonald knew that they had to be stopped. In the American West people had preceded the law. In the Canadian West he wanted the law to precede the people. The existence of the border and of Canadian law had to be asserted. Canadians would not be taken seriously if they failed to crush the smugglers and hoodlums and did not insist on respect for the international border. Sir John needed to send a uniformed force but not an expensive military one. His answer, in 1873, was to found the Mounties, the North West Mounted Police, a small and inexpensive civil force to provide 'rough and ready enforcement of law and justice'. Its senior men would be sworn in as justices of the peace so that they would be arresting officers, prosecutors and judges, the law on horseback, complete and self-contained. Sir John had in mind a Canadian version of the Royal Irish Constabulary.

In the spring of 1874 about 275 officers and men assembled at Fort Dufferin, south of Winnipeg, to prepare for a prairie journey of more than 800 miles out of Manitoba, through Saskatchewan and into Alberta and the foothills of the Rocky Mountains. Their mission was to eradicate the whiskey traders and establish a chain of log forts across the north-west, each flying Canada's flag. In the years to come generations of Canadian schoolchildren would learn that this was the defining expedition of the Mounted Police and the founding epic of the men in scarlet: the legendary March West.

Sir John Macdonald had promised that as a civil force the Mounted Police would wear 'as little gold lace, fuss and fine feathers as possible'. But officers with a fondness for regimental swagger adorned their helmets with plumes and kitted the men with scarlet Norfolk jackets, tan breeches, pillbox caps, white pith helmets, white dragoon gloves and sparkling metalwork and leather.

Old hands, who knew the prairie as a rough and unforgiving place, pointed out that dragoons' rig was unsuitable for the hardships of horseback policing.

On 8 July Colonel George French, aged thirty-two, force commander and military stickler, mounted on a handsome horse he had bought in his native Ireland, turned in his saddle to survey the expedition. For the sake of vanity each of the six divisions was mounted on matching horses. There were greys, dark bays, blacks, light bays, dark browns and chestnuts. The two-mile column had a total of 310 horses, 142 oxen, ninety-three cattle, seventy-three wagons, 114 Red River carts, two nine-pounder field guns, two mortars, ploughs, field kitchens, portable forges and two mowing machines to cut hay. The cart drivers, who were *metis*, men of mixed white and Indian blood, were fighting their hangovers. The temperature was around ninety degrees Fahrenheit and the animals kicked up a cloud of dust. A band played 'Auld Lang Syne'. With reports and stories of hostile Indians in their minds, some of the new Mounties had cut their hair very close in the belief that if it came to a fight and death, Indians would at least be denied their scalps.

French raised his sword and urged the cavalcade westward. A tremendous screeching rent the air. The ox-drawn Red River carts drowned out all other noise. Box-like two-wheeled vehicles, held together with buffalo hide, they were the chief form of land transport on Canada's prairie frontier for three-quarters of the nineteenth century, the equivalent of the canoes and Hudson's Bay Company boats on the lakes and rivers. Originally their wheels were a solid slice sawn from the end of a tree trunk. With more efficient spoked wheels as high as a man they carried loads of 500 pounds. The axles were poplar logs, unlubricated because grease and thick prairie dust combined to jam the wheels. Consequently they moved with a piercing shriek, 'the North West fiddle' it was called, and it could be heard for many miles. 'A den of wild beasts cannot be compared with it in hideousness,' wrote one of the party.

Excruciating as it was, the squealing was the least of the Mounties' suffering. Soon their journey became an ordeal. Heat and lack of water quickly eroded the early jauntiness and the men and their horses sagged. Even after it was filtered the muddy water remained black. Clouds of mosquitoes bloodied faces, necks and hands and, in agony, men rolled themselves into blankets. Mounted men, horse-drawn carts and ox-drawn carts all travelled at different speeds. The two field guns, weighing a ton apiece with their limbers, were a useless burden. The column spread out for miles across the plain. At the end of a day on the trail men camped for the night while their food and cooking equipment were still hours away; so they slept unfed. They suffered the miseries of malnutrition, diarrhoea and dysentery, and their existence became increasingly squalid.

Storms uprooted their tents and pelted them with hailstones the size of plums. The prairie-hardened *metis* driving the ox-wagons had smaller and more practical tents that weathered the storms. They did not depend on fresh meat, as the Mounties did, but thrived on their own dried meat and on pemmican, pounded meat mixed with bone marrow and sweetened with berries.

To the expedition's dismay, the handsome colour-graded horses, raised in Canada's east, could not stomach the prairie grass and began to wilt and starve. The parched and hungry men dismounted and walked to spare them; and the walking began to wear out their boots. The painful progress also began to wear out French's patience. In his journal he catalogued the failings of his subordinates. Always peevish, he thought he saw carelessness and defeatism everywhere. The doctor had to assure him that sick men were not shamming.

After sixteen days and 275 miles on the trail French and his sick and weary men and skeletal horses reached the distinctive wind-eroded rocks at Roche Percee. Today you can see the place where they camped and the daunting view they had of the westward-rolling prairie under a great vault of sky. French faced the fact that his expedition was in danger of turning into a death march. He still had far to go. His horses were collapsing. Broken wagons and dead animals marked the trail.

He divided his force, ordering Inspector William Jarvis, with three officers and twenty-seven men, including all of those who were ill, to take the sickest horses, the cattle and carts, and attempt the 900-mile trek north-west to Fort Edmonton. There was the prospect of some food along the way, in Hudson's Bay Company outposts. French was nevertheless committing Jarvis and his men to a desperate venture.

French pressed on with the remainder. The brave scarlet uniforms were tattered. Mosquito-bitten knees poked through breeches. Saddles and harnesses disintegrated. There were 600 treeless miles to go. Henri Julien, a reporter and artist for the *Canadian Illustrated News*, who accompanied the expedition, wrote of the tedium of the trail: 'The eye dwells on vacancy, tired at glancing at the blue sky above or the brown earth beneath . . . the very labour of talking becomes tiresome and you fall to meditation . . . silence weighs on you like a mechanical power.'

All the Mounties became infested with lice. They drank liquid mud and suffered fresh bouts of diarrhoea. They found and shot bison and suffered agonies after filling their half-starved stomachs with barely cooked meat. The hunting exhausted the tired horses. As they grew more sick some of them stopped drinking; and as their horseshoes wore out their feet became raw. The weather grew bitter and, after five horses died in one night, each man was told to give up one of his two blankets to cover a sick animal and to huddle close to a comrade for warmth. The men bound their worn-out boots with twine; and those whose boots had rotted away wrapped their feet with sacking.

Fifteen-year-old trumpeter Fred Bagley found that his lips swelled so much that he could not blow his bugle. He confided to his journal that French seemed 'very hazy as to where we are. Everyone glum . . .'

'Can't go much farther,' wrote another man.

French considered his ragged men and sick and dying horses. 'I begin to feel very much alarmed for the safety of the Force,' he wrote. He had a compass, a sextant and an odometer to measure the distance covered, but navigation over the trackless prairie was gruelling work.

After resting in the Sweet Grass Hills, close to the American border, French ordered some of his force to take the fittest of the surviving horses and make the punishing return journey eastwards, to Dufferin. He and his deputy James Macleod rode south to Montana to buy horses, food, socks and moccasins. After their shopping they parted company, French to catch up with the

Mounties who were marching eastwards, and Macleod to lead a force westward to the foothills of the Rockies. Reaching their goal, the Macleod contingent chopped down cottonwood trees and built Fort Macleod, the first of a string of Mounted Police log outposts in the Canadian wilderness.

The decisive and doughty Macleod declared the arrival of law and order. Two weeks later his men came upon a gang of American smugglers, driving wagons full of whiskey, guns and furs.

'Halt!' shouted one of the Mounties, with a cry that became part of the reality and romance of the north-west. 'Halt! In the name of the Queen!'

Macleod himself sat in judgement on the whiskey men and shut them in his log jail. Other smugglers went on the run. When winter came Macleod and his men were ill-equipped for it and had to replace their ragged uniforms with boots, coats, trousers and hats made from buffalo skin. The government had promised new uniforms and pay, but sent nothing. Macleod saddled up and with four of his men journeyed 300 miles through blizzards and temperatures of fifty below to Helena in Montana. There he drew his men's pay from a bank and rode back to give it to them.

By the end of the winter Macleod and his men had stamped out the whiskey orgies among the Indians. They had also established a reputation as impartial enforcers of the law. The red coat of the Mountie became in Indian eyes a symbol of protection, while across the border the blue coat of the United States Cavalry represented betrayal and slaughter. The Indians called the border the Medicine Line: to their minds the line of surveyors' cairns stretching along the forty-ninth parallel represented strong medicine, the protection of the Queen, the Great Mother, and of the Mounties who enforced authority and maintained the frontier. Indeed, Indians often addressed officers as Father. Within fifteen years of the March West, however, Indian tribes were in reservations, their free and nomadic existence at an end.

Meanwhile, Inspector Jarvis and his party reached Fort Edmonton after eighty-eight days of marching. The men built bridges over rivers and hauled carts by hand over muddy tracks. When a horse collapsed two Mounties put a pole under its belly and levered it to its feet. The horses, said Jarvis, 'were living skeletons'. So were the men when they staggered into the fort. They owed their lives to Jarvis's leadership. 'We were living together like a family,' wrote one. 'Every heart was beating for Inspector Jarvis.'

The March West nearly ended in disaster but in the telling it was a triumph. The legends multiplied and writers hymned the men of integrity and daring who lived up to their motto: *Maintiens le Droit*, Uphold the Right. Robert Service set the tone in his poem 'Clancy of the Mounted Police': 'Who would wear the scarlet coat shall say good-bye to fear.' Within five years of the March West *The Times* reported that 'in every sense of the word the Canadian North West Mounted Police has been a corps d'élite'. One of its officers in the 1880s was Charles Dickens's son, Francis.

No colonial police force, indeed no police force anywhere, has enjoyed the Mounties' iconic status. Hollywood added to the legends and Mountie films became a genre after the first was made in 1909. Stories and articles established the force as foursquare, manly and moral; and many Canadians felt that these qualities made the Mounties distinctively Canadian. According to the legends, stubbled American sheriffs went for their guns but the clean-cut Mounted Police subdued bad men with the steely gaze of authority.

In one typical story a large force of American cavalry handed a group of 200 disgruntled Indian warriors into the keeping of a few Mounties.

'But there are only four of you,' said the astonished American officer.

'That's so, Colonel,' came the reply, 'but you see, we wear the Queen's scarlet.'

The stories, though, drew on a reality of courage and a tradition of relentless pursuit. The Mounties policed the westward advance of the transcontinental railroad and, when that frontier was tamed, moved to a new one in the Yukon gold rush of the 1890s. In the Arctic they learnt from Eskimos how to travel, hunt and survive. If, on the frontier, many officers were all too human and drank and patronized brothels, they emerged in popular literature as sober incorruptibles and as solid as the Rockies.

The writer Wallace Stegner was five years old when he travelled to Saskatchewan in a stagecoach to meet his father and recalled the impact of seeing his first Mountie. 'He wore not only the scarlet tunic, but yellow-striped blue breeches, glistening boots and a wide campaign hat. Holstered at his belt he had a revolver with a white lanyard, and he was altogether so gorgeous that I don't even remember meeting my father.' He concluded that it was 'to the honour of an almost over-publicized force that having dramatized in scarlet the righteousness of the law it represented, it lived up to the dramatization'.

32

The light of the world

'. . . O thievish Night,
Why shouldst thou, but for some felonious end,
In thy dark lantern thus close up the stars . . .'
John Milton, *Comus*

From 1807 London began to bring new light to its gloomy streets. The flickering and smoky lanterns whose wicks were dipped in oil were gradually replaced by coal-gas lamps. The new illumination in Pall Mall was a small sensation and the cartoonist Rowlandson showed Londoners gawping at the lights and a prostitute wondering where she would now find the shadows appropriate to her profession. Gas lamps were installed on Westminster Bridge five years later and by the 1840s much of the city had the theatrical Victorian gaslight look that seemed to many so romantic and magical and to others, those who were never satisfied, merely lurid. Gaslight, however, could not easily be installed in private houses. Until the 1860s the world after dark was illuminated chiefly by candles and by lamps fuelled by whale oil.

Whales were huge living tankers roaming the oceans, more than sixty feet in length and as many tons in weight. The hard and dangerous business of hunting them was the work of particularly tough and skilled sailors. Very little of a whale was wasted. Whalemen clambered over a corpse and wielded sharp spades to hack off the massive blanket of fatty blubber, up to twenty inches thick, which was fed into pots for boiling down. A single whale yielded as much as twenty-five tons of oil. Whaling ships commonly returned home with more than a thousand barrels, each containing thirty gallons, the raw material for an industry which employed more than 50,000 people.

Apart from the oil whales were the source of other valuable materials. The bowhead whale, also called the right whale, because it was the right whale to catch, had in its mouth a screen of horny fibres which trapped and sieved the creature's meals of plankton and krill. This baleen, known as whalebone, was often more than eight feet in length and in exceptional cases measured fifteen feet. In the years before plastics it was the light and flexible material used in the manufacture of corsets, parasol ribs, furniture springs, venetian blinds, collar stiffeners, hat brims, carriage whips and window gratings.

The sperm whale yielded blubber, a set of eight-inch ivory teeth and 500 gallons of spermaceti. This colourless fluid, given the name spermaceti because early whalers mistook it for sperm, served as a buoyancy aid and was stored in a reservoir inside the huge square head. It could be made into an ultra-light lubricating oil or into the finest wax for candles. From the intestine and rectum of the sperm whale, too, came the waxy grey ambergris, the by-product of an intestinal disease, which matured into a sweet-smelling substance used in the manufacture of perfume and sometimes as a flavour in cooking. Sold by the ounce it was at times worth its weight in gold.

Hunters pursued the immense North Atlantic population of whales to the edge of extinction. As the animals grew scarce in the early years of the nineteenth century, the whalers turned to the almost unexploited Pacific. The first British whaler rounded Cape Horn to hunt in the Pacific in 1786 and returned to London in 1790 loaded with oil and spermaceti, having lost only one man killed by a whale. Thereafter a number of British whalers ran a profitable two-way trade, sailing out to Australia with shipments of convicts, hunting whales for several months and returning to Britain with their holds crammed with barrels of oil. Whales and seals in Australasian waters were slaughtered in great numbers by piratical whaling men, Australian, American and British, sailing out of Hobart and the Bay of Islands in New Zealand, notorious as lawless and drunken places. In New Zealand whalers frequently killed the Maoris; and on several occasions the Maoris seized their ships and ate the men.

It was the New Englanders, however, men from Nantucket, New Bedford, Martha's Vineyard and Boston, who came to dominate whaling in the southern ocean. These Yankee whalers were the major force there for more than forty years after 1820. In their richest years, in the 1840s, they had more than 500 ships at sea, four-fifths of the world's whaling fleet, and their sails speckled the ocean. Pacific whaling was long-haul work. Arctic men, sailing out of Aberdeen, Peterhead, Dundee, Hull and Whitby for Greenland and beyond, were away for months. Pacific crews signed on for voyages of three years or more, rounded Cape Horn and sailed many thousands of miles, ranging to Japan and into the Indian Ocean as far as Kerguelen and Zanzibar.

The whalers were as courageous as they were hard and greedy. Much of the work was tedious and filthy but the thrill of killing whales seemed to make it all worthwhile. As the creatures surfaced and vented their lungs through their blowholes the hunters rowed towards them. There were six or eight men in a twenty-eight-foot pinnace and they were careful to be quiet and to remain out of the restricted vision of their prey: they knew that they were in the death-zone. It took a cool nerve to be at close quarters with a powerful leviathan much larger than the boat.

In the bow the harpooner stood and from a range of about five yards hurled an iron harpoon into the whale's flesh. Then, if he had the chance, he threw another. As often as not the creature accelerated away and pulled the boat at high speed as if it were a sleigh, the bow waves soaring in a plume. Frequently the whale dived steeply and stayed below the surface for half an hour or more as the hunters frantically paid out hundreds of fathoms of line to avoid being dragged under. Sometimes they had to cut it. Sometimes the whizzing line, smoking with the friction, caught luckless men by an arm or leg and jerked them into the sea where they drowned in a moment. Whales often struck back at their tormentors. The right whales smashed men and boats with thrashing blows of their flukes. Sperm whales attacked with open jaws or used their heads as battering rams. A speared whale towed a boat for many hours until it tired; and then the harpooner had his chance to thrust a final lance, aiming to penetrate the arteries and windpipe and flood the lungs with blood.

Moby Dick, the malevolent albino sperm whale of Herman Melville's masterpiece of 1851, was based on Mocha Dick, an aggressive and battle-scarred Pacific monster which killed thirty men. Another sperm whale rammed and sank the Yankee whaler *Essex* in the Pacific in 1820. Her crew launched three boats and headed for Chile. After several weeks at sea they were starving. In one boat the men ate the flesh and heart of a dead shipmate. In another they drew lots to decide who among them should give his life for the others. The captain's nephew, a skinny boy of eighteen, lost the draw and meekly placed his head on the gunwale to receive the fatal blow; and his uncle and the others ate him. Of the twenty men in the *Essex* eight survived.

Arctic whalers, hunting in cold weather off Greenland, relatively close to their home ports, sliced up the blubber of their catches, stored it in casks and sailed it home to be rendered into oil. Pacific whalers operated so far from base that they boiled down the blubber aboard their ships in iron pots mounted on brick furnaces, the try-works as they were called. Their success in hunting was advertised by the glow of the fires, by a baleful column of foul black smoke rising into the sky and the smell of blood carried on the breeze. Ishmael, Melville's narrator in *Moby Dick*, described the 'unspeakable, wild, Hindoo odor' of the funereal smoke. All that was left of the boiled-down blubber, once the oil had been extracted, was the scrap of fatty flesh known as the fritter which was used to feed the furnaces. 'The whale supplies his own fuel', noted Ishmael, 'and burns by his own body.'

Whaling was the heart of Pacific commerce for forty years. As they criss-crossed the ocean the whale-ship captains gathered considerable knowledge of hitherto unknown islands, reefs and atolls and amplified the Pacific charts. They led the way in colonizing the islands where they stopped for food, water and repairs, and they recruited islanders to work for them as labourers, dock hands and servants, drawing them into Western values and the culture of money and prostitution. A new mixed-blood people began to emerge. Many Polynesians – Hawaiians, Tahitians and Maoris, in particular – made their lives as whaling crew and were celebrated by Melville in his character of the tattooed harpooner

Queequeg. In general, whalers steered clear of Fiji because of its reefs and cannibal cookery.

As the whalers spread over the Pacific so did the missionaries. The hunt for whales and souls changed everything. Inevitably, sailors and evangelists brought with them the smallpox, tuberculosis, influenza and measles that laid waste the local populations. There were 35,000 Tahitians in the 1770s, 15,000 at the beginning of the nineteenth century and around 9,000 by the 1840s, while 50,000 Samoans in the early nineteenth century were reduced to half that number by the century's end.

Europeans found it difficult to arrive at a sensible halfway house in their view of Pacific peoples. Captain Cook's astronomer, William Wales, seemed to be exceptional in his prudent judgement of Tahitians, that their character had been deprecated just as their beauty had been magnified. The noble savage of the eighteenth-century Enlightenment, of Joseph Banks's time, was seen by missionaries of the following century as the ignoble, ignorant, idolatrous, treacherous and naked savage, ripe for spiritual rescue. The South Seas evangelists disdained traditions of worship, ceremonial, music, dancing, song and the carvings of gods. They banned them all, put the women into voluminous and uncomfortable clothes and taught both women and men the benefits of toil.

The London Missionary Society acted in the spirit of the Biblical command, 'Go ye therefore and teach all nations', and sent thirty-nine evangelists and artisans to Tahiti in 1797. It took them twenty years but eventually they wore down the islanders and converted them, not through the power of prayer but through the rum and guns they gave the king to persuade him to bring his people to the church. In many other parts of the Pacific, too, missionaries learnt that the most efficient way to convert whole populations was to bribe the chiefs. A visitor to Tahiti in 1820 found that music and dancing had faded away and the islanders wore European clothes. Missionaries noted approvingly that on Sundays 'not a tree is climbed, not a canoe is seen on the water'.

Herman Melville, Robert Louis Stevenson and Mark Twain were

appalled by the missionaries' subtraction of innocent pleasure from islanders' lives. But missionaries were convinced of the rightness of their crusade to bring primitive people to civilization. Samoans and Hawaiians were fairly easily drawn into Christianity by 1830. Protestant missionaries abandoned their efforts to convert the Marquesans and in the 1840s left the French to corral them into Catholicism. In some parts of the Pacific missionaries scrambled like claim-staking gold-diggers, competing bitterly with each other to accumulate congregations. Such a situation persists to this day in Papua New Guinea where rival churches engage in 'sheep-stealing', luring each other's worshippers away.

Missionaries became figures of authority. Many were traders and believed that the gospel and commerce led to civilization. Missionary-explorers were applauded for being 'emphatically imperial' in their work. It was not until the 1850s, after more than twenty years of trying, that preachers began Christianizing the Fijians who were busily engaged in factional fighting. The islanders ate the Wesleyan missionary Thomas Baker in 1869, but only five years later his colleagues in faith converted the new king who ceded Fiji to the British empire. In common with other islanders, Fijians surrendered much of their tradition; but in return the missionaries learnt their language and were the first to write and print their vocabularies and grammar.

Out of the worldwide advance of the Bible in the nineteenth century sprang a prolific missionary art and literature. Pictures, illustrations and the covers of books and journals portrayed missionaries as heroes battling heathenism; and the more fierce the native the more noble seemed the white man. Missionary propaganda depicted Pacific islanders as degraded and violent and their idols as especially monstrous. After the missionary John Williams was murdered in Erromango in 1839 the printer of a picture of his death was instructed to make the natives darker and fiercer and Williams more saintly. Stories of missionary martyrs, spiced with cannibalism, helped to raise money for the cause. The historian Jane Aaron has observed that missionaries were glamorous and haloed figures in Wales. 'Gospel-bearers sent out from Welsh chapels to every corner of the Empire' were 'knights in shining armour and each black convert a pearl that would shine forever in the heavenly crowns' of those chapel-goers who funded the missions. Congregations thought that their missionaries were better than 'English imperial officers'; but the reality was that they worked in areas 'made accessible by British guns and the prestige they enjoyed was largely attributable to their identity as Christian soldiers in the army of the Great White Queen'.

In the 1850s geologists and chemists in Europe and America looked for something better than whale oil. The numbers of whales were dwindling, and their oil was expensive and could no longer supply the demands of the rapidly growing cities. Chemists in Britain, Europe and America refined illuminants from coal, asphalt and turpentine, the coal-oils. A Glasgow chemist, James Young, made paraffin from coal and also extracted oil from Scottish shales. Abraham Gesner, a Canadian geologist, obtained a colourless oil from asphalt in 1854 and called it kerosene, combining the Greek words for wax and oil. By 1859 he had a plant in New York producing 5,000 gallons a day. More than thirty other companies in America were soon manufacturing similar products.

Travelling through Pennsylvania, George Bissell, an impecunious American journalist and teacher, saw villagers in Oil Creek collecting a sticky black fluid which seeped and bubbled from the ground. They spread cloths to sponge it up and sold it to pharmacists who prescribed it for headaches, worms, rheumatism and the sores on horses' backs. It dealt efficiently with constipation. Bissell reckoned that this 'rock oil', as it was called, could be refined and sold as lamp fuel and machine lubricant. He gathered a group of like-minded businessmen to raise money for research and they paid Benjamin Silliman, a chemist, to evaluate the oil. In 1855 Silliman declared it a valuable lighting fuel and backed his opinion by taking shares in the new Pennsylvania Rock Oil Company. All the company had to do now was to find an oil well.

It recruited Edwin Drake, an unemployed railway conductor, to prospect near a timber camp called Titusville in the hills of north-west Pennsylvania. He used a derrick and drill of the kind employed in finding salt. After more than a year he had found nothing, and his masters wrote to him and told him to stop. The letter was on its way when, in August 1859, Drake drilled to just over sixty-nine feet and struck oil. As it was being decanted into tubs and barrels the news triggered the first oil rush. Thousands of adventurers, men and women, hurried to Titusville where so much oil was gushing that there were not enough barrels to hold it. Barrels in fact were more costly than the oil itself. Bissell, the founding figure, was one of the first to arrive. The oil made him wealthy.

In 1860 seventy-five wells were working in America's first oilfield and the product was refined and marketed as kerosene. A refinery at that time was a still in which crude oil was boiled and its products were condensed out. As Daniel Yergin writes in his history of oil, 'the skills required were not all that different from making moonshine, which is why whiskey makers went into oil refining'. The invention in Vienna of an inexpensive lamp with a glass chimney enabled it to burn cleanly and produce 'the light of the age . . . something nearer the clear, strong light of day, a dainty light fit for Kings and Royalists'. Like other technological advances of the century the kerosene lamp seemed to be yet another means of taming time. People could cheat darkness and read and sew as long as they wished. Evenings were extended. 'Kerosene', said a New York chemist, 'has increased the length of life among the agricultural population.'

So much oil was available that the price fell in a few months from $10 a barrel to ten cents. The capacity of an oil barrel was standardized at forty-two gallons, the size of a herring cask established by royal command in England in 1482; and it remains so today. The teamsters who drove oil wagons down the muddy tracks out of Titusville demanded big fees. The bosses defeated them by building the first pipeline, a wooden tube, to take the oil away; but not before the teamsters tried sabotage and gunfire in an attempt get their way.

A cargo of Pennsylvania kerosene in barrels was stowed aboard a London-bound sailing brig in 1861. Terrified of explosion, the crew fled. The captain toured the waterfront, collared drunken sailors and bundled them aboard his ship. He sailed before they could sober up and made the first Atlantic oil shipment. Within five years American oil exports to Britain were worth around

$15 million. People danced to the 'Oil Fever Gallop' and sang a silly song called 'Oil on the Brain'. The oil age had begun.

In 1865, John Davison Rockefeller, a pious and ruthless twenty-six-year-old Baptist businessman, became an oil refiner in Cleveland, Ohio. At thirty he founded the Standard Oil Company and had twenty-five refineries. By the end of the 1870s he owned nine-tenths of American refining and all the oil wagons of the Pennsylvania Railroad. He shaped the modern oil industry and was one of the wealthiest men on earth. He was certainly frugal and a favourite meal was bread and milk. But he loved horses and delighted children by balancing a biscuit on his nose. At the age of sixty, and feeling the twinges of arthritis, he turned philanthropist and gave more than $500 million to medical research. He died at ninety-eight.

In 1892 the mighty Rockefeller was stunned by a coup brought off by the trader Marcus Samuel, from the East End of London. Standard Oil shipped kerosene, packed in five-gallon tins, to the Far East. Because the Suez Canal authorities banned its transport, for fear of explosion, Standard sent it in sailing ships by way of the Cape of Good Hope. Samuel, however, saw a way of undercutting Standard. He would buy Russian kerosene from Baku on the Caspian Sea and ship it through Suez, a much faster and cheaper route than the Cape. To do the job he needed a fleet of special ships, not freighters filled with cans of oil, but specially designed tankers safe enough to go through Suez. He commissioned the first of the fleet, *Murex*, with safety features to minimize the risk of explosion and tanks that could be steam-cleaned to permit the transport of rice or sugar on the return voyage.

In spite of Standard's opposition and mutterings in the press about 'Hebrew influence', *Murex* passed the canal authority's tests. It sailed from Hartlepool to Batum on the Black Sea, loaded with 4,000 tons of kerosene brought by rail from Baku and shipped it through Suez to Singapore and Bangkok. It was a triumph. Samuel had already built a network of storage tanks in the East and now he was ready to exploit a huge new market.

He had thought big, but he had failed to think small. He had not considered how his customers would carry the kerosene home. Standard supplied blue oil cans for this purpose and these were ingeniously recycled by people as roofing sheets, saucepans and bowls. Samuel quickly ordered tons of Welsh tinplate to be sent to the East and made into bright red tins. By the end of 1893 he and his brother Sam had a fleet of eleven tankers, all named after seashells in honour of the popular shell-decorated boxes their merchant father sold. In 1897 they founded the Shell Company. By 1902 nine-tenths of the oil passing through Suez went in Shell ships.

In 1851 the Admiralty decided that coal from the four-foot seam in south Wales, which produced great heat and little smoke, was ideal for the new steam-driven warships. The decision dismayed the colliery owners of Newcastle upon Tyne; but Newcastle coal was smoky and less efficient.

The Royal Navy's imprimatur opened a huge world market for the steam coal that lay deep beneath the pretty, wooded valleys of southern Wales. A new tribe of tycoons moved in. In 1855 the Marquess of Bute began to mine the four-foot seam 350 feet beneath the Rhondda valleys. David Davies, who had started out as a country sawyer and made his name as a railway builder, drilled in the Rhondda in 1865 and discovered the richest coal seam in the world at 660 feet. It was such excellent naval fuel that Davies gave his company the name of Ocean. The valleys of Glamorgan and Monmouthshire, the two Rhondda valleys in particular, became the most intensively mined district of the world and a dynamic mining society.

Coal went by rail and canal to the port of Cardiff which became the coal capital of the world. In the 1880s, the historian John Davies observed, at least a quarter of the international trade in sources of heat and energy originated in south Wales. 'The Severn Sea', he said, 'occupied a position in the world economy comparable with that of the Persian Gulf in the 1980s.'

At the turn of the century, however, Admiral Sir John Fisher sounded the first faint knell for naval coal. Fisher had been in the navy since 1854, almost from the time that Welsh steam coal was chosen as the prime fuel for warships, and as a midshipman he had served under one of Nelson's captains. In his sixties, as determined and energetic a modernizer as ever, he campaigned for ships that punched hard, that were as heavily armed and as fast as possible.

He wanted fifteen-inch guns and more than twenty-five knots. 'The first of all necessities is speed so as to be able to fight when you like, where you like and how you like.'

To this end he championed oil. It burned more efficiently than coal, gave 50 per cent more heat per unit of cost and weight, and drove the shafts faster. It gave ships better acceleration and an extra four knots of speed, and, mercifully, it saved the time, dirt and exhausting labour of coaling. It was pumped, not shovelled. 'Oil', said Fisher in 1901, 'will revolutionize naval strategy.'

More conservative admirals were not convinced. After half a century they felt secure with the reliability and availability of the coal that was so abundant on their doorstep. Britain, after all, had no oilfields. As First Sea Lord from 1904 to 1910 Fisher persisted with his campaign and eventually persuaded Winston Churchill, First Lord of the Admiralty, the navy's political boss. In 1912 Churchill ordered the building of five oil-fired battleships and two years later, to secure oil supplies, Britain took a controlling interest in the Anglo-Persian Oil Company.

33

A gulag in Eden

'A big banyan tree has come up. Under its shade one used to take cool breath, after being flogged and tortured.'
Prisoner's reminiscence in *The Cellular Jail Handbook*

The Andaman Islands lie in a jigsaw scatter on the blue-green Bay of Bengal. With their southern cousins, the Nicobars, they stretch for 500 miles. Burma is to the north and Sumatra to the south. They are hot and slow and seem to belong to long ago. Palms curve over white beaches. Along the wharves of Port Blair yellow-funnelled freighters load oil drums and sacks of rice, the air scented with spice and diesel, a Somerset Maugham backdrop. The islands are a Union Territory of India and, although far to the east of the mainland, keep Indian time so that velvet darkness falls at six.

A few hundred of the four tribes of aboriginal inhabitants survive as forest foragers, whittled away by disease, plantations, greedy tree-felling and road-building. They are small negroid peoples and may be related to pygmies, their presence in the islands perhaps a clue to the way the human race spread from its early roots in Africa. About fifty live on North Sentinel Island, the western outrider of the Andamans and one of the world's last mysterious places, protected by rough seas, coral reefs and the inhabitants' implacable hostility to strangers. The tribes were long ago named after Hanuman, the Hindu monkey god, hence the name Andaman; and for years they were demonized as cannibals. In the Sherlock Holmes story *The Sign of Four*, the diabolical killer with a blowpipe is a tiny Andamanese.

History is soon confronted. The exquisite Andamans were for

eighty years a British gulag. On Viper Island a path winds through a coconut grove and up the hill to a substantial red-brick building crowned with a dome. It was an elaborate death house, visible for miles, like a lighthouse, and built to terrify. Here in 1872 a Pathan was hanged for murdering Lord Mayo, the Viceroy of India, on an Andaman beach.

The first prisoners were 200 men who arrived in chains in 1858, transported for life for their part in the Indian Mutiny. For thirty years the islands themselves were jail enough, snake-infested and malarial, remote and impossible to flee. Prisoners lived in barracks and the more recalcitrant of them were shackled in chain gangs, up to 200 men linked together, working and sleeping, on a single chain. The authorities allowed some prisoners to settle and farm the land after they had served part of their sentence. A number were permitted to choose a wife from the ranks of female prisoners. The jailers arranged a parade and each man selected a woman he fancied. The women, it seemed, had no choice in the matter.

Although there were cells on Viper Island the British authorities felt that a proper large jail was required. A government commission reported in 1890 that conditions in the Andamans were not punishing enough and recommended that all prisoners should be kept in solitary confinement for the first six months of their sentence. In 1893 convicts began to build a formidable red-brick jail at Port Blair. Completed in 1906, it housed more than 600 prisoners, one to a cell, many of them transported for long terms or for life. It was constructed according to the most modern ideas of prison design. The cells were in seven three-storey blocks radiating like spokes from a central security tower. It remains by far the largest building in the islands, the architectural showpiece. An iron flogging easel, manacles and a three-man gallows are still to be seen.

The first chief jailer, David Barry, emerges from the accounts of Indian witnesses as a sadist who made the prison a version of Devil's Island. 'There is a God in heaven and a God on earth,' he told newly arrived prisoners, 'and in Port Blair that God is me.' Many of his warders were ex-convicts who had served their sentences and enjoyed their new power to terrorize the prisoners.

The jail was a weapon against the infant Indian nationalist

movement. Outside its walls is a semicircle of statues of some of the political prisoners who died there. Each figure is manacled. The Andamans were the last stop for many of the political troublemakers who opposed British rule, not least those found guilty of sedition. Terrorist bomb outrages increased the sense among the British that the Raj was threatened by mobs stirred by clever words. Under a policy of repressing literature, writers and publishers were arrested, tried and sentenced, often harshly. Their writings in newspapers and pamphlets were regarded as poison likely to spread disaffection among ordinary people, and police and administrators combed obsessively through books, journals, poems and songs in search of inflammatory material. They raided bookshops and seized works like Aristotle's *Politics*, still considered dangerous after 2,300 years, and sent spies to discover what schoolchildren were reading.

The liberal idea of freedom of expression, promoted as a glory of the British way of life and imbibed by educated Indians, sat uncomfortably with the repressions of imperial rule. Later, in his struggle for independence, Mahatma Gandhi would reach for this contradiction and confront the British with their own ideals of justice, fair play and democracy to make the case that they should quit a land that was not theirs. The last political prisoners left the Andaman jail in 1938; and the building is now a national memorial to Indians who opposed British rule. It is a brooding and troubling place.

On Ross Island, the eerie red-brick ruins of the British officers' mess, the governor's residence, the houses and churches built by convict labour, have all been seized by the jungle. Throttling roots and branches and the tentacles of banyan trees wind around the walls, while roots push through the tennis courts where subalterns served. Deer wander the fallen headstones in the rubbled cemetery. When the Japanese invaded the Andamans in 1942, someone put in a bad word for Mr A.G. Bird, a British administrator, and he was beheaded with a sword on Ross Island beach.

The banishment of Indian prisoners to the Andaman Islands began as transportation of British and Irish convicts to the Australian colonies was being abandoned. No criminals were sent to New South Wales after 1840. Transportation to Van Diemen's Land, later renamed Tasmania, ended in 1853. The last convict shipment from Britain landed in Western Australia in 1868. In eighty years, more than 161,000 men and almost 25,000 women were transported to Australia. The average age was twenty-six. A third were Irish.

As a means of dealing with the criminal rabble, exile to the ends of the earth had been popular with the British authorities. There was no system of prisoner reform, and the overcrowded jails and prison hulks needed regular clearing. As slaves and labourers, convicts were a useful workforce in a distant colony. In the 1830s, however, public opinion began to turn against the system. The number of criminals shipped out declined after the peak of 1833. New laws called for the building of prisons within a civilized penal system, while Australians protested against the use of their territory as a human rubbish dump.

A parliamentary inquiry of 1837, headed by Sir William Molesworth, twenty-seven and burning with the flame of reform, heard disturbing accounts of the cruel treatment of men condemned to chain gangs and of vicious daily floggings. In Van Diemen's Land the overseers used an especially heavy whip. Few prisoners who fled their tormentors survived. In five years in the 1820s 101 of 116 Tasmanian escapees were hanged, shot by soldiers or died in the forests. A few days after breaking free in 1822 eight men were reduced to seven when they killed one of their number and ate his heart and liver before cutting up the rest of the body. A second man was murdered and boiled, then a third, until at last only one was left. He gave himself up and told the story but his confession was not relied upon. He escaped again with a companion and was found with his friend's body on which he had several times dined. He was hanged in Hobart.

In 1837 Captain Alexander Maconochie, who had been the first secretary of the Royal Geographical Society, arrived in Van Diemen's Land as private secretary to Sir John Franklin, the new Governor. Maconochie looked at the convict colony through the filter of unusual experience. As a naval officer during the Napoleonic Wars he had been a prisoner of the French for two awful years. Before sailing for the colony he had agreed to examine the convicts' conditions for the Society for the Improvement of Prison Discipline. His report sharply criticized the 'extreme severity' of officials who ran the convict system, castigating them as the cause of vice, crime and degradation, as slavers profiting from the labour of helpless men. His critique was a sensation in the London newspapers and delighted the foes of transportation. It added to the impact of the Molesworth inquiry which reported to Parliament in 1838 that transportation was inefficient as a deterrent to crime and remarkably efficient in corrupting prisoners.

Maconochie's observations set him on the path to becoming what the historian Robert Hughes called 'the only inspired penal reformer in Australia throughout the history of transportation'. Maconochie believed in punishment, but also in reform. A criminal had a past but needed a future. Prison could avenge and restrain, he thought, but should offer more than despair. A prisoner needed to be given responsibility for himself. Maconochie devised a system in which a man could play a creative part in the management and length of his own sentence. Confinement began with a short spell of hard labour. Thereafter, by being well-behaved and doing approved useful work, a man earned marks which were a form of currency. They bought food, tobacco and clothing. Maconochie believed that reliance on prison rations reduced men to dependence on the institution and made them the prey of prison guards. Bad behaviour cost a man marks. By accumulating enough marks a prisoner could buy his freedom: 6,000 marks were the equivalent of a seven-year sentence.

Maconochie was many years ahead of his time. In 1840 he was sent to try out his system in the horrible penal colony of Norfolk Island, five miles by three, a thousand miles east of Sydney. This was the final resort of the hard cases. It had been designated by Thomas Brisbane, a Governor of New South Wales, as a place without hope, and had been run by a series of sadistic governors. One of them, James Morisset, ensured that floggings were spaced so that lacerations from the first fifty lashes were only partly healed before the next fifty tore them open, and so on. The Molesworth inquiry heard that official brutality in Norfolk Island was so harsh that men welcomed a death sentence. They committed murders so that they would be hanged.

Into this hell, with its population of 1,200 prisoners, went Maconochie the reformer. One of his first acts was to invite everyone to a celebration of Queen Victoria's birthday. He allowed all the astonished convicts to roam free, swim and play games, talked to them as they ate a lunch of fresh barbecued pork and joined them in a toast to the Queen, drunk in rum he paid for himself. In the afternoon they watched a comic opera. All that Maconochie asked was that at the sound of a bugle they should return quietly to their quarters. It was a risk, but it worked. Maconochie made every man feel that he belonged to the wider world.

The Sydney authorities criticized him, but he pressed on. He tore down the gallows intended to cow the men, built two churches, allowed space for a synagogue, gave prisoners allotments to grow vegetables and ensured that convicts' graves had headstones. Three years into his reign the Governor of New South Wales, a sceptical observer, paid a surprise visit, no doubt expecting to find pampered prisoners. Instead he found 'respectful and quiet' men who had not been brutalized. He was impressed. Maconochie was, more or less, vindicated. But London had already decided that his reform ideas were suspect and sent a letter recalling him. On the ship carrying the official dispatch came Major Joseph Childs, his replacement, a martinet and flogger of the old school who restored the rule of terror.

Maconochie's experiment, however, had worked. In his four years he had discharged 920 prisoners and only twenty of them, 2 per cent, reoffended. On his return home he wrote a book about prison reform that in his lifetime was mostly ignored but became influential after his death. His old boss, the affable Sir John Franklin, was sacked as Governor of Van Diemen's Land in 1843, not least because hardliners thought him soft on convicts. Ordered home, he looked for an achievement to expunge the disgrace. In the Arctic, he believed, in the search for the North West Passage, he would find redemption.

34

Faster than witches

'Steaming through metal landscape on her lines
She plunges new eras of wild happiness
Where speed throws up strange shapes, broad curves
And parallels clean like the steel of guns.'
Stephen Spender, 'The Express'

On the crowded Oval Maidan in Bombay cricketers play matches that interlock like Olympic rings, wickets pitched yards apart, the balls whizzing this way and that, so that an inattentive fielder might catch out a batsman in a neighbouring game. Two of Sir Gilbert Scott's Gothic majesties loom over the scene, the university with its tall umpire of a clock tower and the High Court with its mischievous carvings of wolves in lawyers' gowns and a monkey robed as a judge. Many more such buildings rise regally in the city's heart, Gothic and Indo-Saracenic with French and Venetian allusions, pinnacled, gabled, domed, turreted, fretted, medallioned and gargoyled.

Among the colossi is the Victoria Terminus. Frederick Stevens gave free rein to exuberance when he designed it. Its sandstone frontage stretches 1,500 feet and the grand entrance is sentried by a stone lion and tiger symbolizing Britain and India. The interior is a palace of marble columns, polished Indian blue stone, tessellated floors, a soaring vaulted ceiling and wrought iron galleries. Light falls through stained glass as if through the windows of a cathedral. High on the dome presides a fourteen-foot figure of Progress and on the main gables stand figures representing Engineering, Agriculture and Commerce. No railway station in the world matches it in theatricality. It is much more than the largest public building in Bombay and more than a measure of the city's importance. Opened in 1887, the year of Queen Victoria's golden jubilee, it asserts the confidence of an imperial power in its prime. It celebrates the railways of India built by British engineers and millions of Indian craftsmen and labourers, a monument to the steam-powered empire.

No other invention changed the world so suddenly and profoundly. There was the world before the steam locomotive and then there was the transformed world dominated by the railways. Within twenty-two years of the opening in 1830 of Britain's first commercial railway all of the country's main lines had been constructed. Hundreds of thousands of people were accustomed to travelling at speed and regarded a fast journey almost as a right. They revelled in velocity.

Throughout the second half of the nineteenth century railways were the precursor and proof of civilization in the newly won and half-explored territories and colonies, the framework for trade, administration, control and policing.

They were also the means by which many countries, among them Germany, Italy and Argentina, stitched together their various regions, making one unified garment from disparate provinces. In the United States and Canada, too, the railroads stretching from coast to coast proclaimed nationhood and rugged frontiering qualities. The railway locomotive was the avatar of progress. A railway country was a growing country. The nineteenth-century railways were the largest engineering projects of history, the greatest since the Romans' roads, bridges and aqueducts imposed order and sense upon their empire.

Apart from being a phenomenal and daring struggle, the construction of the Indian railways was a grinding together of cultures. In order to build efficiently, the small teams of British engineers had to learn how to manage labouring armies, to discover, like George Stephenson, that engineering great quantities of iron and stone and timber was one thing, but to engineer a mass of men was another.

Some of the British thought it would be too difficult and costly to build railways across the fearsome mountains and roaring rivers of India. They also wondered, in their imperious way, whether Indians would be 'attracted from the bullock cart to the rail', and be 'persuaded to pay a fare rather than meander without any sense of time'.

Lord Dalhousie, the reform-driven and imperious Governor-General of India, felt no such doubt. At the Board of Trade in the 1840s he had worked throughout the British railway boom. Now he was committed to the modernization of India with a unifying network of roads, canals, telegraphs and railways. He also annexed eight Indian states on the grounds that British rule was better than Indian. In all he did he showed scant regard for local feeling. He was also ill, urged on by a sense of time running out. 'The creative and comprehensive nature of his work', wrote the historian Percival Spear, 'surpassed that of all his nineteenth-century peers. He was the apostle of westernized India with all an apostle's zeal.'

Dalhousie approved the first railway east of Suez in 1853, the twenty-one-mile line between Bombay and Thana. By the time he died in 1860, worn out at forty-eight, he knew that hundreds of miles of track had been built and that the railways he had fathered were pushing out from Bombay to Calcutta, Delhi and Madras. More lines were snaking from Karachi to Lahore, from Bombay to Ahmedabad, from Calcutta into Bengal. The onrush of technique, technology and innovation contributed to the anxieties about British rule that were part of the background to the Indian Mutiny. Dalhousie's contempt and disregard for Indian feelings were a factor in the disquiet. As Percival Spear concluded: 'He lacked the intuitive knowledge of how his measures would appear to those they affected. For this he paid the penalty of the Mutiny disaster.'

One effect of the Mutiny was to spur the British to forge ahead with building railways over which they could send troops quickly in case of trouble, and stations which could double as fortresses. Lahore railway station was built in 1864 in the form of a medieval castle, with musket loopholes and iron gates.

Many of the Indian railways were completed after titanic battles. Engineers and contractors assembled masses of timber, rails, girders, bricks, stone and lime, and struggled to ensure that they arrived at construction sites on time. The timber for sleepers and bridge trestles had to be steeped in creosote to proof it against rot and termites. Each sleeper on the usual five foot six inch gauge was ten feet long and a foot wide. Engineers ransacked the forests; but even so, much timber was imported from Britain and Australia. A mile of track required 1,700 sleepers as well as 115 tons of rails. Rails and girders, prefabricated sections of bridges, fish-plates, spikes, nuts and bolts and locomotives were shipped from Britain and generated tens of thousands of jobs at British ironworks. Ian J. Kerr notes in his study of the building of the railways across India that by the end of 1863 more than 2.7 million tons of material had been sent from Britain to India in 3,571 ships. Each mile of railway built in the 1860s required a shipload of 600 tons; and occasionally the cargoes were lost in shipwrecks.

As the lines progressed Indian brick-makers set up kilns and baked tens of millions of bricks for bridges, workshops and station buildings. The sleepers and rails were laid on millions of tons of stone and pebble ballast, especially costly because of the lack of stone ballast in the plains. The foundations and piers of hundreds of bridges were deep and robust to withstand the formidable rushes of water during the monsoon rains. In the early years, before the engineers knew the power of their enemy and how to defeat it, many bridge towers were undermined by the scouring water. The construction of each major bridge created its own amazing spectacle. A temporary town mushroomed on the river bank and up to 20,000 swarming workers toiled day and night to build the foundations and piers and rivet together the iron and steel bridge sections. These had been prefabricated in Britain, dismantled and shipped out. The railways cost £18,000 a mile instead of the £8,000 envisaged by Dalhousie.

The British recruited labourers, load carriers, diggers, brick-makers, stonemasons, scaffolders and regiments of carters, amounting over the years to millions of men, women and children. Almost everywhere in India construction labouring and road-building is work for women as well as for men. Thousands of tons of earth and rubble are shifted in baskets carried on the heads of thin and spare men and women moving in seemingly endless files. Mounds of stones are hammered into smaller chunks by squatting women; and children still labour in brickworks. The railway armies of the Raj, tramping from site to site, formed an almost permanent body of migrant families who were joined by thousands from the surrounding countryside as they built the big bridges, embankments and cuttings. A tribe of quarrymen worked in their traditional way, building fires against rock to make it expand and burst. Then they pounded the loosened debris into rough cubes. Almost all of the manual labour was done by Indians. Exceptionally, sixty-five miners experienced in tunnelling and fresh from building the Severn tunnel, were recruited from Wales to drive the Khojack tunnel through dangerous rock in Baluchistan.

Heavy casualties were inevitable on any difficult section. Many men died in explosions in cuttings and tunnels and in falls from bridges. The seething railway camps, bereft of sanitation and clean drinking water, full of poor and ill-fed people, succumbed easily to cholera, smallpox, malaria and dysentery. Typically, disease killed a third of the labourers in a camp and then spread rapidly through the land as infected people fled.

The evident dangers in laying a track through the steep hills of the Western Ghats, the 2,500-foot ridges which bar the way from Bombay to Madras, prompted an official to warn that 'We must be prepared for numerous casualties. It would be idle to expect that we should overcome the physical difficulties we have in making railroads in such a country as India without heavy sacrifice of life.' The ascent of the Bhore Ghat was one of the bloodiest episodes. Around 40,000 people worked on it and built twenty-two bridges and twenty-five tunnels. More than 10,000 were killed by disease, falls and explosions. Solomon Tredwell, the British contractor commissioned to build the line, fell ill and died after only two weeks in India. His widow Alice took the job on and saw it through, completing it after eight years in 1863.

The British engineers who arrived in the 1850s knew little of India, its people and the challenges of landscape, geology and climate. They often faltered. Bridges collapsed, embankments were washed away and misunderstandings dogged their relationship with the people they recruited. Experience enabled them to build faster year on year. They learnt to manage men better and found that people worked more efficiently when they had decent housing and clean water and were not oppressed and thrashed by drunken British overseers. They found they could not ignore the subtleties of caste and faith in hiring and dealing with people. The best railwaymen adapted to local conditions.

By 1900, when the railways showed a small profit for the first time, the major part of the system, 24,000 miles of it, was complete. It seemed the most natural thing to travel by rail for days. Indians had absorbed the construction skills and had become the main subcontractors and track layers. As if they had always existed, the railways became part of Indian life and of every Briton's idea of India, complete with class divisions on trains and in stations. The newsagents stocked the stories written by Mr Kipling in Wheeler's little railway editions; and travellers could telegraph to a station up the line for a meal to be ready in a basket.

The railways grew into India's story. Pilgrims used them for their long devotional journeys. In spite of having a first-class ticket the young lawyer Mahatma Gandhi was thrown off a train by a white man in South Africa, an event which crystallized his thinking about civil rights; and in India he made the railways an instrument of his campaigning for independence, travelling everywhere, third class, to spread his message.

Leaving Moscow on an overnight train is an adventure and one of the best ways to meet Russians. In the hours spent jolting over uneven tracks through the forests of silver trees there is warmth, intimacy and the discovery of common ground, shared drinks and picnics, and tea from the samovar. The train encourages talk. A Russian train journey always had its particular romance. Tolstoy made the train a tragic presence in *Anna Karenina*, and Dostoyevsky set memorable scenes of *The Idiot* in railway stations. The first Russian railway was a short line built in 1837 from St Petersburg but railways really began after the Tsar, in 1857, urged the construction of lines to enable Russia to benefit from its 'abundant gifts' of nature. In the following year a Siberian railway linking Moscow to Vladivostok on the Pacific coast was proposed; but this greatest of Russian railways and the longest line in the world, stretching almost 6,000 miles, was not started until 1891 and opened in 1903.

Its strategic purpose was to consolidate the Russian empire's grip on Siberia and to open up far-off lands and mines that had been accessible only by foot, sledge, sleigh and horse-drawn mail carts.

A mass of labourers with wooden shovels, Russians, Chinese, Koreans and thousands of exiles and prisoners from the Siberian punishment camps, hacked out the line and built the bridges. In winter the workers froze and in summer they writhed under the vampirism of insects. In Manchuria tigers came to eat them and bandits to rob them. As the line was being built the Manchurian village of Harbin grew into a lively railway town which, when the trains started running, imported such European pleasures as striptease and jazz. In its pre-revolutionary years the trans-Siberian was certainly romantic, its green-and-gold carriages hauled by polished locomotives, its dining-car luxurious. In St Petersburg the director of communications proudly saw off the train to Moscow every day, raising his hat and standing to attention. Stations had fine buildings and excellent buffets set amidst gardens. At the end of the nineteenth century expresses left Paris and reached St Petersburg in forty-six hours. The city's Astoria Hotel displayed a large map of the London Underground and had a substantial English library. Private enterprise built most of the Russian railways. Savva Mamontoc, who drove the line to Archangel, was a renowned patron of the arts, Samuel Polyakov funded schools with his railway profits, and the widow of the magnate Karl von Meck dipped into her late husband's rail fortune to provide thirteen years of financial support to Peter Tchaikovsky. All that time they corresponded intimately but never met.

Australians built their first railways in the 1850s but excitement evaporated when they realized there would be no jackpot. The problem was that the tracks went nowhere. To make any money a railway had to connect two centres of population, and in the middle of the century Australia's chief cities, Sydney, Melbourne and Adelaide, were too distant, both from each other and from the sheep towns of their hinterlands, to be connected efficiently and profitably by rail. Ships were better. 'In Australia,' the historian

Geoffrey Blainey pointed out, 'a thirty-mile railway could cost as much as ten steamships.'

The first working railway, laid in 1854, was a mere tendril that ran two and a half miles from the heart of Melbourne to the city harbour. A writer in the *Melbourne Argus* called on Aborigines to heed its whistle. 'Here comes Christian England', he said, 'to absorb your hunting-grounds, destroy your game and inoculate you with her vices and show her Christian spirit by dooming you to extirpation ... who, while destroying you, cants in her churches about doing to others as she would be done by! Rejoice, you dark-skinned savage, at the advent of your kind and most Christian brother!'

The true birth of the railways was in the 1860s when lines reached out from Melbourne to the gold towns of Ballarat and Bendigo. In New South Wales the heroic struggle was the building of a track from Sydney to Bathurst along a zigzag route up the steep ridges of the Blue Mountains. The journey could be terrifying, as W.G. Grace discovered during the England cricket tour of 1873–4. 'We were painfully conscious', he recalled, 'that the snapping of a coupling would send us down the precipitous slope to certain death.' At the time of that tour there were 1,600 miles of Australian railways. When Grace returned for his second tour in 1890–1 the growing population had access to more than 10,000 miles. Battalions of railway navvies were a familiar sight. Grace was delighted. The new railways 'enabled us to dispense with the tedious coach drives through the bush', he wrote, and also with the rough steamship voyages 'which had been the bugbears of our tour in the seventies'.

Australian railways were built largely by state governments and financed chiefly by money borrowed in London. In the boom of the 1880s the rival cities of Sydney and Melbourne built lines of differing gauges towards each other, a distance of 600 miles. The incompatible lines, of course, could not mate. Rather, they coincided on opposite sides of a platform on the border. Neither could their promoters agree on what to wear at the joint banquet to celebrate completion. Sydney wore dinner jackets while Melbourne chose morning suits and spats. Passengers and goods had to be carried from one train to another, but for all the delay this entailed the railway soon proved twice as fast as the ship, although a ticket cost much more.

The 715-mile Sydney-to-Brisbane railway opened in 1888, but to many minds the significant achievement of the railway builders was to open up the interior. Farming regions prospered as the rails penetrated the plains that lay beyond the coast. Distant wheat lands in New South Wales, South Australia and Western Australia became profitable when harvests could be sent cheaply by rail down to the ports. A railway from Port Augusta, north of Adelaide, was built up through South Australia to Marree in 1884 and then on to Alice Springs. Railways also encouraged an Australian iron industry, wagon works, coal mines and timber mills.

The whistle of the locomotive, however, was not thrilling to everyone. To Ned Kelly it was the sound of intrusion. He grew up wild in an untamed stretch of Victoria and became a cattle-thief, bank robber, furious hater of Aborigines, outlaw-hero to those who shared his resentments, and eventually, in 1878, the murderer of three policemen. With the telegraph and the police, the new settler-farmers and the law courts, the railway was part of the apparatus of modernity encroaching on the unfettered life of bush people like himself. It also seemed to be an old oppression transferred across the world, an English threat. Like the horsemen of the later American West, Kelly construed the locomotive's smoke as a cloud over his own freedom.

In the end the railway played its part in the drama of his downfall. He lured a train, full of police, to the town of Glenrowan and he and his gang tore up a section of track to derail it. They adjourned to a pub to await the crash. In the small hours of the morning they heard the train's whistle. Kelly thought it was the policemen's knell; but it turned out to be his. A Glenrowan schoolmaster stopped the train before it reached the broken line and told the police of the plot. They surrounded the pub and opened fire. Kelly emerged in his famous suit of iron armour and fell with twenty-eight bullet holes in his legs and other unarmoured parts. The police carried him to the railway station and eventually to Melbourne. There, in November 1880, laconic to the end and murmuring 'Such is life,' he was hanged.

The imperial mission, as Queen Victoria distilled it, was to 'protect the poor natives and advance civilization'. By the 1880s there was a feeling in Britain that railways, bridges, missionary schools and decent motives more than vindicated their colonial progress. In East Africa the British saw the appalling crimes of Arab slave traders who continued, as they had for centuries, to lead long lines of captives from the interior and sail them in dhows to the markets of the Persian Gulf and the East African coast. The great junction of the trade was the island of Zanzibar and its market where buyers prodded and squeezed the naked slaves and selected boys for emasculation and the eunuch life. To the traders and their clients it was an ancient and respectable trade, to the British abhorrent.

In 1882 Britain occupied Egypt and, needing to control the Nile and the East African interior, made a protectorate of what is now Uganda and Kenya. It proposed to suppress the Zanzibar slavers, to develop the East African territory and keep the Nile from French and German interference. The strategic instrument to facilitate these ambitions was a 600-mile railway from the Arab seaport of Mombasa to Lake Victoria through deserts and mountains that were barely explored. It seemed a pointless project to some British politicians and it was labelled 'a lunatic line'.

George Whitehouse, a young English railway engineer with experience in India, South America and South Africa, was commissioned to build it and began work in 1896. He constructed a new harbour at Mombasa and assembled shiploads of sleepers, rails, bridge girders, locomotives, wagons and carriages. From India he recruited more than 2,000 labourers and craftsmen, the first of more than 36,000 who would come to work on the railway over the next six years. They found the going very hard. The hills were steep and the jungle as thick as wire wool. Tribesmen attacked the labourers. Every drop of water had to be hauled up from the coast. Half of the construction army contracted dysentery and malaria, and scores of men died in the first few months. Tsetse flies wiped out most of the transport herds of 1,850 bullocks and mules.

The first great battlefield was the Tsavo River, 133 miles from the coast, which had to be bridged. Quite apart from the engineering difficulties the area was notorious for its lions which attacked and ate the workers. As Colonel J. H. Patterson, the engineer in charge of building the bridge, described in *The Man-eaters of Tsavo*, two lions menaced his men for ten months. They killed twenty-eight Indian workers and perhaps more than 150 Africans. Many labourers preferred to take to the trees at sundown. Anyone who has passed the night in a tent in the African bush will never forget the intestinal bassooning rumble of the hunting lion and the explosive roar which, even if a mile or more away, seems three feet from your prickling neck. Patterson, keeping a tense and nervous vigil with his rifle, heard the roars and the labourers' screams many times; and in the morning found the crunched bones of the victims. He finally shot one of the lions and then had his showdown with the other. He sat in a tree and waited for it to emerge. As it prowled in the moonlight he shot it twice and it ran off. At daybreak he and his Indian gun bearer followed its trail. It charged from a bush and Patterson shot it twice more but failed to stop it. He only just made it up the tree where his fast-moving bearer had taken refuge with another gun. It took three more shots to finish off the lion; and Patterson was at last able to get on with building the bridge.

In May 1899, 327 miles from Mombasa and more than 5,000 feet up, the 18,000-strong railway army arrived at a stream called Nairobi, meaning cold water, and a frontier town grew up. Whitehouse built the first brick dwelling. Beyond Nairobi the line reached the precipice of the Kikuyu escarpment and the engineers gazed at the volcanoes and lakes of the Great Rift Valley 2,000 feet below.

Somehow the railway had to descend the almost perpendicular cliff to the valley floor. In the first major engineering job of the twentieth century the ingenious Whitehouse built a series of funiculars and loaded his railway trucks on special wagons hauled up by a cable and a steam winch assisted by the gravity of descending wagons. 'The subjugation of this escarpment', said a construction expert, 'constitutes one of the finest examples of engineering skill in Africa.' From the Rift floor the line bridged deep ravines on the climb to the highest point, Mau Summit, at 8,350 feet, 493 miles west of Mombasa. On the descent towards

Lake Victoria Sudanese troops fought off Nandi tribesmen who attacked survey parties and labourers with spears.

At the end of 1901 the last spike was driven home on the lake shore at Kisumu. A sixty-two-ton steamer was already there. It had travelled under its own power from Glasgow to Mombasa where it was dismantled, crated, put on the railway and finally carried by porters to the lake for reassembly. It was the link with Kampala. Whatever the critics said, the railway was a tremendous feat. Whitehouse and his men fought malaria, lions and spears to build it. The dangerous six-month journey from Mombasa to Kampala was cut to five days. The railway opened Uganda and Kenya to settlement and, as the posters put it, made the highlands of British East Africa 'a winter home for aristocrats'.

During his first visit to Africa in 1896, when he was twenty-one, William Ewart Gladstone Grogan saw a vulture feeding on a man's corpse. He shot the bird and ate it. In this self-reliant and apparently nerveless young man the master imperialist Cecil Rhodes spied a kindred spirit. 'Give yourself to Africa,' he advised Grogan. 'You will never regret it.' Ewart Grogan gave himself wholeheartedly and became Rhodes's disciple. 'The urge and surge of British blood', he said, 'must carry our people willy nilly into the last attainable confines of a finite earth, there to persist, absorb, dictate, boss and impose our Will.'

Grogan believed in Rhodes's dream of a trans-Africa railroad, from Cairo to Cape Town, and in 1898 he decided to prove its feasibility by walking the route from south to north. No colonial yarn would be more ripping. Grogan made his journey a romance, a Victorian version of a knightly quest to win a maiden's hand. It is probable, however, that, given his adventurous and arrogant nature and commitment to Rhodes's view of Africa, he would have set off anyway. But the pretext was his love for Gertrude Watt, whom he had met in New Zealand. Her stepfather refused to let him marry her and in an extreme and melodramatic gesture, Grogan promised that he would prove himself worthy by walking the length of Africa.

He set off, not from the Cape but from Beira in Mozambique with his friend Arthur Sharp. In their baggage they packed three British flags, copies of Shakespeare and a large quantity of Worcestershire sauce. Their medical supplies were primitive and so was Grogan's doctoring. When a venomous snake bit a porter's finger, Grogan gave the man whisky, slit the finger, sprinkled some gunpowder in the wound and ignited it. This was the treatment recommended by Francis Galton, the Africa explorer, in his work *The Art of Travel*, first published in 1855. As an alternative, Galton suggested cutting the bitten area with a knife and burning it out with a rifle ramrod 'heated as near a white heat as you can get it'. Grogan's gunpowder remedy did not work and he did not attempt the hot ramrod method. Instead he gave the man more whisky and plunged the finger into boiling potassium permanganate. This appeared to be successful. At least the man did not return for more treatment.

Grogan and Sharp travelled to Blantyre, canoed up Lake Nyasa and walked to Lake Tanganyika. They boarded the small steamer *Good News* which had been carried there in pieces and assembled by British missionaries in 1884. At Ujiji, where Henry Stanley had met David Livingstone twenty-seven years before, Sharp nearly died of fever and Grogan became seriously ill. Fevers tormented them both throughout their long march and they were often carried, soaked in sweat, shaking and semi-conscious in hammocks slung on poles.

Between the bouts of desperate illness there were interludes of terror. They came upon the aftermath of a tribal massacre and walked among thousands of corpses. In a cannibal kitchen they found a partly eaten baby, a pot of human soup and an opened skull with a spoon in the brain. In his arch and studiedly casual way Grogan described an encounter with cannibals. 'As I turned a corner, I found myself confronted by half a dozen gentlemen of anthropophagic proclivities on supper intent. The unexpected apparition of a white man checked their rush and, dodging a spear, I got my chance and dropped one with a shot through the heart . . .'

At Lake Edward Grogan fell ill again, screaming in his delirium while Sharp dosed him with quinine. In camp one evening, the porters became drunk and thirty of them confronted Grogan, brandishing spears. Grogan shot the most threatening one and the others retreated. After more incidents and bouts of malaria, after their camp was washed away in a storm and their once large caravan was reduced to seven men, Sharp mentioned to Grogan that, after sixteen months of walking across Africa, he really ought to be getting back home. He proposed to tramp across what is now Uganda and Kenya to Mombasa and from there get a ship to Europe.

Grogan's upper lip was as stiff as Sharp's as he shook hands in farewell.

'Damn good of you to have given up your time,' he said.

Grogan recruited more porters and pressed on to Lake Albert and the Upper Nile. His clothes rotted away and he was furious when a porter absconded with one of his two remaining shirts. At the end of December 1899, embarking on a crossing of 400 miles of swamp and wilderness, he lightened his load by tossing away his camp bed. He encountered tall, ash-smeared Dinka tribesmen who escorted him through a maze of lagoons, but, later, more than a hundred Dinkas attacked his party, spearing one of his porters through the heart and smashing the skulls of three others with clubs. Grogan shot four of the attackers, and the rest of them fled. Marching on and finding no water he and his men were reduced to sucking handfuls of mud for moisture. They all thought they would die but Grogan saw a flight of birds and followed their direction until he came to a waterhole.

Grogan was a scarecrow when he stumbled into the camp of a British army doctor on a hunting expedition south of Sobat on the Nile.

'How do you do?' asked the doctor.

'Oh, very fit, thanks,' said Grogan.

In the comfort of the doctor's boat he sailed to Fashoda, 'towards clean shirts, collars, friends, all that makes life a joy'. He reflected that to appreciate 'the exquisite flavour of bread and butter, the restful luxury of clean linen, the hiss of Schweppes, one must munch hippo meat alone and drink brackish mud for days'.

He reached Cairo in February 1900 and kept a promise to the three porters who had travelled much of the way with him. He showed them the sights of Cairo and sent them back home by steamer. He telegraphed Gertrude: 'My feelings just the same. Anxiously await your answer. Make it yes.' She did.

In London Grogan presented Queen Victoria with a flag he had carried all the way and told her the story of one of the last great adventures of her reign.

He returned to Africa and, famously aggressive and energetic, was one of the founders of the British Kenya colony. He had a number of mistresses and extramarital children, and in 1923 caused uproar when he publicly flogged three black men who had jolted his sister's rickshaw. He was sentenced to a month's imprisonment in Nairobi and later acquitted. Untroubled by doubt, he died in 1967 aged ninety-two.

35

All aboard for the Pacific

'Do you wish to make the mountains bare their head
And lay their new-cut forests at your feet?
It is easy! Give us dynamite and drills!
Watch the iron-shouldered rocks lie down and quake.'
Rudyard Kipling, 'The Secret of the Machines'

When a few visionary men, with scant knowledge of what they were talking about, proposed in the 1870s that a railroad should be built across Canada to the Pacific they were scorned for their 'insane recklessness'. Alexander Mackenzie, the Liberal leader, said it was 'one of the most foolish things that could ever be imagined'. It did not help that the railroad's most passionate advocate, Sir John Macdonald, the Prime Minister, was a legendary consumer of bottles of port, so much so that the newspapers from time to time kindly reported that he was 'indisposed'. It helped even less that there was not a single dollar to finance the ribbon of steel with which he proposed to stitch the component parts of Canada together. In any case, as many agreed, the mountains were impenetrable, the engineering difficulties insuperable and the railway therefore impossible.

Canada came into being by way of evolution rather than revolution. It has no independence day or founding struggle against oppression. A group of separate British colonies grew up in a frieze of territory along the forty-ninth parallel, sandwiched between the solemn tundra to the north and the Stars and Stripes to the south. As we have observed, Sir John Macdonald brought about their confederation in 1867 and thereafter longed to bring into the new Dominion of Canada the west coast colony of British Columbia, four times the size of Britain. Its inclusion would make Canada one united nation stretching 4,000 miles from the Atlantic to the Pacific; and, Sir John cunningly argued, it would be another safe route to India for the British. To persuade British Columbia to join he promised in 1871 that a railway would be driven through the mountains 'to connect the seaboard of British Columbia with the railway system of Canada'.

Underlying all this was his fear that if he did nothing the Americans would simply ooze into Canada, as a fat man fills a chair. Some Americans were already talking of the 'irresistible doctrine' of advancing beyond the forty-ninth parallel. Americans had a vigorous idea of their destiny; but Canadian national sentiment was not strong. To Sir John's mind the railway was vital to the invention of Canada; and it had to be Canadian, every mile of it on Canadian soil, a steel assertion of Canadianness.

The Americans, too, saw their own railways as unifiers, forming a web that bound together an immense country with profound cultural divisions. The railways had an immediate economic benefit, played a major part in the North's victory over the South in the Civil War and, with the development of insulated wagons, carried the Hudson River ice that made possible the supply of meat, butter, fruit and fish over long distances. The first transcontinental line was completed in May 1869 with the joining of the Union Pacific and Central Pacific lines at Promontory Point, Utah. More railroads linking east to west were completed in 1881, 1885, 1889 and 1909. The pioneering railway promoters were in the public imagination the heroic agents of the American destiny, constructing the line 'from sea to shining sea'. It was some years before these knights of the American romance fell in public esteem and were despised as venal ratbags.

Sir John knew that his Canadian Pacific would be the greatest and most expensive railway in the world, hundreds of miles longer than the transcontinental line completed by the Americans in 1869. He was adamant that until it had its railroad Canada had no backbone and was little more than a geographical expression. In 1871, upon his guarantee of a railway, British Columbia opted to join Canada.

Sir John made the astonishing, and some thought absurd, promise that the line would be started in 1873 and completed in 1883. A scheme that would have bankrupted Midas needed the drive of a ruthless, daring and flamboyant arm-twister. Sir John was such a man and he staked everything on his railway. As it turned out, its construction was a proper nation-building feat, a chronicle of death and dynamite and, like the Royal Canadian Mounted Police, a Canadian legend. Within a few years the Canadian Pacific's romantic posters of trains threading through river canyons and traversing the prairies became the national picture book.

The first task was to search out a route for the track that would be laid westward for 2,000 miles from the north shore of Lake Superior to the mountains of British Columbia and thence to the ocean. A tribe of indefatigable pathfinders and surveyors began their patient measuring of forests, mountain ranges, lakes and rivers. Some parts of the wild lands they walked and canoed had never been crossed even by Indians.

It was extreme and bloody labour, and for many surveyors only large measures of whiskey eased the fatigue and the loneliness. In six years they surveyed and staked out 46,000 square miles of country and buried thirty-eight of their number. They were thoroughly accustomed to living rough, whether it was freezing or burning. In winter they sheltered in skin tents or brush shelters, felt the temperature sink to forty and fifty below, and heard the trees crack as the sap froze. A scarf over the mouth could freeze to a man's face and half-suffocate him. A bare hand that touched a metal instrument could burn or leave a strip of skin behind. An eye could freeze fast to a theodolite eyepiece. Summer brought swarms of mosquitoes and the flies that maddened horses.

Many surveyors emerge from the journals and reports of the day as not only tough but also obsessed, seeing their work as a personal struggle with vertiginous gorges, glaciers and torrents. Marcus Smith, in charge of surveying in the mountains in 1873, found his way barred by a roaring river and crawled for fifteen miles over a glacier on his hands and knees to reach a crossing place. He was nearly sixty.

Those who complained that Canada could not afford the enormous amounts of money needed for the railway were right. The surveys alone were beyond the country's pocket. Sir John, who was not to be thwarted, persuaded British and American capitalists to bear the cost in exchange for large tracts of prime virgin land in the West. The uncovering of corruption and kickbacks brought his government down in 1873, but he was back five years later declaring the opening up of the prairie and full steam ahead for the Canadian Pacific Railway.

In 1881, after taking over the existing railways of eastern Canada, the CPR bosses started building the track from Lake Superior. Seven hundred miles of it were laid across the bottom of the Canadian Shield, the wilderness of grey and red granite pitted with dark lakes. The track builders used tons of nitroglycerine to blast the rock, pouring it from cans into drilled holes, and frequently blowing themselves up in the process. The explosive was too dangerous to carry on wagons so that labourers had little choice but to transport it in ten-gallon cans on their backs: they were pedestrian bombs, a spark away from obliteration.

After crossing the brooding Shield they drove the line over more than 300 miles of muskeg, deep and sticky bogs. As they filled them with mountains of gravel they swore that many of these horrible swamps had no bottom. They must have felt they were stuffing the holes with dollar bills. The blasting, felling, digging and shovelling were done by a multinational army of thousands of engineers and navvies and rough regiments of sub-contractors, English, Scots, Irish, Scandinavians and Americans, who followed in the trailblazers' wake, bending their backs to the most exhausting labour and fuelling themselves with liquor.

Cargoes of sleepers and telegraph poles, ballast gravel, rails, spikes and tools were hauled by hundreds of teams of horses and wagons and stacked beside the advancing line. As the men moved westward they threw up sprawling towns of tents and shanties, riproaring Gomorrahs soaked in liquor and brawlers' blood. The most notorious was Rat Portage, a bedlam of 3,000 hellions. A priest sent out to save their souls found them so incorrigible that he wished himself back among the quiet and amenable Indians.

After the ordeal of the Shield and the muskeg the construction teams broke out on to the seemingly endless bald prairies. They laid 900 miles of track from Winnipeg across Saskatchewan and Alberta to the mountains of British Columbia. Even by the standards of North America these are massive bulwarks. The railroad men had to build 450 miles of track through the Front, Main and Western ranges of the Rockies, then the Columbia Mountains, the Selkirk Range and the Monashees, and at last the North Cascades and the Coast Mountains.

When the railway was started no one knew how and where it would wind through these avalanche-prone mountains or whether such construction was an engineering possibility. In common with almost every other Canadian, Sir John Macdonald had never seen the mountains.

Major A.B. Rogers, an American railway surveyor, had not seen them either, but he was hired to find a way through. He was fifty-two, small, confident and pugnacious, determined to put his name, literally, on the map. He had eyes of piercing blue, white Dundreary moustaches stretching to his lapels and a tongue so foul that he was nicknamed 'the Bishop'. His meagre diet of raw beans and biscuit would have left a squirrel hungry and he expected his men to feed as sparingly. He derided those who 'made gods' of their stomachs. 'Effeminate,' he growled, chewing away at his tobacco plug. For their part the men feared he would starve them; and sometimes he nearly did because his Spartan outlook and meanness led him to cut his survey parties' rations. Hunger forced them to retreat from their expeditions: the skinflint major's economies cost his masters more money.

The pressure on Rogers to discover a way through the mountains was very heavy when he decided, with no great certainty, that the spectacular and barely surveyed Kicking Horse Pass was the key. He knew that an earlier surveyor had concluded that it offered no way through. But Rogers trusted his own instinct. The first critical step was to discover a passage piercing the fortress walls of the Selkirks. His bosses offered him a reward of $5,000 and a gold watch if he could find it. His initial expedition failed because, typically, he did not pack enough food; but on his next attempt he breached the Selkirks and achieved his dream. His name is on the map, the Rogers Pass.

By now the railway was under the dynamic generalship of William Cornelius Van Horne. Unlike the nibbling Rogers, he enjoyed eating, drinking and smoking cigars; and with his beard and bulk he looked rather like the Prince of Wales. The mountains were his ultimate test. By the end of 1883 the line had reached the eastern end of the Kicking Horse, and early in the following year contractors advertised in the newspapers for 12,000 'good, able-bodied steady men' to join in the great assault for $1.50 a day.

The enemies were chasms, rock falls, snow slides, forest fires, terrific torrents, rickety platforms and tenuous scaffolding from which men tumbled to their deaths. Dynamite killed a man a week and left many maimed. The camps seemed to cling to the cliffs. Labourers felled thousands of trees, built timber bridges, blasted tunnels and hacked out steep supply roads on which horses hauled food and materials. The campaign in the Rockies was a long siege and at the end of May 1884 the citadel fell. The sound of a locomotive whistle floated hauntingly through the Kicking Horse Pass.

Meanwhile another force of labourers toiled from the coast and into the mountains, laying the railway eastwards to meet the westbound steel. These were mostly Chinese men who endured not only hardship and danger but also the hatred of the white men with whom they worked. They came from the Pearl River region of southern China and crossed the Pacific to California to escape unrest and economic oppression. In 1858 some of them migrated from California to the gold mines of British Columbia and, although attacked and shunned by Europeans, made money in the niches of the frontier economy. They ran restaurants and laundries, drove wagons and reworked gold claims abandoned by whites. They prospered, as a tax inspector remarked, because of their 'more frugal habits and greater industry'. A member of the Canadian Parliament observed that 'the only labour employers can depend on is the Chinese'.

Nevertheless the Anti-Chinese Association opposed their recruitment to the railway; and although they lacked any knowledge of them, newspapermen scorned 'the beardless and immoral children of China'. The fact was, though, that more than 10,000 labourers were needed to build the railway up from the coast and there were simply not enough white men. The railway bosses hired thousands of Chinese to clear the way for the track and to lay the sleepers: it was that or have no railway at all.

At a dollar a day, the Chinese were cheaper, faster and more sober than Canadians and other whites. Between 1881 and 1885 about 17,000 of them came to Canada and many went to work on the railway. They were so vital to the construction that in 1882 the railway chiefs chartered ships to bring a further 6,000 from Hong Kong.

More than 600 Chinese died on the railway, blown to bits, drowned in rivers, crushed by rocks, murdered by whites. They died of scurvy, too, because of their thin diet, and suffered terribly in the freezing winters. A newspaper in British Columbia reported that in a three-month period of 1880 there were no deaths on the railway construction; but Chinese deaths were not officially counted, an echo of Mark Twain's observation on racism in *Huckleberry Finn*:

> 'We blowed out a cylinder-head.'
> 'Anybody hurt?'
> 'No'm. Killed a nigger.'
> 'Well, it's lucky; because people sometimes do get hurt.'

In 1885 Canada passed a law which obstructed Chinese immigration and imposed a $50 head tax on each migrant and a strict limit to the number of them a ship could carry. By 1890 the Chinese had found ways of raising money for the tax, and migration resumed. The law, however, institutionalized prejudice. Not until 1947 were discriminatory rules repealed and the Chinese admitted to the legal and medical professions and allowed into swimming pools.

The Canadian Pacific was completed in November 1885 and its engineers had their triumphant scene at the last of the construction camps, Craigellachie in the Monashee Mountains. Every Canadian knows the country's most celebrated photograph, the picture of the top-hatted Donald Smith, the Canadian Pacific's president, hammering home the railroad's last spike. When the Americans completed their transcontinental railroad the final spike was made of gold. Canada's was of humble iron, a metaphor both for the toughness of the struggle and a certain Canadian modesty.

Maybe the railway was a birth-of-a-nation drama but William Van Horne took the unromantic view that 'the Canadian Pacific was built for the purpose of making money for the shareholders', and for no other reason. In 1886 Sir John Macdonald and his wife travelled across Canada by rail. For exhilaration's sake the sporty Lady Macdonald rode for a few miles on the front of the locomotive, perched upon the cowcatcher.

The Canadian Pacific Railway was a plough turning virgin earth. It did not go from town to town, for there were none to speak of. It needed to create a country, to develop settlements and attract migrants who in turn would build a market. A publicity drive in Britain and Europe eulogized the Canadian prairies and took care to be economical with information about the harshness of the climate. The campaign urged tourists to see the magnificence of the Canadian West and to enjoy the sumptuous railway hotels, a pleasure to this day.

The railway planted 800 towns like seeds as it went, among them Winnipeg, Regina, Moose Jaw, Calgary, Revelstoke and Kamloops, giving Canada its modern pattern. It took years, but gradually settlers caught the train to make their lives as farmers and traders beside the track. Winnipeg, on the banks of the Red River, an uproarious and debauched drinking town in the 1870s, was a frenzied Eldorado in the early 1880s. Thousands of real-estate dealers, gamblers and half-crazed immigrants scrambled for property, and one successful English speculator invited his friends to watch him enjoy a bath in champagne.

The long-haul trains drew the strings of the country together. It took a week to cross from Montreal to the Pacific. Passengers could mark their progress by counting the days or the hours or the meals. Toronto to Winnipeg was a thirty-six-hour, five-meal journey. Another leg took four breakfasts, three lunches and three dinners. Long before pre-packaged food the train cooks cut meat from a swaying side of beef in a refrigerator room. They knew that if they were to make a delicate pie or custard they would have to wait for a particular smooth length of track to do so.

If they went all the way across the continent passengers arrived in the Pacific terminus chosen by William Cornelius Van Horne. They found themselves in a watercolour of blue mountain slopes, green islands and limpid sea, breathing the sweet salt air of the Gulf of Georgia. In Van Horne's day it was a little one-pub lumber town on the Burrard Inlet called Coal Harbour. It was a name too grubby for Van Horne. With his ear for euphony and his eye on history he called it Vancouver.

George Vancouver had surveyed this coast between 1791 and 1795, searching for any outlet that might be related to the North West Passage. Van Horne built his own North West Passage, of steel rails, and launched Vancouver as Canada's chief Pacific port. The industrious Chinese prospered, too, and in Vancouver created one of the great Chinese communities of North America.

36

Specimens

'And much it grieved my heart to think
What man has made of man.'
William Wordsworth, 'Lines Written in Early Spring'

Joseph Banks orchestrated the first grand botanic collections through his worldwide network of dedicated hunters and gatherers of exotic plants; and the thousands of specimens he garnered formed the foundation of that British Eden, the Royal Botanic Gardens at Kew, the greatest plant collection in the world. For forty-four years from 1841 the work of collection and research at Kew was commanded by Sir William Jackson Hooker and his son Sir Joseph Dalton Hooker who succeeded him as director.

In the distant forests, mountains and deserts their collectors sought out the rarest plants. These men and women were missionaries, military officers, colonial civil servants and travellers, devoted to extraordinary botanical quests. In the nature of things they were sometimes obsessive and eccentric. For little or no financial reward they frequently submitted to hardship and danger in pursuit of some shy jungle orchid or elusive mountain bellflower. No doubt most of them accepted Joseph Banks's stern advice that 'Collectors must not take upon themselves the character of gentlemen', and knew that botany meant roughing it. Thirty-five-year-old David Douglas, whose collection of conifer seeds from the Rocky Mountains had a dramatic impact on the British landscape, went plant-hunting in Hawaii in 1834, fell into a pit excavated to trap wild cattle and was gored to death by the animal that had preceded him.

The young Joseph Hooker learnt the naturalist's craft while an assistant surgeon aboard HMS *Erebus* which, with HMS *Terror*, took part in Sir James Ross's expedition to the Antarctic and the waters of the southern ocean from 1839 to 1843. In 1847 Hooker voyaged to India with Lord Dalhousie, the new Governor-General, and led a botanical expedition into Sikkim in the Himalayas. As he wrote in his account of it, his party was fifty-six strong, mostly coolies carrying tents, rice, flour and scientific equipment, for the going through dense forests was too hard for horses and everything had to be loaded on men's backs. Boys were hired to climb trees for plant samples. As he travelled Hooker was in rhododendron heaven. His spectacular booty included more than forty species which started the Victorian rhododendron mania and changed the appearance of gardens all over Britain. Seeds of the shrub were sent from Kew to nurseries and botanic gardens in Britain, Ireland, France, Italy, America, the West Indies, Australia and New Zealand.

Sir William Hooker, who died in 1865, saw to it that his son succeeded him as director at Kew. Like Banks, both of the Hookers believed strongly in economic horticulture, the international transfer of useful plants; and Joseph certainly did not see Kew as a place for idle saunterers. Just as, under Sir William's aegis, the cinchona tree had been transplanted from the Andes to India to provide a source of quinine, so his son oversaw the removal of rubber plants from Brazil for cultivation in Ceylon and Malaya. Robert Cross gathered 600 plants and samples from swamps, made his way out by canoe and loaded his haul aboard a ship. The vessel was wrecked on a Jamaican reef and, while other passengers left in boats, Cross gallantly stayed with his plants and was rescued by the Royal Navy. Transported in heavy glass Wardian cases, the plants arrived safely in Kew and Cross returned to the Amazon forest for more.

The flow of plants into Kew and other gardens grew into a flood. Robert Fortune, a Berwickshire-born gardener, first went to China in 1843 to collect plants for the Horticultural Society's garden in London. He was glad that he took his shotgun: he needed it to beat off an attack by pirates. After three years he returned with knowledge of the art of bonsai and twenty-four Wardian cases of plants. In 1848 he went back to China on assignment for the East India Company, gathered thousands of tea plants and smuggled them over the mountains into India. He also took several Chinese who knew how to tend them.

After some years of experiment, a tea industry took root and prospered.

Ernest Wilson went to China in 1899 to gather seeds for the Veitch Nurseries and returned after three years with a rich harvest, including the kiwi fruit and 400 other new plants. In his next expedition to China he found many new rhododendrons and roses; and then became a renowned collector for the Arnold Arboretum in Boston, Massachusetts. George Forrest and Frank Kingdon-Ward were also great rhododendron hunters, the latter making twenty-two expeditions in forty-five years. In this rugged, largely masculine, world of the botanical explorers Lucy Bishop stands out. A courageous collector of plants in North America and Persia, she became, in 1892, the first female Fellow of the Royal Geographical Society.

Travellers collected everything. Treasures, sculptures, pottery, weapons and all kinds of hardware were shipped home. Also animals: the royal menagerie in the Tower of London exhibited lions, leopards, bears and elephants. In the thirteenth century a polar bear on a collar and chain was permitted to walk from the Tower into the Thames to catch salmon. A rhinoceros once gave rides on Ludgate Hill. Bears were commonly tormented for entertainment.

People were collected, too. A group of Eskimos were brought to Bristol in 1501. King Henry VIII met a Brazilian imported for his inspection. In 1578 Sir Martin Frobisher presented Queen Elizabeth with an Eskimo and the Queen allowed him to spear some of her royal swans from his kayak. Matoaka, an Indian girl, came to London in 1616 with her English husband John Rolfe. Her nickname was Pocahontas and whether or not she saved the Englishman John Smith from execution in Virginia, she became a noble savage of art and legend.

The British buccaneer William Dampier returned from the East Indies in 1691 with 'the famous prince' Gilolo who was 'exposed to public view every day in the Blue Boar's Head in Fleet Street', before he died of smallpox. Men of the Hudson's Bay Company sent an Indian girl to England as an oddity in 1716. Ten thousand people jostled to see Cherokee Indian chiefs in London in 1762. Crowds craned their necks to glimpse the Eskimos brought to London in 1773. At Plymouth, while they were on their way home, most of them were wiped out by smallpox. The only survivor, a woman, carried the disease to Labrador and her entire tribe died of it.

All such visitors were from other worlds, exotics, objects of curiosity. The outermost space was the Pacific. An islander was taken to Paris by the circumnavigator Louis de Bougainville in 1769 and charmed everyone as a perfect example of the noble savage, the natural man celebrated by the philosopher Jean Jacques Rousseau. Joseph Banks, in the Pacific with Captain Cook in 1769, invited Tupaia, a Tahitian priest, to sail back to Britain, well aware that he would be a social sensation. Banks reflected that Tupaia would cost him less to look after than the lions and tigers kept by some of his English neighbours. Tupaia performed valuable service as an interpreter during the voyage but died on the homeward journey.

Banks was therefore delighted in 1774 to be presented with a young man from the Society Islands brought to England in HMS *Adventure*. This was Omai who was personable and had exquisite manners. Banks spoke to him in Tahitian. Omai spoke some English but had the Polynesian difficulty with English sounds, so that he pronounced Cook as Tootee and Banks as Opano and addressed King George III, in the story of their meeting, as 'King Tosh'. The King, like everyone else, was charmed and gave Omai an allowance and a sword. He also recommended inoculation against smallpox and Banks saw to it.

Indeed, Banks took great care of his guest, dressed him in good clothes and installed him in lodgings in Warwick Street. Fashionable London was captivated by him. Banks took him to the theatre to see David Garrick perform and to the House of Lords to see the aristocrats, ditto. He gave him dinner many times at the Royal Philosophers' Club, and escorted him to dine with Dr Johnson and his companion Mrs Thrale, who were both impressed by his breeding. Omai was the social lion of London. He melted the ladies with his attention. 'How much more nature can do without art,' cooed the novelist Fanny Burney, who sat next to him at dinner.

Omai spent Christmas with Lord Sandwich, another of his patrons, and went to Yorkshire with Banks. He learned to shoot and also to skate.

He famously sat for Sir Joshua Reynolds, the society artist, who painted a masterpiece with Omai a handsome and confident classical figure set in a Tahitian landscape, although wearing a decidedly unTahitian robe and turban. It was the idealized South Sea islander that many people in that age of enlightenment and discovery wanted to see. Less well known is the portrait of Omai painted by William Hodges, the artist on Cook's second voyage, which is much more true to life and certainly not idealized.

Omai had his uses. To the romantic and optimistic London mind he was the primitive and innocent man in harmony with nature; and his friendliness and bearing seemed the proof of a natural nobility. Sceptics scorned this as pure sentiment and pointed to the dark side of island life, wanton sex and human sacrifice. Far from being noble the savage was just that, a savage. Evangelists certain of the superiority of Christianity saw little that was admirable in paganism and deplored the fact that while Omai was fêted in London nothing was done to save his soul.

Satirists, meanwhile, used Omai as their mouthpiece, depicting him as the impartial visitor observing the greed and vanity of London society. Their Omai commented waspishly on the 'pilfer'd wealth' and 'bloated bishops' supping turtle while curates starved; and coolly observed the bellicose and butchering European imperialists always ready to 'cross o'er the seas, to ravage distant realms'.

During his third voyage, in 1777, Captain Cook returned Omai to the Pacific. Of course, the lion of London no longer belonged. He was not welcomed in Tahiti for he was not high-born and the people were only interested in the red feathers he had brought. Cook landed him on Huahine, an island a hundred miles off, and set him up in a house with his sword and suit of armour and London clothes. He wept on Cook's shoulder at their leave-taking. In 1785 a gorgeously costumed pantomime, *Omai, or a Trip Around the World,* was staged at Covent Garden. It told how Omai, a prince of Tahiti, sailed to London and wooed Londia, an English girl. It was a theatrical sensation and demonstrated once more the power of the South Seas to inflame the imagination. By that time, however, Omai had died of fever beneath the palms of his island.

In 1792, at the end of his term as Governor of New South Wales, Captain Arthur Phillip sailed for London. His party included a popular member of his household, a Sydney Aborigine called Bennelong, who volunteered to make the journey. The first native Australian to go to London, he was celebrated for a while and people remarked on his courtesy and politeness. He returned to Sydney after three years, but by then he was neither one thing nor the other, distanced from his old tribal world and unable to be a part of the white one. He took to rum, shed his jackets and trousers, and lost his dignity and the respect of people both black and white. He lived in a hut at what became known as Bennelong Point and died in 1813, aged about forty. The point that bears his name is where the Sydney Opera House now stands.

At least poor Bennelong did not suffer the degradation heaped on a South African woman in 1810. William Dunlop, a British ship's doctor, persuaded Saartjie Baartman, aged twenty-one, to sail with him from Cape Town to London. There, led around by a keeper, she was 'exhibited like a wild beast' on stages in Piccadilly and Haymarket, naked so that crowds of spectators could see her pronounced buttocks and vulva. She was called the Hottentot Venus. After four years she was taken to Paris for exhibition and on her death in 1815 a surgeon extracted her skeleton, brain and genitalia for display in a Paris museum. At a moving ceremony in 2002 her remains were returned to her people and taken back to South Africa for burial, a dignity after a life of inflicted indignity.

In 1830 Captain Robert FitzRoy took three young men and a girl from Tierra del Fuego, the territory near Cape Horn, to England. Deeply religious, he believed that the four could be schooled in English and the Bible to become interpreters and Christian bridgeheads in their native islands. As he himself explained it, FitzRoy was 'concerned for the eternal souls of people who had never heard God's word'.

He originally seized them as hostages after one of his boats was stolen, and he whimsically called the boys Jemmy Button, York Minster and Boat Memory and the girl Fuegia Basket. She was about ten or eleven. At Falmouth he had them vaccinated against smallpox by the method pioneered by Edward Jenner; but Boat Memory died of the disease. FitzRoy dressed the three survivors in English clothes and took them to Walthamstow in London to be taught English and the rudiments of the Christian faith. Jemmy was a sunny boy of sixteen who liked his clothes and his shiny shoes and often preened before a mirror. Fuegia, likewise, was amiable and made a good impression when the three of them were presented to King William IV and Queen Adelaide. The Queen gave her a bonnet. York Minster was older, about twenty, a taciturn scowler moodily attracted to Fuegia.

After eleven months of schooling the three Fuegians boarded the small survey brig HMS *Beagle*, commanded by FitzRoy, and at the end of December 1831 sailed for Tierra del Fuego. There, FitzRoy intended, they would set about civilizing their own people. Aboard the *Beagle*, too, was the twenty-three-year-old naturalist Charles Darwin, a drinking-and-shooting graduate who had just scraped his degree at Cambridge. The stroke of fortune that led to his spending five years in a cramped cupboard of a cabin aboard the ship in its circumnavigation of the world shaped his life. All his experiences and observations in South America, the Pacific islands and Australasia, his discoveries of new species of animals and plants, formed the storehouse of his knowledge. Influenced by his reading of Thomas Malthus on population he developed the chief work of his life, published in 1859, *On the Origin of Species by Means of Natural Selection or the Preservation of Favoured Races in the Struggle for Life*. It exploded in the Victorian consciousness.

Darwin grew to know the three Fuegians on *Beagle*'s southward voyage. He did not believe that any good would come of FitzRoy's experiment. Dressing two young men and a girl in smart clothes and immersing them briefly in civilization would not turn the tide for an aboriginal people living a primitive life in one of the world's harsh places. FitzRoy landed the three at their home. When he returned a year later he was horrified to see Jemmy's wild appearance. 'I could almost have cried,' he said. Darwin noted Jemmy's matted hair and skinny form. 'Instead of the clean, well-dressed stout lad we left,' he recorded, 'we found a naked thin squalid savage.' Still, Jemmy came aboard the *Beagle* for dinner and, washed and dressed, ate and talked with all his old good manners. But he was determined to stay with his people in Tierra del Fuego.

Over the next decades missionaries and others came to colonize the islands, and the native people were whittled away by disease and murder. In 1889 seven Fuegians were exhibited in London, dressed in animal skins, sitting around a stove for warmth. As Darwin wrote after observing the Aborigines in Australia: 'Wherever the European has trod, death seems to pursue the aboriginal.'

Robert FitzRoy was Governor of New Zealand from 1843 to 1845. He became meterological officer at the Board of Trade in 1854 and pioneered the drawing of weather charts for forecasting, indeed coined the word 'forecast'. In the 1860s he sent gale warnings by electric telegraph and devised the system of storm cones hoisted on poles. He perfected a barometer to measure the sudden drop in pressure that precedes stormy weather. Forecasting was an aspect of his concern for safety at sea; and he was wounded by the scoffing that followed inaccurate predictions.

Nothing, however, could have astounded him more than the controversial work of his former shipmate, the young naturalist of the *Beagle*. Charles Darwin's theories, positing a natural rather than a divine origin of species, the survival of the fittest, seemed to FitzRoy as to any other Christian who believed that the Bible was literally true, a heresy. It was an article of faith that the Book of Genesis was sacred, that God had created the world, that the creatures of the world had been saved aboard the Ark and that nothing had changed since.

There was a showdown at the meeting of the British Association at Oxford in 1860, a debate between the Creationists and the Darwinists, although Darwin himself was not present. The main battle, before 700 people, was between Samuel Wilberforce, Bishop of Oxford, on the side of orthodoxy, and, in Darwin's corner, Thomas Huxley and the botanist Sir Joseph Hooker, assistant director of Kew Gardens. The Bishop attacked Darwin for being sensationalist; Huxley seized on the Bishop's prejudice and ignorance. In the uproar Robert FitzRoy rose with a Bible in his hand and cried that the truth lay in the very book he was holding. He was shouted down. Five years later, in failing health and despondent over criticism of his weather forecasting, he took a razor and cut his throat. He was not forgotten. When meteorologists needed to rename the shipping forecast sea area of Finisterre in 2002, they chose the name of FitzRoy.

In 1866 a woman called Trucanini stared into the lens of a camera and the shutter clicked. The photograph was an epitaph, for she was the last of the aboriginal Tasmanians. Her people painted their bodies with red ochre, lived on wallabies, shellfish, snakes and berries, and spent much time in dancing and singing. They were affectionate and showed no hostility. British settlers hunted them as vermin, caught them in man traps, enslaved and raped them, and sought to erase them all, a process that many Christians explained as the inevitable fate of the savage in the advance of civilization. The last fifteen of Trucanini's people were photographed by the Bishop of Tasmania in 1858 and the pictures were exhibited in London. She herself died in 1876 but found no peace in the grave. Her bones were exhumed and her skeleton displayed in the Tasmanian museum.

The American impresario Phineas Taylor Barnum preferred his specimens alive. A pioneer of mass entertainment, he had a genius for concocting sensation. His shows responded to public curiosity, and pandered to it also, with displays of freaks, disfigurement and exotic indigenes. Into his repertoire he crowded 'mermaids', and 'bearded ladies', the original Siamese twins, the extremely diminutive 'General Tom Thumb', 'the Wild Moslem Nubians', a group of so-called cannibals, 'the Wild Man of Borneo' and suitably fierce Zulus.

In 1882 his agent R. A. Cunningham acquired nine Aborigines in Queensland and shipped them to the United States. Advertised as uncivilized savages they toured in a circus with, among other attractions, the original Jumbo the elephant. Londoners saw the show at the Crystal Palace in 1884. One of the Aborigines was Jimmy Tambo, aged about twenty-one, and another was his wife. Tambo did not last long. He died after a year on the road, and his wife and four of the others died shortly afterwards. Death, however, did not diminish Tambo's value: his body was sold to a museum, mummified and put on show. In 1993 his remains were found in an undertaker's basement in Cleveland, Ohio, and his descendants took them back to Australia for burial.

To gather human material for his 'Great Ethnological Congress of Curious People From All Parts of the World', Barnum sent agents to make 'a collection in pairs or otherwise of all the uncivilized races in existence ... to astonish and instruct' the American public. In this he was supported by anthropological institutions eager to measure and categorize the skeletons. The Natural History Museum in London, for example, has the bones of 17,000 people.

The American explorer Robert Peary rounded up six Eskimos in 1897 and shipped them to New York where they were displayed as living specimens in the American Museum of Natural History. In the first two days of the exhibition 30,000 people queued to see them. The youngest was seven-year-old Minik. Peary, who took no responsibility for the fate of his captives, said Eskimos were 'much like children and should be treated as such'. The museum plainly thought they were not really human and confined them in a cellar. When Minik's father died, the curators fooled the boy by burying a body-sized log while the body itself was stripped to a skeleton; and, in time, Minik had the shock of coming face to face with the skeleton which was mounted in a glass case. The museum always refused to give him the bones for burial. In the story of Minik's life, *Give Me My Father's Body*, Kenn Harper described how, after twelve years in America, the boy returned home, to learn to hunt and to speak his native language. Caught between two cultures he did not really fit and went back to America where he died of influenza in 1918. In 1993 the bones of Minik's father and others of Peary's Eskimos were returned to Greenland and buried under a plaque saying: 'They have come home.'

37

Your worship is your furnaces

'This extraordinary metal, the soul of every manufacture, and the mainspring perhaps of a civilised society.'

Samuel Smiles on iron, 1884

During the summer of 1802, a brief interlude in the wars against Napoleonic France, Lord Nelson took a holiday with his mistress Emma and her husband Sir William Hamilton. The battered naval hero, the vivacious redhead and the scholarly ex-ambassador travelled as a genial ménage in a carriage from their home at Merton in London and across England and Wales to Milford Haven. On the way they stayed in Merthyr Tydfil.

Merthyr then was a gaudily violent place of gushing furnaces, well on the way to becoming Thomas Carlyle's 'vision of Hell, sootiest, squalidest and ugliest . . . all devoted to metallic gambling'. Nelson, who had come to see iron forged for his cannons, the Royal Navy's fists, must have been reminded of the scorching heat and swirling smoke of a gundeck during battle.

The furnaces were fired by coke, the wonder fuel developed early in the eighteenth century by Abraham Darby at Coalbrookdale. Before he discovered how to burn the impurities out of coal to make coke, ironmasters heated their furnaces with charcoal and plundered forests for timber. Coke was the great technical advance in fuel, and Darby increased his iron production tenfold.

Richard Crawshay, a Yorkshireman who ran the Cyfarthfa works at Merthyr and had a pyramid of cannonballs on his coat of arms, was one of Britain's first millionaires. He showed Nelson how the navy's guns were made. In 1760 John Guest, a Staffordshire coal merchant who knew the secret of coke, rode to Wales with his servant, both of them on one horse, to take over the management of the Dowlais works. He, too, made his fortune from armaments. The once-empty hills drew people in their thousands, and then tens of thousands, and the furnaces roared like dragons through the day and night.

John Hughes was born in 1814, the son of a Merthyr engineer, and grew up in that lurid frontier world before becoming manager of an ironworks in Newport, Monmouthshire, and then a shipbuilding director in Millwall, London. After 1860, when the Royal Navy moved from oak to iron construction, he designed armour plating for warship hulls and mountings for the new heavy guns.

His innovative work attracted the attention of the Russian government which invited him to start an iron industry from scratch in Russia. The Russians had come to the right man. Hughes was iron to his fingertips. For more than thirty years Welsh foundries had been the major suppliers of rails to the world, and the names of the chief ironworks were stamped on bridges and rails everywhere. The Merthyr-born historian Gwyn A. Williams observed that the tragic Anna Karenina who, in Tolstoy's novel, threw herself under a train in Russia, would surely have had the word Dowlais stencilled on her body.

The Russians, however, wanted to manufacture their own rails and asked Hughes to visit southern Russia, what is now the Donetsk region of the Ukraine, a bleak virgin steppe peopled by peasants and wandering Cossacks. It had plenty of coal, for it marked the eastern end of the folds of coal reserves that stretched from Ireland through south Wales, Shropshire, the Pennines, Kent, northern France, Belgium, the Ruhr, Czechoslovakia and Silesia, the geological foundation that helped to shape Europe. The coal was similar in composition to the high-grade coal of Wales but the people lacked the technical skill to extract it. In 1869 Hughes made an agreement, approved by the Tsar, and formed the New Russian Company to develop mining and iron-making. He built a smithy and a small ironworks with equipment and materials shipped from Britain and brought from the port by ox-cart. This was Russia's industrial revolution.

With the completion of so many railways in Britain and the general fall in demand for iron rails, craftsmen were leaving Wales to find jobs in America, Brazil and elsewhere. In 1870 seventy Welshmen joined Hughes in Russia and smelted the first iron in the following year. Within three years the foundries were making 8,000 tons of rails a year and miners were working seven coal seams. The town that grew up around these burgeoning industries was called Hughesovka, Russified to Yuzovka. Just like any British ironmaster, Hughes ruled as a baron, bred a dynasty with his sons John, Arthur, Ivor and Albert, and became a substantial landowner. His company paid dividends of nearly 100 per cent. He rewarded his skilled workers well and built good red-brick houses for them; but he paid the labourers meanly and housed them in dark earth-floor dugouts with one stove for every six families. *Sobachevka*, the workers called them, the kennels.

As any British industrialist did, Hughes worked his people twelve hours a day and more, and used female and child labour. He reduced wages to keep paying good dividends and was backed by stern anti-strike laws. He built a free hospital, Russian and English schools, a church, several taverns and the Great Britain Hotel. During his twenty-year reign Yuzovka grew into a town of more than 20,000, with 6,000 people making iron and steel, and 2,000 working in the mines. After his death in 1889 his sons ran the business successfully, and by 1910 Yuzovka was producing half of Russia's steel and most of its coal. The town had a Welsh flavour into the twentieth century, and many Welsh workers were absorbed into Russian life. Williamses, for example, became Vilyams. The rule of the Hughes family ended in the Bolshevik seizure of power in 1917. In 1924 Yuzovka became Stalino and in 1960 Donetsk.

Like those who went to Russia many of the Welsh craftsmen who migrated to the United States in the second half of the nineteenth century had valuable skills for sale, experience in puddling iron, forging steel, rolling tinplate and mining coal. They settled in the ironworks city of Pittsburgh and the mining towns of Scranton and Wilkes-Barre, all in Pennsylvania. There, for some years, they formed an elite.

As Richard Edwards wrote home: 'The Welsh get the jobs of foremen.' Certainly their knowledge helped the metal and mining industries of America to find their feet and prosper. Their expertise, however, was overtaken by new American techniques. Few sons of the pioneers followed their fathers into the coal mines because the work generally became less skilled and the jobs were in any case taken by the new wave of Italian, Hungarian and Polish immigrants, prepared to work for lower wages. The newcomers were also attractive because they were less union-minded and less inclined to strike than the Welsh, English, Scots and Irish, and easier to manage.

In an experiment at Chalk Farm in London in 1867 a railway engineer laid iron rails on one side of the track and steel rails on the other. The iron rails needed renewing twenty-three times before the first steel rail showed signs of wear. It was a decisive test.

In 1850 there were 18,000 miles of railway in the world. Twenty-eight years later there were 206,000 miles. Steel was by then the commanding material, and British ironworks were badly hit by the declining demand for iron rails. Railways were for many years the main consumers of steel and used almost half of British and American production. The Victorian steel age began in a tentative way in 1856 with Sir Henry Bessemer's introduction of a converter and, later, with the open hearth process developed by William Siemens. As with any other advance, there were many ancestors. Steel is iron combined with carbon. It is exceptionally hard, and, unlike iron, it can be given a sharp edge. For centuries it was the expensive specialist product of cutlers and sword-makers in Rome and, later, in Toledo and Sheffield. Around 1740 Benjamin Huntsman, a Sheffield clockmaker, invented a crucible for casting steel at high temperatures, the first time that steel had been melted after production, but for more than a hundred years steel remained a small sideline.

Bessemer looked for a method of producing cheap bulk steel, particularly for artillery, and developed his converter in Sheffield. It forced compressed air through molten iron to burn out carbon impurities; and a precisely measured amount of carbon was added afterwards. It was soon found, however, that ores containing phosphorus produced steel that was brittle and of poor quality.

There were non-phosphoric iron ores in Britain and Spain but, clearly, the phosphoric ores in America and in the Lorraine region which Germany had taken from France in the war of 1870 could not be used for high-grade bulk steel.

In 1876, the cousins Percy Gilchrist and Sidney Gilchrist Thomas found a way forward. Percy was a chemist at the Blaenavon ironworks in Wales, Sidney a court clerk in London. Every night Sidney worked on the phosphorus problem in a bedroom laboratory at his home; and every weekend he travelled to Wales to work on more experiments with his cousin. In two years they demonstrated that a furnace lined with special limestone and silica bricks eliminated phosphorus. It was the making of modern steel. The cousins earned their fortune but Sidney died in 1880, aged thirty-five, his lungs ruined by the fumes produced in his experiments.

The immediate result of their work was that the phosphoric ores of America and Lorraine could be exploited by the great steel makers: Andrew Carnegie in Pittsburgh and Alfred Krupp in Germany. Carnegie declared that 'Moses struck rock and brought forth water', but Sidney Gilchrist Thomas 'struck the useless phosphoric ore and transformed it into steel, a far greater miracle'. By 1890 the United States was making more iron and steel than Britain and at the turn of the century made more than Britain and Germany combined. 'Farewell, then, Age of Iron,' wrote Carnegie, 'all hail King Steel.' He himself was the king and he foresaw the myriad possibilities of steel.

Iron ships were built from the 1830s and in larger numbers from the 1850s. Steel ships offered the advantage of more cargo for the same draught as an iron ship but it was only from 1877 that shipbuilding in steel became general. By 1890 nine-tenths of British vessels were of steel. In that year the Forth Bridge, in Scotland, the first major steel bridge, was completed. The design, by Benjamin Baker and John Fowler, replaced that of Thomas Bouch, whose Tay Bridge blew down in a gale in 1879 killing more than seventy people. After that catastrophe the public wanted a bridge over the Firth of Forth that would withstand all the violence of nature. The engineers chose steel, which had just been cleared for construction work, and built a bridge which endures as a symbol of formidable strength.

38

In all things be men

'If you've got a bit of muscle,
And enjoy a manly tussle,
Then go and put your flannels on and let the fun begin!'
Sedbergh School song

'If we abolished Games, as certain unhealthy people would have us do, and put in their place mere brain-work, we should soon cease to rule the sea and land and deserve to lose our Empire.' No one set out more fervently the case for the rugged sport ethic in Britain and its colonies than Eustace Miles, an assistant master at Rugby School and at one time the amateur tennis champion of the world. The best test of a man's ability to rule in India, he declared in 1902, was the captaincy of a football or cricket team. He noted approvingly that parents who wanted their sons to 'govern our colonies' sent them to be trained by men who had scarcely any qualifications 'apart from the fact that they are gentlemen and athletes and not absolutely ignorant'.

The British way of ruling, as he described it, was a superior moral mixture of stern humanity and Christian resolve combined with the positive qualities of games and the open air.

'When we take some new place,' he wrote of the process of colonization, 'we do not merely set up fortifications, government buildings and churches, we begin a cricket ground.' Rather than subjecting colonial people to an iron rule, 'we admit them to our own life and say "Come and play football" or "Have a try at cricket". Every one will admit that the Empire is ours by right of conquest, that it is very large and a mine of wealth. We had far better justify our imperial rule by its results, by showing that we are a good, healthy and fair nation, not bullies. Not only have Games helped us to gain an Empire, they have made the rulers fit to rule, healthy and honourable.'

Games, he said, also helped to train boys for war, for 'a bold dash by a soldier is almost exactly the same as a bold dash by a huntsman or a football-player. Games teach pluck. Englishmen are the pluckiest people in the world. If little Tommy is to become a real man he must be made to play Games.'

Miles also warned against the dangers of too many books in a boy's life. He preferred the rugby scrummage to the academic crammer. It was better for a boy to watch football than to listen to a lecture in a badly ventilated room or spend the afternoon at a music-hall or theatre. 'In a hot lecture-room and in an average church we notice that people's veins sometimes stand out with dark and poisoned blood, not bright red blood rich in oxygen. I would sooner have Jack a healthy athlete than a weedy pedant.'

The greatest of cricketers, W.G. Grace, thought along similar lines and once remonstrated with a player he found reading a book in the dressing-room. 'How', he demanded, 'can you make runs when you are always reading?'

Like many others, Eustace Miles saw public schools as integral to the imperial ethos, hothouses where viceroys, governors and district officers sprouted from seed. 'It would be terrible', he said, 'if our public schools were swept away or if Games were swept away from our public schools.'

In Britain, he pointed out, phrases like 'fair play' and 'play the game' had real meaning; and what made Tom Brown a man was a public-school education with its steely emphasis on honour and games. He did not mention cold baths.

Theodore Pennell would have agreed. He himself entered Eastbourne College as a delicate boy but left as 'a well-proportioned, spare and yet muscular man of six feet two inches'. In 1892 he went as a medical missionary to the North West Frontier of India, setting up in Bannu where he went on his rounds by bicycle. Utterly fearless, he stood up to preach the Christian message at the Bannu horse fair and was greeted by a hail of stones and clods of earth. He was undeterred. Among people notoriously well-armed he never carried a gun on the grounds that no true warrior would be cowardly enough to attack an unarmed man. Pennell cut a striking figure in the yellow robes of a holy man or the picturesque clothing of a Pathan, a costume he once wore to add drama to a lecture he delivered in Cardiff.

Pennell sought to change Pathan traditions by opening a school at Bannu to steep the boys in the British public-school ideas of *esprit de corps* and honour, quite different from the Pathan code of honour based on murderous vengeance and eternal vendetta. He inaugurated the new school swimming bath by diving from the high board, led the way in the compulsory morning swim, summer and winter, and dealt with boys who overslept by throwing them into the pool.

He was in the mainstream of those evangelical missionaries convinced of the value of cleanliness, rigour and games. Many men like him went out to the empire with soap and cricket balls. He himself played enthusiastically in the football and cricket teams he started at Bannu to encourage his pupils towards a nobler life. It was not easy, however, to get Pathans to accept the decisions of referees and umpires. It was harder still to get them to see the benefits of 'playing the game'. In the early days the football tournaments often erupted in 'hand-to-hand fights and in feuds nearly as bitter as the tribal vendetta'. But under Dr Pennell's patient tutoring the Bannu Mission players 'often showed a truly British spirit of magnanimity to their opponents'.

It nevertheless took some time for them to grow accustomed to the idea that because they were Pennell boys they were expected to set an example to others, often to their own disadvantage. Keen to see his students emerge like gentlemen from 'character-forming' defeats and humiliations, and to develop self-control, he was delighted by an incident when Bannu played a British army team at football. An officer violently threw one of the Bannu boys to the ground but 'not one of the Pathans lost his temper'. Pennell had his disappointments, of course, and awarded a black mark to the Peshawar football players who went into a huff after their defeat by the Bannu Mission and boycotted the hearty post-match sing-song.

One of Pennell's great adventures was a football tour. He led the Bannu boys, who played in pink shorts and blue shirts, on a rail trip to Bombay, Hyderabad and Calcutta. He was 'justly proud' of the dignity with which his hot-tempered Pathans bore 'unmerited defeat' as well as the 'unjust decisions' of referees.

Dr Pennell also took a party to visit Canon Cecil Tyndale-Biscoe's mission school at Srinagar in Kashmir. Athletic, passionate and driven, Tyndale-Biscoe took charge of the school in 1890 and for half a century devoted himself to turning Indian boys into strong and responsible men through a regimen of frequent washing, games, moral instruction and social work. The motto he chose for the school was 'In All Things Be Men'. Each pupil had a 'character sheet' on which the headmaster scored not only performance in history, mathematics and English, but also the qualities of pluck, 'colour of heart' and 'absence of dirty tricks.'

Robert Baden-Powell, founder of the Boy Scouts and a strong believer in public-school sport and manliness, praised Tyndale-Biscoe as 'a man who could put backbone into jellyfish'; but at first Tyndale-Biscoe had difficulty in persuading his pupils even to kick a football. A leather ball was taboo to high-caste Hindus and they refused to play with it. The headmaster was determined. He grasped a riding crop, armed his teachers with sticks and drove the reluctant players on to the pitch, ordering them to kick off. It was a bizarre scene. Many of the boys in that place of early marriage were bearded husbands and fathers, and they wore long gowns under which they carried fire pots, the traditional way of keeping warm in winter. At last they submitted to the threats of the menacing teachers and began to play. The boy who kicked the ball first was banned from his own home for being polluted by the unholy leather and had to live with relatives. For some time after that a few rebels carried needles to puncture the ball. It took Tyndale-Biscoe years to overcome the taboo and get the boys to enjoy football. He also persuaded them into trousers.

Both Pennell and Tyndale-Biscoe found that their scholars had their own interpretation of the games ethic. Pennell's pupils prayed for goals, runs and wickets. Tyndale-Biscoe discovered that one football squad had arrived with a priest to cast a spell on their opponents' goalkeeper to make him butter-fingered. A visiting tug-of-war team seemed invincible until Tyndale-Biscoe saw that they had tied their end of the rope around a heavy stone garden ornament. At cricket matches, opponents slyly moved the boundary back when fielding. On the other hand, a contented Pennell watched a match and saw that 'the technicalities of the game are observed with as much punctiliousness as in England, the white flannels show off well under the bright Indian sun, and but for the boys' dark faces and bare feet one might imagine he was watching a public school match in England'.

Pennell died at Bannu in 1912, aged forty-four. Khan Abdul Ghaffar Khan, who was born in 1897, was educated on the Frontier by British missionaries like Pennell. For half a century Ghaffar Khan was the outstanding leader of the Pathans, one of the figures in the forefront of the Indian independence movement in the 1930s and 1940s, and he was labelled 'the Frontier Gandhi'. He was almost ninety when he told me that as a boy he was so impressed by the missionaries' commitment to serving others that he decided to spend his own life trying to improve the conditions of his people. Like the missionaries, he sought to persuade Pathans to renounce their traditions of vendetta and violence. He, too, failed, but he is remembered for his ideals and integrity.

Belief in the virtues of the British public school and the religion of games was deep and widespread among colonial administrators. In India during the 1870s several schools were established to educate the sons of princes so that they could assist in managing the empire. The idea was to make them gentlemen-feudatories, familiar with the English language, the customs and mores of the British, and especially the manly sports. One of the first schools was the Rajkumar College in Gujarat, founded in 1870, a piece of transplanted Victorian England with its Gothic buildings, quad and large cricket field. The statue of J.W. Coryton Mayne, who was headmaster here for twenty years, suggests a man who knew the value of discipline, punctuality and a straight bat. At Mayo College in Rajasthan, another Indian Eton, founded in 1870, the boys were made to play cricket every day to improve their moral condition.

Lord Lugard, Governor-General of Nigeria from 1914, believed that just as the Romans civilized the barbarians of Britain, so British imperialism could educate, civilize and regenerate Africa; and he was convinced that the public-school system, with its emphasis on character-building games rather than 'mere book-learning', was the perfect means of doing so. He expected young Africans to learn respect and to know their subservient place in the new civilization. He did not think the schools needed to nurture inquiring minds.

Just as they transported their religion, the British took all of their games to the world, football, rugby, hockey and golf, and the lesser ones of tennis and badminton. But of all their games the pre-eminent, the most sacred, beautiful and satisfying, was cricket.

'Wherever men are English and the flag's unfurled,' sang the boys of Sedbergh School in Cumbria, 'You will there find cricket/And the willow and the wicket . . .' The boys of Tonbridge School in Kent could answer with their own anthem: 'The game of old England, the link with our home,/Wherever we wander, wherever we roam.'

Soldiers, administrators, teachers and traders – they all packed their bats as they set off for the imperial territories. The Calcutta Cricket Club founded in 1792 was the first outside Britain. The Indians sat and watched the strange spectacle. No one expected that they would have any interest in playing; and in any case the British for many years thought of cricket as a game into which they alone could withdraw and for a few hours get away from India, or Africa, Malaya or China. Watching cricketers running about on broiling days in Shanghai, the Chinese shook their heads and called the game 'imperial buffoonery', a description the British probably relished. Cricket's rituals and rules were for them a reminder of home. Buckling on their pads, shaping up to the fast ball, they distanced themselves from the people they ruled.

Indians, however, gradually took up the game. It suited them and, over the years, they embraced it. When India became independent some nationalists thought that the people would, or should, reject the game as a legacy of colonial rule. It was the vainest of hopes. The first Indians to play cricket were the Parsi merchants and civil servants in Bombay who started a club in 1848. The British, whose clubs excluded both Indians and socially unsuitable whites, were dubious about this development. For their part the Indians saw that the sahibs were mere men who could be beaten at their own game. When the Parsis defeated the England tourists in Bombay in 1889 a British officer wrote that the crowd only knew that 'the black man had triumphed over the white and they ran hither and thither, gibbering and chattering'.

By the 1890s there were dozens of Indian clubs. Lord Harris, Governor of Bombay, looked down his nose and remarked that batting was 'easier for the phlegmatic Anglo-Saxon than for the excitable Asiatic'. To his mind cricket in India was best between white teams, but he conceded that the British had introduced 'a manly game open to poor as well as rich'. The cricket historian Ramachandra Guha has revealed the heroic character of Palwankar Baloo, a low-caste bowler who took more than a hundred wickets on a tour of England in 1911, having learnt his skill as an Englishman's personal bowler, serving up thousands of deliveries.

The glamorous Indian cricketer K.S. Ranjitsinjhi, a product of the Rajkumar school in Gujarat, went to Cambridge in 1891, played for Sussex and England, and was co-opted as an honorary Englishman. 'Ranjitsinjhi,' went the song, 'All the way from Inji ... Though foreign his name in this English game he's certainly one of we.'

In cricket Indians found an enduring passion. W.G. Grace, who toured Australia twice and made missionary cricket expeditions to the United States and Canada, was convinced of the game's value as international cement. Cricket tours, he said, 'helped to deepen British interests in our colonies and to bind us in closer harmony with other nations ... the good fellowship born on the cricket field has done more than is recognized to knit together the various sections of the British Empire and to advance the cause of civilization'.

Many felt, with him, that cricket was a blessing that the British spread around the world. The Colonial Office in late Victorian times had a strongly cricket-minded outlook. Sir Ralph Furse, who served for more than forty years in the Colonial Office, drew up a list of British works and achievements: 'The abolition of slavery, the suppression of cannibalism and tribal warfare, the campaign against disease and want, the example of justice and fair play, cricket and the rule of law.'

The public-school sporting ethos, however, had its critics. Herbert Branston Gray wrote a book in 1913 attacking the public schools that produced 'a useless drone trained only to wield a willow or kick a bladder'. Kipling derided the 'flannelled fools at the wicket' and pilloried the cult of games:

> How is this reason (which is their reason) to judge a
> scholar's worth,
> By casting a ball at three straight sticks and defending the
> same with a fourth?

Such critics, however, seemed to be a small and carping minority to those who believed that the empire was best run by straightforward, clubbable games players who inoculated themselves against the dangers of being sucked into local cultures, of going native, with regular doses of cricket, tennis, golf and games at the club. Of course, there were always men who slipped into bad ways and stayed at home reading books and writing poetry. John Lawrence, who ruled the Punjab in the 1850s, heard that one of his officers had a piano. He vowed to smash it and made the man move post six times in two years.

Wherever the British set foot, racecourses, cricket grounds and golf courses were laid out. Tennis and squash courts were, like the churches, part of the standard kit of military cantonments. Indian boys learnt to be excellent squash players by playing, the very hard way, with the shattered balls discarded by British players. Army officers had ample leisure for sports of all kinds, for riding, jackal-hunting, polo and shooting as well as cricket. And in the club and the mess, too, there were rough, rowdy, masculine games.

Philip Mason, in *The Men Who Ruled India*, wrote that the men of the Indian Civil Service, the 'guardians' of developing India, knew that the country would one day be free, just as Macaulay, Queen

Victoria and Gladstone had said. But in the meantime there was a need for guardianship.

> It was to peace and unity rather than freedom, that the effort in India was directed, to equal justice for all, roads, railways, canals, bridges. That was the mixture, very good for the child, to be given firmly and taken without fuss. And to give it was a corps of men brought up in a rigour of bodily hardship to which no other modern people have subjected their ruling class, trained by cold baths, cricket and the history of Greece and Rome, aloof, superior to bribery, discouraged from marriage until they were middle-aged . . .

Men who endured lonely lives as district officers, he said, accepted the idea of hard work and hard play and believed that rugged exercise kept them sane and balanced. In his view, 'the game of games' was pigsticking, spearing wild pigs from the back of a horse. 'Pigsticking was purgation. The danger and excitement, the ferocity thus harmlessly given an outlet, sweetened men who might otherwise have been soured by files and hot weather and disappointment. Good pigstickers were usually good officers. The district officer who spent all his time with his nose in his files did not always have his district in good order.'

Games, hard exercise and a macho culture presumably filled many a need among celibate and family-less men. Clive Dewey, in his *Anglo-Indian Attitudes*, identified the small cadre of mandarins in the Indian Civil Service as a ruling class who were intellectuals 'yet pretended to be men of action to escape the stigma attached to cleverness by the late-Victorian middle class. "Character" counted, not brains.'

39

Something in the air

'It is good for man
To try all changes, progress and corruption, powers, peace
and anguish, not to go down the dinosaur's way
Until all his capacities have been explored: and it is good for
him
To know that his needs and nature are no more changed in
fact in ten thousand years than the beaks of eagles.'
Robinson Jeffers, 'The Beaks of Eagles'

A century after the Great Exhibition of 1851 the British steel industry published a boast-book saluting 'the merchant adventurer spirit' of British civil engineers and chronicling the achievements of 'constructional engineering in the service of world civilization'. Its list of great bridges and railways lent substance to Rudyard Kipling's suggestion that 'transportation is civilization'. Pictures of telegraph stations, dams, docks, tunnels and harbours displayed the empire of nuts and bolts reality rather than the empire of ostrich-plumed surreality. Viceregal pomp and tight trousers vanished. The bridges endure.

Even a small selection from the list demonstrates the reach and ambition of British engineers and the opportunities for investors. In 1853 the Ganges canal opened, work started on the Victoria bridge in Montreal and the first Indian railway was built. Calcutta's streets were lit by gas in 1857, a harbour constructed at Cape Town in 1858 and railways in Cape Colony in 1859. At this time British capital went abroad at the rate of £30 million a year. Engineers built a railway from Colombo to Kandy in 1863 and started the

Central Argentine railway in 1864. Two years later, telegraph cables connected Rangoon, Singapore, Sumatra, Malacca, Australia and China. The Saõ Paolo railway was built in 1867 and more bridges and irrigation canals were completed in India. Karachi harbour was finished in 1873, Colombo harbour started and the first railway in Mexico completed. Engineers began work on the Canadian Pacific Railway. Major sanitary works were undertaken in Buenos Aires in 1878, and the Admiralty dockyard was begun in Hong Kong in 1882. In the 1880s the Kaisar-i-Hind bridge was built over the Sutlej, the Dufferin over the Ganges at Benares, the Lansdowne over the Indus at Sukkur, and the bridge over the Bio Bio in Chile. The Bangkok–Paknam railway was completed in 1893, the Periyar dam finished in south India in 1895 and the Uganda railway started. In Egypt in 1898 work started on the first Aswan dam, one of the mightiest of all British constructions, designed by Benjamin Baker, engineer of the Forth Bridge. The Uganda railway reached Nairobi in 1899 and in 1904 engineers began the Victoria Falls bridge, siting it dramatically as Cecil Rhodes commanded, so that railway carriages would shine in the spray thrown up by the thundering Zambezi. There was much more. The adventure and romance of Victorian engineering so inspired Frank Hornby, born in 1863, that in 1901 he invented the Meccano construction kits that became popular with generations of British schoolboys.

Frederick Braby and Company, of London, Liverpool, Glasgow and Bristol, was typical of firms which specialized in iron and steel construction abroad. In 1863 it built all the railway stations on the Tudela-to-Bilbao line in Spain. It shipped great quantities of its 'Empress brand' iron sheets, cast-iron columns and steel frames to build pioneering colonial structures all over the world. It built water-towers, footbridges, railway workshops, stations and warehouses for grain, cotton, rubber and sugar. Its construction teams erected offices and staff bungalows at gold mines, tin mines and tea and coffee plantations. Planters and district officers drank their sundowners on Braby verandas. Telegraph operators tapped out their Morse in Braby telegraph and cable buildings in the Pacific islands and Australia, and soldiers lived in Braby iron huts in South Africa. Braby fashioned the copper dome of the Bengal legislature, zinc roofs for docks in Gibraltar and

Buenos Aires, temple cupolas and the seventeen-foot terminal of the Namirembe cathedral in Kampala, a copper globe surmounted by a cross.

The company also built churches which marked the march of missionaries across Africa, and mission schools which educated many young Africans. These were usually the only available gateway to literacy, and many twentieth-century political leaders, Kenyatta, Nkrumah, Banda, Mugabe and Kaunda among them, learnt English and British history in their classes. Nelson Mandela always acknowledged the influence of British culture and ideas; and recalled that when he attended a mission school 'Britain was the home of everything that was best in the world'.

At the beginning of this book I imagined a person living through the nineteenth century, the witness of the extraordinary evolution of the sinews of empire reaching across a world that in his youth was largely unexplored. Even in 1850 there were still huge blank spaces beyond the charted coastlines. By 1872, however, there were transport links in place to enable Thomas Cook, from his office at 98 Fleet Street, to arrange the first round-the-world package tour, a Victorian dream turned into reality. That year a French newspaper serialized Jules Verne's *Around the World in Eighty Days*. Cook's tourists took three times longer than Phineas Fogg. Cook, who called himself 'the Conductor', led the adventure himself.

On the first leg the tourists crossed the Atlantic from Liverpool in the new Belfast-built White Star liner *Oceanic*, 2,200 tons and as graceful as a yacht. Its comforts included a grand saloon, coal fires and a 'narcotic paradise' of a smoking-room. It had electric bells for summoning stewards, oil lamps rather than candles, the wonder of baths and the luxury of steam heat. 'No work', said the brochure, 'is too great for the giant, Steam. He warms the child's berth; he weighs the anchor; he turns the barber's brush . . .'

From New York Cook's tourists took the train to Chicago, St Louis and Denver, enjoying George Pullman's sleeping and dining cars, watching Indians riding on the prairie and experiencing the American phenomenon of mingling with all classes, what the actress Fanny Kemble, on another occasion, called the 'most painful proximity to coarse and unpolished brethren'. The train ascended the Rocky Mountains and descended to San Francisco where the travellers boarded a paddle steamer for the twenty-four-day voyage to Yokohama. From Japan they went by P&O steamer to Ceylon and India, and by P&O again from Bombay to Egypt.

P&O steamers were comfortable enough but still relatively small and lit by candles. In 1881 the company launched its first vessel of more than 5,000 tons and three years later fitted a ship with electric lighting. At the same time Cunard installed electric light in its new Atlantic liners *Umbria* and *Etruria*. Electricity brought the benefits of refrigeration which, at last, foreshadowed the end of the ocean farmyards of hapless cows, sheep, chickens and ducks that provided fresh meat. Electricity reduced the risk of fire. Safety at sea was also improved by the campaigns of Samuel Plimsoll in the 1870s and 1880s. Against the opposition of shipowners he pushed measures through Parliament to outlaw the 'coffin ships' so cynically overloaded by their owners. It took a long time for owners to improve living conditions: in 1914 the Port of London medical officer said many seamen lived like prehistoric cavemen. Passenger steamers, however, became steadily more opulent, with electric light, heating and better lavatories. First-class passengers enjoyed spacious mahogany-panelled cabins, swanked in mirrored saloons and havered over a breakfast choice of steak and onions, mutton chops, omelettes, herrings, porridge and Irish stew.

In the last two decades of the nineteenth century Europeans filled in the gaps and completed much of their exploration of the world. Scientists, surveyors and missionaries helped soldiers, traders and administrators to secure cultural dominion over hundreds of millions of people. In the Scramble for Africa, Britain, France, Germany, Italy, Portugal and Belgium behaved as if participants in a grotesque board game, half-crazed by ambition, seizing new territories to thwart their competitors. Years before, David Livingstone had described his vision of an Africa civilized by trade and Christianity. For twenty years or so from the late 1870s empire-builders marched in Africa with an evangelical conviction that they were taking light into pagan darkness and might find great wealth along the way. They carved up the continent and gave it its modern shape.

In those years Africa and much of the rest of the world were gridded, charted, wired, railed, garrisoned, divided, tamed, managed and made productive. A quarter of the world's territory and a quarter of its people, 372 million subjects of the Great White Queen, belonged to the British Empire. A drawing in *The Graphic* in June 1897 showed Queen Victoria, on the morning of her diamond Jubilee, sitting in a wheelchair and pressing a button to send a telegraph message. So thoroughly was her empire wired that the Queen's words of thanks and blessing to her 'beloved people' were soon being read across the globe. They helped to give a sense of coherence and meaning to the largest empire in history. It was an immense jigsaw of territory, much of it acquired haphazardly, but usually for raw materials and profit. Jawaharlal Nehru thought that the British represented mighty forces which they themselves hardly understood; but in 1897 the empire seemed to many in Britain and in her distant colonies to have a shape and meaning as a structure nobly governed by a chosen people, improvers, God's civilizers, By Appointment.

Seven years before, the United States Census Bureau had announced the closing of the American frontier. Around that time six new western states, Montana, Idaho, Wyoming, Washington, North Dakota and South Dakota, entered the Union. Only Utah, Oklahoma, Arizona and New Mexico waited to join the contiguous United States. There were still wild places, but the wilderness had largely been explored and settled, the manifest destiny fulfilled. In 1804, when Lewis and Clark set out into the great uncharted, the land west of the Missouri was empty of white men. At the end of the century 30 million people were living there, a third of the American population.

Guns, it was said, had 'won the West', Samuel Colt's revolvers and Oliver Winchester's rifles among them. 'Buffalo Bill' Cody wrote to Winchester that 'for general hunting or Indian fighting' the Winchester '73 was the boss. 'While in the Black Hills Mr Bear made for me and I am certain had I not been armed with one of your rifles I would now be in the happy hunting grounds. The bear was not thirty feet from me when he charged, but before he could reach me I had eleven bullets in him, which was more lead than he could comfortably digest.'

Guns played a part, but it was wire that brought order to the frontier and resolved the difficulties of enclosing the prairie. In the wooded east of the country farmers had plenty of fencing material. But on the treeless plains there was neither wood nor stone to make boundaries and fences to protect stock and crops from the cattlemen's roaming herds. Fencing cost almost as much as land itself. A dollar spent on livestock had to be matched by a dollar for fences. In 1871 fencing costs equalled the national debt. Quarrels over boundaries and encroachment raised tensions on the frontier.

A hedge of osage orange thorn was a partial answer, but it grew slowly, was costly, and harboured vermin and insects. Better were the varieties of barbed wire that imitated thorn hedges; and best was the wire developed by three men at De Kalb, Illinois, in 1873. Mass-produced, cheap and long-lasting, it made fortunes for its manufacturers, enabled farmers to protect their homesteads from the cattle herds and opened the Great Plains to sensible and affordable agriculture. Cattle raisers, who saw themselves as noble pioneers, complained that fencing eroded their freedom, by which they often meant their tyranny. After some years of strife peace came and, by 1890, most western land was in private ownership. Barbed wire ended the romance of the open range, the freewheeling cattleman, the cowboy and the long cattle drives; another of history's details that made a difference.

So, too, did refrigeration. In 1879 the first American beef reached Britain in the refrigerated compartment of a ship. The year before, Argentina shipped eighty tons of frozen mutton to France. Two Scots, Andrew McIlwraith and Malcolm McEacharn, launched the Australian trade in frozen beef, mutton and butter in 1880, installing a cold chamber in the steamer *Strathleven*. A dinner held aboard the ship when she docked in London was the talk of the town: in more than one sense frozen meat had arrived. In 1882 the *Dunedin* carried New Zealand's first frozen meat to Britain, 5,000 carcases of mutton and lamb, thereby transforming the colony's economy and securing the small farmer's future. By 1890 South Africa started shipping fresh fruit to Britain. The first load of bananas came from Jamaica in 1896.

Although steam was everywhere driving sail from the seas, there were still routes where sailing ships prospered until the First World War. We have seen how ships like the *Cutty Sark* sailed Australian wool and wheat to Britain until the steamers took the trade. Sailing ships, however, were fast and economical for the transport of coal from Newcastle in New South Wales 6,200 miles across the Pacific to Chile. Driven by the dependable winds of the Roaring Forties, 2,000-ton sailing vessels could reach Valparaiso in thirty days. They sailed back to Australia to fetch more coal or loaded cargoes of Chilean nitrates and sailed by way of Cape Horn to deliver them to explosives and fertilizer factories in Germany. Sailing ships also made profits where steamships could not by carrying timber from Western Australia to South America, Africa, India and China; and also to Britain where hardwood blocks paved city streets.

Even in the 1920s and 1930s there remained a tenuous niche for ocean sail. On the edge of war, in 1939, thirteen three- and four-masted barques gathered in Spencer Gulf, South Australia, to load wheat for Europe, eleven of them sailing by way of Cape Horn. It was the last grain race. Eric Newby, who took part in it as one of the twenty-eight crew aboard a Finnish ship, wrote dramatically of the hard and dangerous work aloft on the yards, of setting and re-setting sails, the heaviest of which weighed one and a half tons. His ship was first home, running to Queenstown in ninety-one days.

It was Britain's maritime might that attracted twenty-two-year-old Guglielmo Marconi to London in 1896. Britain possessed more than three-fifths of the world's shipping tonnage, fleets of liners steamed to schedule and hundreds of tramp steamers were directed by the worldwide telegraph network to pick up cargoes. For Marconi Britain was where the money was.

Marconi was the son of a wealthy Italian landowner and an Irish mother who was a member of the Jameson whiskey family. When he arrived in Britain he carried two bags containing the simple wireless transmitter and receiver he had built in the loft of his home in the countryside near Bologna. He was not the inventor of radio; rather, he was its inspired and determined developer.

Marconi was not well-educated, had been regarded at school as less than bright, was no scientist or academic and failed his university entrance examination. He was, however, interested in electricity, and the course of his life was set in 1894 when he read the obituary of the German physicist Heinrich Hertz.

Hertz had built a battery-powered transmitter and receiver to demonstrate the existence of electromagnetic waves. In doing so he proved the theories of James Maxwell, a Scot who gave a mathematical explanation of Michael Faraday's ideas about electricity. Other physicists, Oliver Lodge, William Crookes and the New Zealander Ernest Rutherford, the father of atomic physics, also worked on the principles of wireless telegraphy. Marconi had an engineering mind and was encouraged in his trial-and-error experiments with electromagnetic waves by the physicist Professor Augusto Righi. He worked doggedly to extend the range of transmission and in 1895 sent a signal over a mile and a half. Failing to interest the Italian navy in his experiments he went to London. He believed that wireless telegraphy had a future in shipping.

There he met William Preece, chief engineer of the Post Office, who saw the possibilities. Preece had brought the first telephones, made by Alexander Graham Bell, from the United States in 1877. He arranged for Marconi to send a signal from the Central Telegraph Office in St Martin's le Grand which rang a bell at the Post Office bank in nearby Carter Lane. In further tests Marconi increased the distance. In 1897 he sent signals over the Bristol Channel from Penarth to Flatholm Island and to Brean Down near Weston-super-Mare, and from the Isle of Wight to Bournemouth and Poole. A shrewd publicist, he also set up a wireless link in 1898 between the Prince of Wales's yacht at Cowes and Queen Victoria in Osborne House. The Prince sent bulletins to his mother on his knee injury.

In 1899 Marconi radioed across the English Channel from South

Foreland to Wimereux. The first radio distress call was made from the East Goodwin lightship in the English Channel asking for help for a ship that had run aground. In 1901 a wireless transmitter was fitted into a merchant ship for the first time.

Building more powerful transmitters, larger aerials and better receivers, Marconi increased the distance and showed that the earth's curvature was no barrier to radio waves. In December 1901 he went to Heart's Content, Newfoundland, where his receiver picked up the Morse letter S, three dots, sent from Poldhu in Cornwall. In later experiments he conclusively demonstrated to sceptics that transatlantic wireless telegraphy was a fact.

Two events secured wireless in the public imagination. In 1910 Dr Crippen, having poisoned and dismembered his wife, Belle, in London, assumed the name of Robinson and sailed for Canada aboard the liner *Montrose*. With him was Ethel Le Neve, dressed as a boy and posing as his son. The liner's captain, Henry Kendall, who had read of the murder, saw the 'father and son' holding hands. He sent a wireless message to Liverpool. Chief Inspector Walter Dew of Scotland Yard boarded a ship and set off in pursuit. It arrived at Quebec ahead of the *Montrose* and Dew went aboard and confronted his quarry. 'Good morning, Dr Crippen,' he said. Crippen and Ethel Le Neve, reported *The Times*, were 'encased in waves of wireless telegraphy as securely as within the walls of a prison'.

It was a sensation. Two years later there was a greater one. Distress calls from the *Titanic* summoned the *Carpathia* to the place where the great liner sank and she picked up hundreds of survivors. Without Marconi's wireless, *The Times* said, 'the ship might have passed from our human ken, her fate for ever unknown'. No one now doubted the importance of radio.

'For some years', wrote Wilbur Wright in 1900, 'I have been afflicted with the belief that flight is possible to man. My disease has increased in severity.' At Kitty Hawk on 17 December 1903 he and his brother Orville tossed a coin to decide which of them should make the first powered flight in the biplane they confidently called the Flyer. Orville won the toss. His historic hop of 120 feet in twelve seconds was bettered that same day by Wilbur's flight of

852 feet in just under a minute. They flew in suits, starched collars and ties.

The Wright brothers were modest bicycle mechanics who became self-taught aeronautical engineers and pioneered both the reality and the science of aviation. With their short flights machine transport entered its third dimension. Their swoops, however, were witnessed by only five people and certainly did not make newspaper headlines. Much more definitive as a beginning of aviation was Wilbur's flight in France, five years later, when, in front of a crowd, he gave a display of controlled flying. In that year he made a hundred flights, including one of seventy-seven miles. The following year Louis Blériot flew the English Channel to Dover in thirty-seven minutes, a timely shower preventing dangerous overheating of the engine.

In the romantic infancy of aviation the British were daring and enthusiastic pilots. From 1919 Royal Flying Corps veterans flew passengers to Paris and Brussels, and the fare was the same as a first-class rail and steamer ticket. Pilots wore leather coats and goggles and relished the open cockpits and the wind in their faces. They navigated by railway, and the station roofs had names painted on them to help pilots find their way. One aircraft made twenty-two forced landings on a flight to Paris. In 1919 the Mancunian John Alcock and the Glaswegian Arthur Brown flew a Vickers Vimy bomber across the Atlantic from Newfoundland to Galway; and later that year the Australian brothers Ross and Keith Smith flew a Vimy from England to Australia by way of India in twenty-nine days.

Imperial Airways was founded in 1924 with a flight to Paris and a bitter quarrel over pilots' pay. Passengers sat in wicker chairs in the unheated cabin. Not all aircraft had a lavatory. Regulations, introduced to improve safety, insisted that aircraft carry a wireless operator and, for flights of over a hundred miles, a crew member proficient in navigation. A new and stiffer medical test cost a one-eyed war veteran his pilot's job. As a stunt to promote air travel, an airliner and the Royal Scotsman left London for Edinburgh at the same time. The idea was that the air passengers would be waiting on the platform when the train steamed in. As it turned out, the train passengers waited for the aviators.

To live up to its name Imperial Airways was determined to forge the route to India. A furrow was ploughed across the desert from Amman to guide pilots to Baghdad. In case of forced landing, emergency petrol was placed in underground caches, locked to foil desert tribesmen who, although they had no use for the petrol, shot off the locks to get at the prized cans. Since there were many forced landings, the planes carried water and rations, and a survival book advised passengers not to drink their urine.

In 1929 the service reached Delhi. From the 1930s aircraft linked most of the empire, flying to Egypt, East Africa, West Africa, South Africa, Australia, India, Burma, Singapore and Hong Kong. In a fairly short and leisured age Imperial Airways flying-boats carried the citizens and servants of the empire to their homes and duties. In the nature of things the service was not always efficient and punctual, but the style was grand, more like Imperious Airways, and an extension of British sea power. Long flights with overnight stops seemed civilized and romantic, but such costly travel was for a fortunate minority. I had a taste of it once, an eccentric adventure in which I flew in a Catalina flying-boat, built in 1944. We set out to follow the old Imperial route through Africa. We took off from the Nile at Cairo and passed the Pyramids which flashed gold in the sunset, followed the ribbon of the Nile and landed at Luxor for the night. Next day we flew to Aswan and stayed in the Old Cataract Hotel, built by Thomas Cook in 1889. The Catalina made its swan-like amerrissage on the Nile at Abu Simbel and then flew to Khartoum and across Sudan and northern Kenya to Lake Turkana, a jade sea in a red land, first seen by white explorers in 1888. From there we flew on to Nairobi and then to Arusha where an engine breakdown put an end to the journey.

During the 1960s the jet airliners became supreme and carried more people across the oceans than the ships. Heathrow airport and JFK superseded Southampton and Manhattan as the gateways to the Atlantic. The twinkling traffic of American film stars switched swiftly from sea to sky, disembarking from the *Queen Elizabeth*, the *Queen Mary* and the SS *United States* and boarding the jumbo. As a glamorous social institution and indicator of rank the captain's table faded away.

The breadth of the Atlantic shrank from five days to eight hours of apparently hanging in the seamless sky. In 1840, Bostonians had hailed Samuel Cunard and his new Atlantic service and had exultantly drunk a toast 'to the memory of Time and Space, famous in their day and generation, but now annihilated by the Steam Engine'. How astonished they would have been to see time telescoped by the jet engine.

Quite soon the fleet of liners dwindled, their majesty swept into the past. The sea itself diminished in its significance, no longer a place of prolonged terrors and monstrous mobile mountains. It could be crossed during a night's sleep and not even seen, an old ordeal transformed into a pleasant slumber. An airline pilot told me that he regarded his flight deck as his office, the controls and instruments as his desk. Once, though, just before take-off for New York, a captain announced that we were just about to embark on 'a voyage into that great ocean of air we call the sky'.

He must have been a romantic. In the earliest days of aviation there was some poetry about the freedom of flying and soaring above the earth; but not much and certainly not now. It is, as ever, the sea that fills the imagination and anthologies. Crossing the North Atlantic under sail in the summer of 2001 I saw in my mind's eye the thousands of people flying five or six miles above us and knew that some were gazing at the immense disc of the ocean, for, in ways the sky can never be, the awesome sea is the focus of contemplation.

From my place among the running waves it seemed that the tiny spearheads of the airliners encapsulated everything that many of the men and women in my story could only imagine and ache for: the exhilaration of velocity, the subjugation of distance, the world explored, known, collated and charted, the mysteries plumbed. The skyfarers had superseded the seafarers from whom they sprang and it was the blue sky, not the dark blue ocean, that was scored with white wakes. The jets flew on the pinnacle of knowledge, on the minute coral-like accretions contributed by all those scientists, engineers, navigators and seekers, the finders of longitude, the developers of metal and power, the makers of the sinews. Thus the planes incorporated the spirit and achievement of Cook, Harrison, Watt, Stephenson, Brunel, Faraday, Marconi, the Wrights, Whittle and myriad others; and symbolized the dreams and quests of our common human adventure story.

In a few hours the travellers above us would be dining in London or New York. We vagrants had weeks to go before we would see a tablecloth and a knife and fork that did not slide from port to starboard. In the meantime we had the compass and the company of whales and wheeling birds, the uncertainties of the Atlantic, the extraordinary vividness of ocean dreams, a hope for flying fish and auspicious gales, the old struggle with the ageless sea where it all began; and the gift of time.

Acknowledgements

I owe warm thanks. Editors and foreign editors sent me to many parts of the world and encouraged and indulged my ideas. Grant McIntyre at John Murray was the genial mover behind this book. Gail Pirkis edited with her customary grace. As ever, Penny, my wife, was the heart of things.

In years of reporting I have mined many books. Rather than a formal bibliography, I am listing works I have found informative and enjoyable. Unless otherwise noted, they were published in London.

Especially useful were Jan Morris's masterly *Pax Britannica* trilogy; Lawrence James's *The Rise and Fall of the British Empire*; five of Alan Moorehead's books, *The Blue Nile*, *The White Nile*, *The Fatal Impact*, *Cooper's Creek* and *Darwin and the Beagle*; and E. J. Hobsbawm's *The Age of Revolution*, *The Age of Capitalism* and *The Age of Empire*.

For Africa I referred to Hugh Thomas, *The Slave Trade*; Thomas Pakenham, *The Scramble for Africa*; John Hatch, *The British in Africa*; Elspeth Huxley, *Livingstone*; Charles Miller, *The Lunatic Express* (New York); William Kingston, *Great African Travellers*; Brian Roberts, *Kimberley: Turbulent City* (Cape Town); and *The Autobiography of H.M. Stanley* and Dr Emyr Wyn Jones's critical dissection, *Sir Henry Stanley: The Enigma* (Denbigh).

For the American West I referred, among others, to Robert Hughes, *American Visions*; John Hawgood, *The American West*; Thomas Schmidt and Jeremy Schmidt, *The Saga of Lewis and Clark* (California); Mark Sufrin, *George Catlin* (New York); Philip L. Fradkin, *Stagecoach* (New York); and Marquis Childs, *Mighty Mississippi* (New York). *The New Encyclopaedia of the American West* (New Haven) is addictive. For the Welsh Indians I consulted Professor David Williams's remarkable lecture of 1948 to the Honourable Society of Cymmrodorion, London; Gwyn A. Williams, *Madoc: The Making of a Myth*; and Emyr Humphreys, *The Taliesin Tradition.*

For Australia I studied Geoffrey Blainey's *The Tyranny of Distance*, Manning Clark's *History of Australia* and Robert Hughes's *The Fatal Shore*. Also helpful were Strathearn Gordon and Barnett Cocks, *A People's Conscience*; Bryce Moore, *The Voyage Out: 100 Years of Sea Travel to Australia* (Fremantle); Michael Cigler, *The Afghans in Australia* (Melbourne); Doris Blackwell and Douglas Lockwood, *Alice on the Line* (Alice Springs); Nicolas Baudin, *Terre Napoleon: Australia Through French Eyes, 1800–1804* (Sydney); Geoffrey Dutton, *Australia's Last Explorer: Ernest Giles*; H.H. Wilson, *Westward Gold!* ; John Clay, *Maconochie's Experiment*; and Hans Mincham, *The Story of the Flinders Ranges* (Adelaide). *The Explorers*, by Tim Flannery, is an excellent anthology of Australian journeys.

For Canada and northern exploration I consulted Samuel Hearne's work of 1795, *A Journey from Prince of Wales's Fort in Hudson's Bay to the Northern Ocean*; Barry Lopez, *Arctic Dreams* (New York); Hugh Brody, *Living Arctic* (Vancouver and Toronto); *Arctic Animals* (Northwest Territories Renewable Resources); *Frozen in Time* and *Dead Silence*, both by Owen Beattie and John Geiger; Fergus Fleming, *Barrow's Boys*; Roland Huntford, *The Last Place on Earth*; Glyn Williams, *Voyages of Delusion: The Search for the Northwest Passage in the Age of Reason*; *Company of Adventurers*, Peter C. Newman's tremendous chronicle of the Hudson's Bay Company; *The Last Spike* (Toronto), Pierre Berton's epic of the Canadian Pacific Railway; William M. Baker (ed.), *The Mounted Police and Prairie Society, 1873–1919* (Regina); S.W. Horrall, *The Pictorial History of the Royal Canadian Mounted Police* (Toronto); and Wallace Stegner, *Wolf Willow* (New York).

Books on India included Percival Spear, *The Oxford History of Modern India*; O.H.K. Spate, *India and Pakistan*; Mildred Archer, *Early Views of India: The Picturesque Journeys of Thomas and William Daniell, 1786–1794*; Antonio Martinelli and George Michell, *Oriental Scenery*; Vidya Dehajia, *Impossible Picturesqueness* (Ahmedabad); Michael Jacobs, *The Painted Voyage: Art, Travel and Exploration*; Peter Quennell (ed.), *The Memoirs of William Hickey*; *William Russell Special Correspondent of The Times*, introduced by Max Hastings and edited by Roger Hudson; William Russell, *My Diary in India*; Clive

Dewey, *Anglo-Indian Attitudes*; Dennis Kincaid, *British Social Life in India, 1608–1937*; William Dalrymple, *White Mughals*; Lord Carver, 'Wellington and His Brothers' (Southampton); Ian J. Kerr, *Building the Railways of the Raj*; John Noble Wilford, *The Mapmakers* (New York); Matthew Edney, *Mapping an Empire* (Chicago); and John Keay, *The Great Arc*. For the broader story of cartography there were Denis Cosgrove, *Apollo's Eye* (Baltimore); and Peter Whitfield, *New Found Lands: Maps in the History of Exploration*. I was guided by A.N. Porter (ed.), *The Atlas of British Overseas Expansion*.

For the Pacific I used J.C. Beaglehole's monumental *Life of Captain James Cook*; Alan Villiers, *Captain Cook: The Seaman's Seaman*; Richard Hough, *Captain James Cook*; A. Grenfell Price (ed.), *The Explorations of Captain James Cook in the Pacific* (New York); Hugh Paget, *To the South There is a Great Land* (Sydney); and *Captain Cook's World*, John Robson's wonderful book of charts. Bernard Smith's *European Vision and the South Pacific* (New Haven) is a towering and vital work. In his discursive *Passage to Juneau*, Jonathan Raban writes perceptively about Captain George Vancouver. I also referred to Steven Roger Fischer, *A History of the Pacific Islands*; Jonathan Lamb, Vanessa Smith and Nicholas Thomas (eds.), *Exploration and Exchange*; Keith Sinclair, *The History of New Zealand*; and John Harrison, *Where the Earth Ends*.

I tripped over Joseph Banks everywhere. Among others I drew on the biographies by Hector Charles Cameron and Patrick O'Brian. For more botany I used Henry Hobhouse, *Seeds of Change*; Maggie Campbell-Culver, *The Origin of Plants*; and Clements Markham, *Peruvian Bark*.

My chief sources on animals were Major Arthur Glyn Leonard's work of 1894, *The Camel*, engaging and magisterial; the two-volume *History of the Royal Army Veterinary Corps, 1796–1961*; and the Marquess of Anglesey's splendid five-volume *History of the British Cavalry*.

John Falconer, Curator of Photographs in the British Library's Oriental and India Office Collections, educated me in the story of photography in India when I interviewed him for a newspaper article and I have referred to that as well as to his *India: Pioneering Photographers, 1850–1900* and *India Through the Lens, 1840–1911* (Washington, DC). I also referred to James R. Ryan, *Picturing Empire* (Chicago), and drew on Samuel Bourne's own vivid account of his photographic odyssey for the *British Journal of Photography.*

On mails and telegraphs I read John Sidebottom, *The Overland Mail* (Postal History Society); Howard Robinson, *Carrying British Mails Overseas*; Krishnalal Shridharani, *The Story of the Indian Telegraph* (Government of India Press); Hugh Barty-King, *Girdle Round the Earth*; John Steele Gordon, *A Thread Across the Ocean*; Laszlo Solymar, *Getting the Message*; Hari Williams, *Marconi and His Wireless Stations in Wales* (Carreg Gwalch); and James Hamilton, *Faraday.*

From the vast literature of the sea and ships I consulted, among others, N.A.M. Rodger, *The Wooden World: An Anatomy of the Georgian Navy*; Douglas Phillips-Birt, *A History of Seamanship*; Ian K. Steele, *The English Atlantic, 1675–1740*; Paul Butel, *The Atlantic*; Ronald Hope, *A New History of British Shipping*; John Malcolm Brinnin, *The Sway of the Grand Saloon* (New York); David Howarth and Stephen Howarth, *The Story of P&O*; Ewan Corlett, *The Iron Ship*; Robert K. Massie, *Dreadnought*; Basil Greenhill and Ann Giffard, *Steam, Politics and Patronage: The Transformation of the Royal Navy, 1815–1854*; *The Nelson Dispatch*, the journal of the Nelson Society; Colin White, *The Nelson Encyclopaedia*; Andre Charbonneau and Andre Sevigny, *Grosse Ile: A Record of Daily Events* (Parks, Canada); Duncan Crewe, *Yellow Jack and the Worm* (Liverpool); Alan Villiers, *Cutty Sark*; and Eric Newby, *The Last Grain Race.* I also drew on Basil Lubbock's *The Colonial Clippers, The Western Ocean Packets, The Log of the Cutty Sark, The Down Easters, The China Clippers, The Opium Clippers, The Nitrate Clippers* and *The Arctic Whalers*, all published in Glasgow.

Ships, railways and bridges overlapped in the biographies of Isambard Kingdom Brunel by L.T.C. Rolt, Adrian Vaughan and R. Angus Buchanan. At hand were Asa Briggs, *The Power of Steam*; George Basalla, *The Evolution of Technology*; Nicholas Faith, *The World the Railways Made*; Judith Dupre, *Bridges* (New York); and J.R.Harris, *The Copper King* (Liverpool). For the impact of coal, iron and oil I also used John Davies's grand *History of Wales*; William D. Jones, *Wales in America: Scranton and the Welsh* (Cardiff); Stephen Howarth, *A Century in Oil*; and Daniel Yergin, *The Prize* (New York).

For the fight against disease I referred to Sheldon Watts, *Epidemics and History* (New Haven); Andrew Spielman and Michael d'Antonio, *Mosquito*; and Clements Markham, *A History of Peru.*

For sport I consulted the autobiography of Theodore Pennell;

J.A. Mangan, *The Games Ethic and Imperialism*; *A Corner of a Foreign Field*, Ramachandra Guha's fine history of Indian cricket; Simon Rae's tremendous *W.G. Grace*; and David Rayvern Allen's delightful *A Song for Cricket. The Noonday Sun*, Valerie Pakenham's account of the British at play in India and elsewhere, and Edmund Swinglehurst's books on Thomas Cook, *The Romantic Journey* and *Cook's Tours*, are entertaining and absorbing.

Among sources on London I used Peter Ackroyd, *London: The Biography*; Roy Porter, *London*; A.N.Wilson, *The Victorians*; Peter Hoffenburg, *An Empire on Display* (California); Louise Purbrick (ed.), *The Great Exhibition of 1851* (Manchester); and Yvonne ffrench, *The Great Exhibition, 1851*.

I was helped by library staff in Adelaide, Alice Springs, Sydney, Kimberley, Cape Town, Barberton, Bridgetown, Bangalore, Delhi, Port Blair, New York, the RCMP Museum, the London Library, the National Maritime Museum, the National Library of Wales and the Royal Signals Museum, and by Sheila Markham at the Travellers Club. Ken Jones, Brecon historian, told me about the Brecon nabobs; R.M. Jones about H.M. Stanley's antecedents; and Eryl Rothwell Hughes showed me the Copper King's final resting-place. In Marree, Karen Burk told me of the Afghan cameleers. Jose Petrick guided me to historic places in Alice Springs. Professor Karl Miller helped with the story of Barbados. Roy Anderson took me on the Mounties' trail. Jane Aaron helped me with her work on missionaries; and Michelle Harrison with the manuscript of her 'King Sugar: Jamaica, the Caribbean and the World Sugar Industry'. David Rattray of Fugitives Drift widened my knowledge of South African history.

John Ridgway, rugged and thoughtful seafarer, widened my horizons in a different way, generously inviting me to help sail his ketch *English Rose VI* across the North Atlantic from Scotland to Cape Cod and back again.

Index